Sampling and Assessment of Contaminated Sites

World Scientific Series on Environmental Remediation

Series Editor: Ravi Naidu *(University of Newcastle, Australia)*

This book series focuses on one of the most challenging problems confronting humanity – the need to clean up the nearly 10 million contaminated sites globally, which are directly linked to (a) 13 million deaths per year, (b) declining male fertility, and (c) 75% of babies born with pollutants in their blood. In their present form, contaminated industrial sites are leaching their poisons into the air we breathe, the water we drink and the food we eat. Only a small fraction of these have undergone an attempted clean-up, and not all remediation efforts have succeed. This book series responds to the EU Action Plan: "Towards Zero Pollution for Air, Water and Soil".

Published

Vol. 1 *Sampling and Assessment of Contaminated Sites*
edited by Ravi Naidu, Junaid Sadeque and Alvin Lal

World Scientific Series on
Environmental Remediation
Volume 1

Sampling and Assessment of Contaminated Sites

Editors

Ravi Naidu
CRC for Contamination Assessment and Remediation of the Environment and the Global Centre for Environmental Remediation, University of Newcastle, Australia

Junaid Sadeque
AECOM, USA

Alvin Lal
CRC for Contamination Assessment and Remediation of the Environment and the Global Centre for Environmental Remediation, University of Newcastle, Australia

World Scientific

NEW JERSEY • LONDON • SINGAPORE • BEIJING • SHANGHAI • HONG KONG • TAIPEI • CHENNAI • TOKYO

Published by

World Scientific Publishing Europe Ltd.
57 Shelton Street, Covent Garden, London WC2H 9HE
Head office: 5 Toh Tuck Link, Singapore 596224
USA office: 27 Warren Street, Suite 401-402, Hackensack, NJ 07601

Library of Congress Control Number: 2025030389

British Library Cataloguing-in-Publication Data
A catalogue record for this book is available from the British Library.

World Scientific Series on Environmental Remediation — Vol. 1
SAMPLING AND ASSESSMENT OF CONTAMINATED SITES

ISBN 9781800617834 (hardcover)
ISBN 9781800617841 (ebook for institutions)
ISBN 9781800617858 (ebook for individuals)

For any available supplementary material, please visit
https://www.worldscientific.com/worldscibooks/10.1142/Q0525#t=suppl

Desk Editors: Nambirajan Karuppiah/Gabriel Rawlinson/Shi Ying Koe

Typeset by Stallion Press
Email: enquiries@stallionpress.com

https://doi.org/10.1142/9781800617841_fmatter

Preface

The increasing prevalence of environmental contamination—whether from industrial, agricultural or urban sources—poses significant challenges to public health, ecological integrity and environmental sustainability. Contaminated sites are often difficult to assess and manage due to the complex, heterogeneous nature of contaminants, the wide range of potential exposure pathways and the diversity of environmental matrices involved. As such, effective sampling and assessment are pivotal for the identification of risks, guiding remediation efforts and ensuring the safety of impacted communities. This book is designed to provide both foundational knowledge and advanced insights into the strategies, techniques and emerging technologies used in environmental contamination assessment.

The 11 chapters in this volume span a broad spectrum of topics, from traditional methods to cutting-edge innovations in environmental monitoring. The contributions herein reflect a collaborative effort by experts across multiple disciplines, bringing together a wealth of experience from the fields of environmental science, risk assessment, soil and water chemistry, toxicology and computational modelling.

In Chapter 1, the foundational principles of sampling and assessment in contaminated environments are discussed, setting the stage for the more specialised topics covered in subsequent chapters. Chapter 2 explores risk-based sampling strategies, emphasising the importance of tailored sampling approaches that consider site-specific contamination profiles and regulatory frameworks. Chapters 3 and 4 discuss soil sampling and analysis of contaminants, focusing on herbicides and metallic components, respectively.

One of the more innovative contributions comes from Chapter 5, where cryogenic core collection and high-resolution sample analysis are presented as novel techniques for subsurface remediation efforts. This is complemented by Chapter 6, which delves into the integration of machine learning models to enhance contaminated site assessments, allowing for more accurate predictions and optimised sampling strategies.

Chapters 7 and 8 offer insights into the contamination of water resources. Groundwater–surface water exchange measurement methodologies, essential for understanding hydrological dynamics and contamination transport, are explored in Chapter 7, while Chapter 8 specifically focuses on arsenic contamination in the Ganges–Brahmaputra–Meghna River delta, a critical area of water security concern.

Chapter 9 addresses the emerging challenge of per- and polyfluoroalkyl substances, with an in-depth examination of their chemical behaviour, isomeric complexity and implications for site remediation and conceptual site modelling.

Finally, Chapters 10 and 11 invite the readers to an in-depth discussion on health risk assessments of contaminated sites with multiple case examples. While Chapter 10 outlines methodologies for preliminary health risk assessments, Chapter 11 delves into a detailed case example of health risks associated with industrial contamination at the Bor industrial sites in Serbia.

Each chapter not only provides the technical depth required for professionals and researchers in the field but also addresses practical considerations for those involved in site assessment and remediation projects. The interplay of theory and practice in these chapters will aid in developing more effective and efficient environmental management strategies.

Moreover, this endeavour contributes to the mission of the Food and Agriculture Organization's International Network on Soil Pollution to provide advocacy for site assessment globally and, in particular, in developing nations. By bringing together the latest research, methodologies and real-world case studies from around the globe, *Sampling and Assessment of Contaminated Sites and Sediments* aims to foster a deeper understanding of the complexities involved in environmental contamination. We hope that this book will serve as a valuable resource for environmental scientists, engineers, policymakers and students who are engaged in the sampling and assessment of contaminated sites.

Ravi, Junaid and Alvin
Editors

https://doi.org/10.1142/9781800617841_fmatter

About the Editors

Ravi Naidu has been a leading researcher in environmental contaminants, bioavailability, and remediation for over 30 years. He has authored or co-authored more than 1,000 research articles and technical publications, 13 patents, and co-edited 16 books and over 100 book chapters in the field of soil and environmental sciences.

He holds a master of science degree in soil chemistry from the University of Aberdeen, UK, and the University of the South Pacific, Fiji. He completed his doctoral studies at Massey University, New Zealand. From 1985 to 1989, he served as a senior lecturer in chemistry and the dean of the School of Pure and Applied Sciences at the University of the South Pacific.

In 1989, Professor Naidu joined CSIRO Land and Water in Adelaide, where he was swiftly appointed chief research scientist, research group leader and leader of the Remediation of Contaminated Environments Programme, as well as coordinator of the Restoring Contaminated Environments Component. During his tenure at CSIRO, he led the Sodic Soils Programme and the Soil Contamination and Remediation Programme for the Cooperative Research Centre (CRC) for Soil and Land Management.

He is the founding managing director and chief executive officer of crcCARE, Australia, and leads the Global Centre for Environmental Remediation at the University of Newcastle, Australia. His outstanding contributions to environmental research have been recognised through fellowships with the American Association for the Advancement of Science, the New Zealand Society of Soil Science, the American Society of Agronomy, and the Soil Science Society of America.

Junaid Sadeque earned his master of science degree in petroleum geology from the University of Bergen, Norway, and his PhD in geology (sequence stratigraphy) from the University of Texas at Dallas, USA. He is a recognised subject matter expert in sequence stratigraphy and sedimentology, with extensive experience in both the petroleum and environmental sectors.

As a senior stratigrapher at AECOM, USA, Dr Sadeque plays a pivotal role in applying sequence stratigraphic concepts to predict preferential flow paths in groundwater aquifers and to develop contamination remediation strategies. Over the past eight years, he has led several major projects focused on developing conceptual site models for the US Department of Defense (DoD).

His prior decade-long experience as a clastic stratigrapher at Chevron, USA, has enabled him to bring innovative geological perspectives to the environmental domain. His work has been featured at both national and international conferences, and he has delivered numerous professional training courses on contamination investigation and remediation. Dr Sadeque has co-authored several key technical publications across the petroleum and environmental sectors.

Alvin Lal is the research director of crcCARE and a research associate at the Global Centre for Environmental Remediation at the University of Newcastle, Australia. He has more than 7 years of research and academic teaching experience.

Dr Lal earned his master of science degree in geophysics from the University of the South Pacific (USP), Fiji, and his PhD in groundwater modelling from James Cook University, Queensland, Australia. His research expertise spans groundwater remediation, 3D groundwater numerical modelling (finite element and finite difference), groundwater quality monitoring and assessment, monitoring network design, and artificial intelligence and predictive data analytics.

He was a recipient of the prestigious James Cook University Dean's Award for Research Excellence in 2020. Before joining the University of Newcastle, Dr Lal served as a postdoctoral research fellow at the University of Twente, the Netherlands. From 2020 to 2021, he was employed as a lecturer at USP, where he developed the Civil Engineering programme and established courses in hydrology, water resources, and coastal engineering.

Dr Lal is currently involved in several groundwater modelling and remediation projects in collaboration with researchers from Australia, the Netherlands, Bangladesh, Indonesia, the US, Fiji, Vanuatu, and Kiribati. He has published several book chapters and journal articles and has presented his work at numerous international conferences.

https://doi.org/10.1142/9781800617841_fmatter

List of Chapter Reviewers

Note: Chapter reviews represented the reviewers' experience and opinion and were not necessarily made on behalf of their organisations. Also, some reviewers who preferred to remain anonymous are not in this list.

Alan Fryar, University of Kentucky, USA

Bill DiGuiseppi, Jacobs, USA

Charles J. Newell, GSI Environmental, USA

Charles J. Werth, University of Texas at Austin, USA

Daniel Montiel, Geosyntec Consultants, Inc., USA

Frank Swartjes, RIVM National Institute for Public Health and the Environment, the Netherlands

Gaurang Parekh, crcCARE, Australia

Girish Choppala, Global Centre for Environmental Remediation, Australia

James Graham, Geosyntec Consultants, Inc., USA

Jason Brodersen, Tetra Tech, Inc., USA

Mezbaul Bahar, Global Centre for Environmental Remediation, Australia

Nemanja Barać, University of Belgrade, Serbia

Paul Santosh, Global Centre for Environmental Remediation, Australia

Prerona Das, North-Eastern Hill University, India

Robin Boyd, AECOM, USA

Ross McFarland, AECOM, Australia

Scott Warner, BBJ Group, USA

Snežana Šerbula, University of Belgrade, Serbia

Srinivas Nowduri, Bloomsburg University of Pennsylvania, USA

https://doi.org/10.1142/9781800617841_fmatter

Acknowledgements

This book would not have taken its present shape without the meticulous peer review of each chapter by the mentioned (see page of reviewers) and anonymous subject matter experts of their respective topics. Their guidance, in both the envisioning of this book and the writing of the chapters, vastly improved the content and its delivery. We are also grateful to the Global Centre for Environmental Remediation (GCER), University of Newcastle, Australia, and the Cooperative Research Centre for Contamination Assessment and Remediation of the Environment (crcCARE), Australia, for their logistical support, including chapter formatting. AECOM, USA, also provided valuable consultation in producing this book.

https://doi.org/10.1142/9781800617841_fmatter

Contents

https://doi.org/10.1142/9781800617841_0001

Chapter 1

Sampling and Assessment of Contaminated Sites and Sediments

R. Correll

Rho Environmetrics, Canberra, Australia

rho.environmetrics@bigpond.com

Abstract

Sampling is a cost-effective method used to gather useful and representative information that might otherwise be unattainable or unaffordable. Before planning a sampling, the objective and intended use of the sampling data should be documented. In particular, the way in which the sampling results will be used to influence decision-making should be clearly defined. That information should be recorded in a sampling and analysis plan (SAP).

The initial phase of a study must include a desktop investigation. That needs to include a delineation of the study area or population and site history. That phase will also provide information on the chemicals of potential concern as well as their likely distribution.

The cost of the SAP should be estimated before field sampling commences. This will include costs associated with planning, the cost of obtaining the samples, laboratory expenses (including replicates and blanks), and costs related to statistical analysis and report writing.

Most regions have guidelines for sampling contaminated sites; these guidelines can include the number of samples and the method of collection. There must be sufficient samples to meet the guidelines.

Systematic sampling has the advantage over random sampling in that it is easier to implement and well defined. It is also possible to guarantee that all points in the population being sampled are within a given distance of a sampling point. Although there are theoretical problems in obtaining an estimate of the variance, there are methods available to obtain a useful estimate.

Field sampling variation needs to be assessed by including field replicates. The data quality assurance requires the inclusion of blanks and spiked samples. There must also be a documented chain of custody for the samples.

Typically, samples are analysed in a laboratory, and laboratory errors are another source of variation. Subsampling errors may be eliminated by adjusting the size of the field sample to meet laboratory requirements. Techniques to ensure that laboratory methods are both precise and accurate are outlined.

The data need to be summarised and presented as means or proportions of detections or the proportion exceeding a nominated level. Recommendations are given for software that provide confidence limits for the key parameters.

1.1 Introduction

Sampling is a cost-effective method used to gather useful and representative information that might otherwise be unattainable or unaffordable. Before planning or undertaking sampling, the objective and intended use of the sampling data should be stated. In particular, the manner in which the sampling results will be used to influence decision-making should be clearly defined. That information should be recorded in a sampling and analysis plan (SAP).

Typically, sampling involves taking a set of elements drawn from a population and analysing them to estimate the characteristics of the population. It is useful to consider the components of that statement:

- In the context of this chapter, "population" refers to the soil or sediment in a defined area. The depth included in that population must be defined with reference to the decision-making resulting from the sampling. The sampling depth may be determined by a distinct layer (e.g., sediment subject to acidification or soil that is subject to disturbance or available to roots). The sampling depth may also be less than the targeted depth due to the limited thickness of the targeted layer. Further refinement may be required within that volume of soil or sediment to determine whether stones over a given size or litter are part of the soil or whether plant roots should be included.

- The characteristics of that population could be the average concentration of analytes, possibly with estimates of their standard error of the mean. Other characteristics may include the proportion of the population where a given concentration is exceeded or defining areas with unacceptable concentrations.
- The sampling elements (samples) typically consist of a volume of soil or sediment that is representative of the population and is sent to a laboratory for analysis. Such samples are sometimes referred to as analytical samples. The set of sampling units is chosen to enable the estimation of the target characteristics of the population.
- Once the sampling elements have been analysed, there is typically some laboratory analysis to provide measurements of the sampling elements. The data from the laboratory analysis of the set of samples are then summarised to produce estimates of the characteristics of the population. Examples of the characteristics of the population may include an arithmetic mean or the proportion of samples exceeding the investigation level (NEPC, 2013).

Details of the above components should be recorded in an SAP.

The importance of sampling at contaminated sites has been recognised by various authorities, and much of the experience has been distilled into guidelines. A useful example for sampling contaminated soil is NSW EPA (2022). Other examples are the Canadian Council of Ministers of the Environment (2013), AS44821 (2005), Malaysia DEM (2024), EU (2014) and USEPA (2014). There are also published guidelines for sediments; for example, the Australian guidelines for dredging (NAGD, 2009) and USEPA (1972). This chapter highlights some principles of sampling and includes some guidance developed from experience.

In practice, the size of the samples is small compared to that of the population being sampled. While this is the case, no finite sample corrections are required. That simplifies the calculations.

1.2 Planning an Investigation

Significant planning is required before sampling of a potentially contaminated site is undertaken. An outline of the process is given in Figure 1.1, but each stage is likely to have several sub-stages. The objectives of each stage must be defined, and those objectives must be recorded in an SAP. There is a broad range of publications concerning planning and investigation, for example, Health Canada (2021) and WA Guidelines (2021).

The first stage is to obtain a map of the site in question; it could potentially come from a deed of title. A detailed aerial photograph of the site should be obtained—possibly using multispectral images from a drone. The site's history should be obtained, possibly in consultation with current and previous owners and workers. Any old photographs that are available should also be considered. A site inspection is then required to record any changes in topography or vegetation, especially any drainage lines. Several judgemental samples could be taken at this stage. If there is evidence of run-off, water or water quality could be assessed from the samples or possibly using passive samplers (Fernandez *et al.*, 2012). Those data, together with the current and potential land uses, are then used to define chemicals of potential concern (COPCs).

The detailed site investigation would initially refine the aims of the sampling and the list of COPCs. It may be practical to subdivide the area into strata that should be sampled differently. Typically, the intensity of subsequent sampling will be determined by the maximum area above trigger levels (hot spots) that it is acceptable to miss. That defines the dimension of any grid sampling system. Other survey methods may be

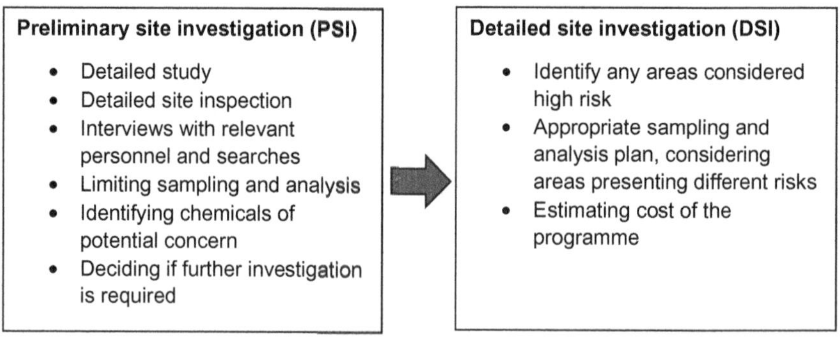

Figure 1.1. Overview of planning an investigation of a potentially contaminated site.

useful, for example, metal detection or ground-penetrating radar (GPR) that can give useful information.

An estimate of the cost should now be calculated. That estimate would include (but not be limited to) the cost of field sampling, cost of transport of samples, laboratory costs (including control samples) and costs of report writing and presentation. The cost of the sampling must be considered in relation to the value of the property.

1.3 Taking Samples

Typically, a sample is a fixed amount taken from a population. In this chapter, we concentrate on confined samples. There are non-confined measures, such as a test well being monitored for gaseous hydrocarbons at a fuel station or soil properties that are measured using GPR.

The actual mechanism of taking a confined sample must be consistent with the aims of the sampling and analysis programme. A bias can be introduced when taking samples, especially when there is a change in the concentration of a contaminant with depth. Taking samples requires:

- using appropriate sampling techniques (for example, corers),
- collecting representative samples that enable inferences to be made about the population,
- collecting an appropriate number of samples, and
- protecting or preserving the samples until they are analysed in a laboratory.

There must be a chain of custody from the time of sampling to laboratory analysis. Pre-labelling sample receptacles is part of that procedure; pre-labelling also saves field time.

1.3.1 *Soil sampling*

The population of soil to be sampled is determined in part by the area being studied. The depth to be sampled will depend on the aims of the sampling and, especially, decisions made based on the survey.

The definition of soil must also be considered in soil sampling. For example, are stones considered part of the soil? If they are sieved out, would that affect the concentration in terms of total soil volume? A further potential source of bias occurs when a soil corer or auger hits a stone; the temptation is then to relocate the core position to where the corer can be more readily inserted. That in effect changes the population being sampled.

Soil sampling should be performed using cores or perhaps augers that are calibrated to sample a particular range of depths. Use of a trowel or spade typically biases the sample towards surface layers. Segregating surface layers from deeper layers may be important in detecting and quantifying contamination at a site from incoming dust. There is a range of situations where this is important. This includes the following:

- Quantifying spray drift, particularly if the active ingredient has a high affinity for soil organic matter and a long half-life.
- Contamination from a smelter—concentration of heavy metals in surface layers could indicate recent contamination—especially if there is a lower concentration in deeper layers (Cartwright *et al.*, 1977).
- Radionucleotide contamination has been reported near some coal-fired power stations. Although the concentration of uranium in coal is low, the consumption of millions of tons can cause significant amounts of uranium to be released. In that case, concentrations may be higher in the surface soil (Habib *et al.*, 2022; Beaza *et al.*, 2011).

In each of the above cases, it is important to carefully segregate the surface layers from the lower layers.

There are also cases where samples at a specified depth may be more important. Two examples are leaking underground tanks or leaking underground pipelines. These examples stress the importance of determining what data are required for decision-making.

1.3.2 *Sediment sampling*

Corers are the preferred method of sampling sediment when feasible. Corers are required when there is an interest in the stratification of the sediments. Obtaining cores may be difficult if the sediment will not stay in the corer. If the act of coring changes the bulk density of the sample, there is a potential for bias when using the results to make inferences about the total sediment. Good results can be obtained by using a piston.

With piston coring, an inner tube contains the sample, which rises as the tube fills with sediment (NAGD, 2009).

Sampling sediment in a wetland can typically be performed using simple corers, perhaps to a depth of 2 m. A core diameter of at least 50 mm may be required to provide sufficient material for analysis from a 100 mm interval.

Metals and metalloids are likely to be concentrated in the fine fraction of a sediment sample. Pore water concentrations have been shown to be a good indicator of zinc toxicity (Kader *et al.*, 2015); however, extracting sufficient amounts for accurate chemical analysis can be challenging. Estuarine sediments typically contain a high level of organic matter, and organic soluble chemicals tend to be concentrated within that organic matter. The WQG (2000, Table 3.5.1) recommends normalising the concentration of organics to a total organic carbon (TOC) level of 1%. A decision to normalise will depend on whether it is the concentration of organics or total contaminants that is of interest, but it is recommended that TOC be measured.

Sampling sediments in a harbour may require using divers to obtain cores. In some cases, the sediment layers may be mixed, especially if the sampling is in the vicinity of shipping lanes or berths. In those cases, a grab sample would suffice.

As in many situations, there would be a compromise between the number of samples and the cost of the sampling programme. The Australian Guidelines (NAGD, 2009) allow for grab samples where it can be shown that there is mixing. More samples could be taken using a grab rather than a corer within the same budget. If the grab samples are biased, increasing the sample number will not decrease the bias.

Monitoring of the dredged sediments is simpler than *in situ* monitoring. It could consist of running composite samples from the spoil. Data from the dredgings may be relevant in assessing the potential effects on a dumping area.

Work health and safety regulations must be followed when taking sediment samples. Risks encountered have included snakes, sharks and crocodiles.

1.4 Simple Random Sampling

When the variations in the characteristics do not relate to subgroups in the population, and the cost of sampling and analysing one unit is approximately the same for all units in the population, the basic sampling design to be used is *simple random sampling*. With this design, objects are drawn completely randomly from the population; therefore, each potential sampling unit has an equal probability of being chosen, and the choice of samples is not affected by previous sample choices. Simple random sampling is assumed in the estimation of many sample characteristics.

1.4.1 *Terminology*

The symbols used in this chapter follow those used by Cochran (1977). Let N denote the potential total number of sampling units in the population, and let n denote the number of units to be sampled. Typically, N is much larger than the number of sampling units n.

The most common measure of the population is the mean. It is estimated using the formula

$$\bar{y} = \sum_{i=1}^{i=n} y_i / n. \tag{1.1}$$

The measure of dispersion among the population units is usually quantified as the *variance*.

An estimate of the variance is given by the formula

$$s^2 = \frac{\sum_{i=1}^{i=n} (y_i - \bar{y})^2}{n-1}. \tag{1.2}$$

It is often easier to visualise the standard deviation, which is the square root of the variance. The units of measurement for the standard deviation are the same as those of x_i and the mean.

The dispersion of the mean is less than that of the sampling units. This is estimated using the formula

$$\mathrm{SE} = \sqrt{\frac{s^2}{n}}. \tag{1.3}$$

Other characteristics of the population may also be required. For example, the proportion of samples with a concentration of an analyte that is greater than some criterion should be stated. Another useful set of parameters is quantiles—for example, the value that includes the top 10% of samples.

1.4.2 *Implementation*

Random sampling is simple in concept but often difficult to implement in practice.

Random soil sampling in a field requires a grid system, with random numbers being used to select X and Y coordinates. This is time-consuming and seldom done in practice. This may be feasible in small fields but not in areas greater than 1 ha. The use of pre-determined positions identified using a GPS can be helpful in locating random positions in a field.

Approximations may also be used; for example, throwing a marker over one's shoulder, or a haphazard selection of sites. Sometimes there is a conscious effort to obtain greater coverage of the field. In practice, complete random sampling has limited use (see, for example, NSW EPA, 2022). There is a dilemma when all samples are clustered and not spread across a site: the temptation is to require a good spread, thus compromising the concept of randomness.

1.4.3 *Number of samples to be taken*

Typically, the larger the sample size, the greater the precision. It may be possible to determine the sample size required to obtain confidence limits that are sufficiently small (say with an acceptable error of D). This is discussed in Section 1.11.2.

Often the budget limits the number of samples. There may be a situation in which the number of samples that can be afforded may be insufficient to yield a useful estimate.

1.5 Stratified Random Sampling

When there is a discrete auxiliary measure that can identify subgroups of the population, a stratified random sample may be appropriate. Stratified random sampling is more effective when there are significant differences among the subgroups. In that situation, a simple random sampling design would overestimate the dispersion in measurement values. Conversely, by removing between-subgroup variations, the estimate of the mean or other characteristics would become more precise.

Soil horizons provide a natural stratification. Other stated definitions could be defined by altitude or soil type.

Subgroups in stratified sampling are referred to as *strata*, hence the term *stratified sampling*. If we assume a total of L strata, we denote the total number of sampling units within stratum h by N_h. The strata may be of different sizes, and so strata weights (w_h) need to be taken into consideration. The strata weights are either formed as $w_h = N_h/N$, where N_h is the number of sampling units in stratum h, or as $w_h = A_h/A$, where A is the total area covered by the population and A_h is the area of the hth stratum.

The mean of a stratified sample is given by the formula

$$\bar{y}_{st} = \frac{\sum_h w_h \bar{y}_h}{\sum_h w_h}. \quad (1.4)$$

The variance of the mean of a stratified sample has to take into account the weights for each stratum:

$$s_{st}^2 = \frac{\sum_h w_h^2 s_h^2}{\left(\sum_h w_h\right)^2}. \quad (1.5)$$

The simplest situation is where each stratum has the same weight, and the dispersion is similar within each stratum. In that situation, the terms w_h can be taken as 1. The number of samples from each stratum (n_h) may not be constant. Ideally, the number of samples from each stratum should be proportional to both the weight of that stratum and the variability within that stratum. The number of samples from within a stratum should be at least five to enable a stable estimate of the variance. A more comprehensive account of sample size determination and allocation methods is given by Cochran (1977) and Lohr (2010).

A simple method of sample allocation to strata is that of *proportional allocation*. This means that the sample units taken from a stratum should be proportional to the size (N_i or A_i) of that stratum. The proportionality constant is simply the stratum weight w_h. The number of samples must be at least $5L$, where L is the number of strata, to obtain a stable estimate of the standard error of $\bar{y}_{st}$.

Historical data may be useful in determining strata at a potentially contaminated site. For example, there may be a greater chance that high levels of contamination will be found near the boundary of the site.

Compared with simple random sampling, stratified sampling will give a smaller standard error of the estimate for the mean when there are differences among strata (Cochran, 1977). In addition, the sampling system gives the means of each stratum, which would enable future sampling to be better focused.

1.6 Systematic Sampling

In systematic sampling, instead of allocating the sampling points completely randomly, a regular sampling system is used. Typically, every *k*th unit of the population is included in the sample. The population can be items in a list or units on a production line. If a random sampling point (between 0 and *k*) is chosen, the mean of the samples is an unbiased estimate of the mean.

Systematic sampling has several practical advantages over random sampling:

- It is easy to describe and implement.
- In field sampling, it could be repeated, so it is auditable.
- On average, studies have shown that it is generally more precise than random or stratified sampling (see Cochran, 1977, Table 8.7).
- With grid sampling, there is a guarantee that every point in the field is within a known distance of a sample.

Systematic sampling also has two disadvantages:

- A problem with systematic sampling is obtaining an estimate of the sampling variance. Methods of obtaining an estimate are given by Cochran (1977, Chapter 8) and Fewster (2011). One method is to use the residuals after fitting a linear trend. Other trends (e.g., quadratic or a low-order spline) could also be used in that situation.
- Systematic sampling is susceptible to cyclic variations, especially if the phase interval is similar to interval *k*. Such a situation may arise while collecting samples at the same time of the day when there are regular daily cycles. One example is the amount and quality of domestic wastewater that has a daily pattern.

Special case 1: Transect sampling

Transect sampling is a design in which transects (lines) are drawn across an area to be sampled. The lines are normally drawn with equal inter-distances, but they can also be randomly spread over the area. The sample size, n, is divided by the number of transects, rendering n_T, the number of sampling points per transect (Figure 1.2).

Special case 2: Grid sampling

A further variation of systematic sampling involves using a two-dimensional grid. The grid defines a set of paths along which systematic sampling can be made (see Figure 1.3). With a regular grid, where rows and

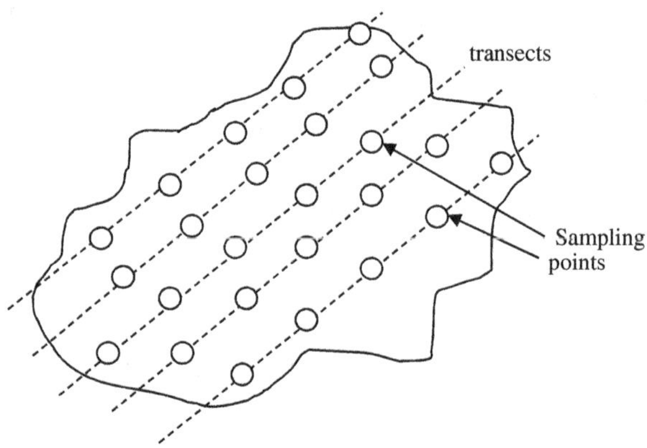

Figure 1.2. Illustration of transect sampling.

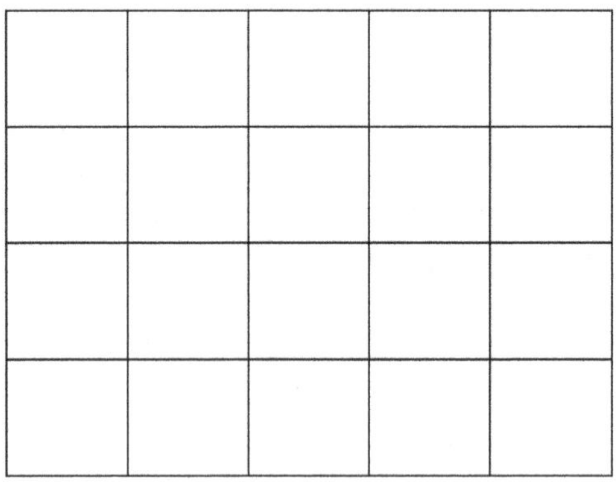

Figure 1.3. Illustration of grid sampling: Samples are taken at the centre of each grid cell.

Table 1.1. Latin square used for sampling.

b	a	c	e	d
a	e	b	d	c
d	c	e	b	a
e	d	a	c	b
c	b	d	a	e

columns are each *s* m apart, there is guaranteed to be a sample within 1/sqrt(*s*) of any point in the field. That property can be useful in writing reports on the extent of contamination.

Special case 3: Latin square sampling
When a grid can be arranged so that there are the same number of rows and columns, Latin square sampling (Cochran, 1977) can be employed. First, the points are labelled as in a Latin square, and samples are taken from each point that has the same letter, for example, all plots with the label "c" in Table 1.1.

1.7 Composite Sampling

Composite samples are formed by aggregating a series of small subsamples to create a composite sample. Composite sampling offers an opportunity to reduce laboratory costs. A summary is given by Boswell and Patil (1987).

Composites within a close radius can be used to obtain a good representation of that local area. Typically, a series of 10 cores from a small area are combined to create a composite. This is particularly useful when there is large local variation. The concentration of an analyte in the composite is comparable to the average of the concentrations in each of the subsamples. The variation associated with the composite will not include variations between the subsamples, so the variability associated with the composite is expected to be less than that of a single sample. Because the concentration in the composite is effectively the average of the subsamples, its distribution will be closer to normality than the distribution of the concentrations among the subsamples.

The number and size of the subsamples can be adjusted so that the size of the composite sample is suitable for analysis without the need for subsampling in the laboratory. The gain in precision must be weighed against the added cost of taking several subsamples rather than a single sample. The gain is typically worthwhile for surface samples but not for deeper samples or where a larger corer is required.

1.7.1 *Detection of an analyte*

Composite sampling is very effective when the aim is to detect an analyte. For this purpose, subsamples are taken from adjacent but separate parts of the field and combined. A detection in the composite indicates that the analyte was present in at least one of the subsamples.

1.7.2 *Detection of hot spots*

Although composite sampling reduces laboratory costs, it also reduces the sensitivity of detection. A further drawback is that when the composite is positive, analyses of the individual subsamples are required to determine which part of the field is a hot spot. This may require additional field sampling and laboratory costs.

Correll (2001) conducted a simulation based on a site that had been sampled using a grid. The use of composite samples in that situation was found to be cost-effective. It was also recommended that a modified investigation limit of IL/sqrt (number of subsamples per composite) be used, where IL is the relevant investigation limit.

Composite sampling is particularly effective when there is a low expected frequency of samples with concentrations of the analyte above the investigation limit.

1.8 Ways of Increasing Precision

Typically, some information is available about a site to be sampled. We denote those data as x variables. The measures may be qualitative variables, rankings or continuous.

1.8.1 *Use of qualitative data*

Qualitative variables can be used to define subgroups of a population. Each subgroup is then sampled separately using random sampling from a systematic sampling. This is further discussed in Section 1.5.

1.8.2 *Use of ranks*

When units in the population can be ranked, rank set sampling (RSS) may be feasible. RSS, introduced by McIntyre (1952), requires taking *m* series of samples which can be ranked within a series, but only one sample per series is measured. Since then, the theory of RSS has been further developed by Ozturk and Kravchuk (2022) and Ozturk *et al.* (2021). The analysis of RSS can be performed in R using the package "*RSSampling*" (Cetintav *et al.*, 2019).

1.8.3 *Use of continuous auxiliary measures*

When a continuous auxiliary variable is available, a regression estimator may be useful in estimating the population mean and its standard error. Details are given in Chapter 7 of Cochran (1977) and in Singh and Mangat (1996). The use of a regression estimator assumes that there is a linear relationship between the analyte of concern and the auxiliary variable. If the auxiliary variable were also a sample, its estimate would have to be unbiased. As long as the correlation between the variables is positive, there is benefit in using a regression estimator.

A special case is when the intercept of the regression is zero—when a ratio estimator gives a more precise result.

1.9 Estimation of Confidence Intervals

Typically, the 95% confidence interval of an estimate is given by

$$\mathrm{CL} = \bar{y} \mp \mathrm{SE} \times t, \tag{1.6}$$

where t is the 95% limit of the t distribution on the appropriate degrees of freedom. Equation (1.6) assumes that the underlying y values are approximately normally distributed; therefore, by the central limit theorem, $\bar{y}$ will be closer to a normal distribution.

When the distribution of y_s is skewed, a logarithmic transformation may be approximately normally distributed. The back-transformed mean is likely to be biased, and consequentially, the resultant confidence intervals will be unreliable. The bias can be corrected if the transformed data are in fact normally distributed and there is a reliable estimate of the variance.

When the assumption of normality is questionable, other methods exist for estimating confidence intervals. The USEPA package uses a variety of methods to construct confidence ProUCL to estimate the upper confidence interval.

Some data will be available as the proportion of samples that had a concentration exceeding a given limit. If the proportion of n samples exceeding the limit is p, approximate confidence limits for confidence intervals is given by Equation (1.7):

$$\text{CL} = \hat{p} \mp 1.96 \times \sqrt{\frac{\hat{p}\left(1-\hat{p}\right)}{n}}. \tag{1.7}$$

Exact confidence intervals can be obtained using the website *Exact Binomial and Poisson Confidence Intervals*, available at https://statpages.info/confint.html (Pezzullo, 2009).

1.10 Laboratory Errors

For a measure to be accurate, it needs to be both precise (repeatable) and unbiased. Random errors affect the precision but may not cause a bias. The lack of precision can be overcome by increasing replication. The random component is indicated by the spread of the concentrations obtained from the same sample.

There are many examples of erroneous results coming from laboratories that have been accredited. Sometimes the errors occur in the laboratory, but they may also occur during sample preparation or sample collection. Strict protocols need to be observed when collecting samples. The types of errors encountered can be grouped into three classes:

- random variations,
- outliers, and
- calibration errors.

Each of these three error classes has different effects.

1.10.1 *Random errors from subsampling in the laboratory*

Subsampling a sample of soil or produce is surprisingly difficult. Many subsampling schemes can cause bias—for example, shaking a sample may result in the less dense fraction being over- or underrepresented.

There can also be biases related to particle size or particle structure. A method that minimises the source of error must be included in the sampling scheme. Two approaches are given as follows:

(i) Adjust the sample size collected in the field so that no subsampling is required. Where core samples are involved, the diameter of the corer or the number of cores in a composite sample is adjusted to obtain a sample size ready for analysis. Replication of field samples may be required for quality control.
(ii) A common method of subsampling involves mixing the sample on a flat surface, dividing it into four sectors and using material only from two opposite sectors. That effectively halves the size of the sample. The process is repeated until a sample of manageable size has been obtained. Subsampling variation still occurs even when this method is used.

There will also be variations in weighing samples and in laboratory instruments. Typically, those errors are small compared to sampling errors.

1.10.2 *Incorrect dilution*

Estimations of chemical concentrations typically involve calculations that include dilutions. Figure 1.4 shows a comparison of total nitrogen

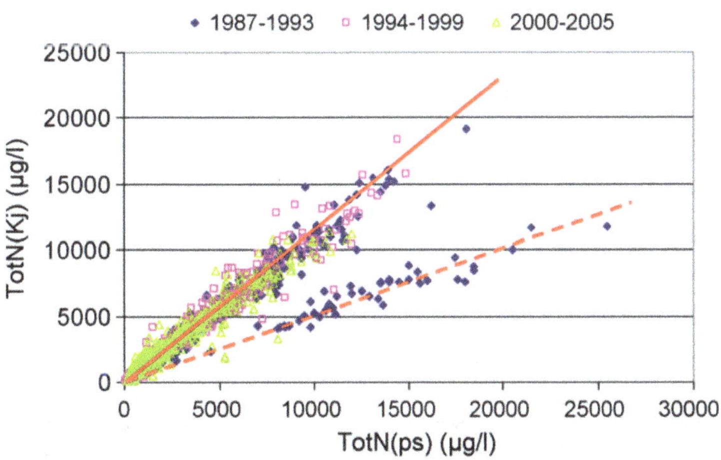

Figure 1.4. Scatter chart of Total N estimated using the Kjeldahl method and Total N estimated using the persulphate digestion method (Wahlin and Grimvall, 2008).

concentrations in 75400 water samples collected in Sweden. Some of the estimates obtained through the Kjeldahl method were half of those obtained through the persulphate method. This implies that there was a dilution or calculation error made in those samples. Unfortunately, such errors do occur.

1.10.3 *Outliers*

Outliers are typically large positive values that do not appear to belong to the underlying population being sampled. The outliers could be either a true reflection of field conditions or, more often, an artefact due to contamination. In many cases, there is no replication; therefore, it cannot be determined if the outlier is a real value or an artefact. Outliers can be either positive or negative, but they should never be recorded as a negative value, as that would not be plausible. Inclusion of only positive outliers therefore creates a positive bias. Depending on the size and frequency of the outliers, this bias can be serious and influence the decision-making process.

Another effect of the outlier is to decrease the precision of the measures—this part of the problem can be overcome by including additional replication. A further nuisance caused by outliers is that, if they do not fit in with the main distribution, there arise a challenge in assessing confidence intervals and a need for performing other statistical tests.

Because of the problems created by the outliers, care should be taken at the technical level to eliminate outliers caused by contamination. If that fails, there needs to be a recognised procedure for their removal from the data set; that may require triplicate field sample measurements. Without replicates, outliers can be detected by an excessive Z score, where Z is the residual divided by the standard deviation. A Z score greater than 3 can be considered an outlier, as it represents a probability of approximately 0.1%. It is important that an outlier should not be simply discarded, as it may have high practical significance.

1.10.4 *Calibration errors*

Figure 1.5 shows a comparison of trends in total nitrogen, as assessed by two methods. While the concentrations based on the Kjeldahl method showed no marked trend, there was a downward trend in the

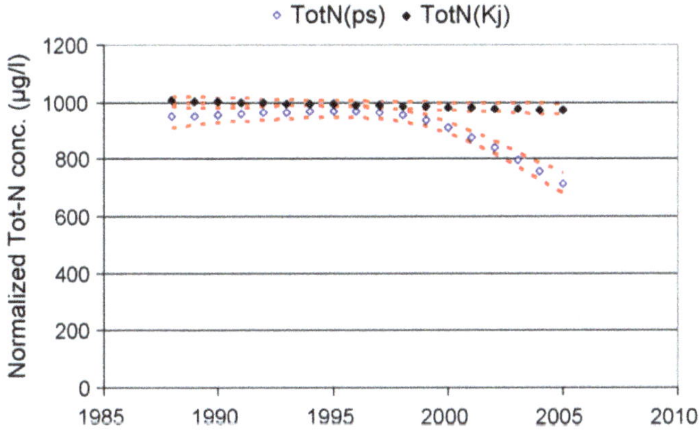

Figure 1.5. Trend lines for total persulphate nitrogen and total Kjeldahl nitrogen in seven Swedish rivers (Wahlin and Grimvall, 2008).

concentrations obtained using the persulphate method. The strong implication was that the calibration of one method had drifted from the correct values.

1.10.5 *Quality control of data*

A range of control samples needs to be included in a sampling programme. These must include field blanks and spiked samples. There should be triplicate samples, with at least one per every 10 sample sites, and the RSD (Standard Deviation divided by the mean) between them should be no more than 50%. Further, recovery rates from the spiked samples should be within the range of 75–125% (NAGD, 2009).

Pre-labelling sample containers has been found to not only reduce sample mix-ups but also save valuable field time. Pre-labelled sample containers should be allocated for field duplicates.

Some reference material should be included in each laboratory run, and the laboratory results should be within the range of 80–120% of the certified value. The reference material may include certified material, but it is also advisable to have an in-house resupply of reference material that is maintained for the duration of the monitoring period. Such a sample must be stored in a cold room so that it does not undergo any change during storage. Use of such a standard should warn of laboratory drift, as shown in Figure 1.5.

A scan of the data by an appropriately experienced worker will often detect obvious outliers. Under laboratory conditions, there is more opportunity to re-analyse outstanding samples.

1.11 Sample Number Determination

1.11.1 *Number specified in guidelines*

The number of samples may be determined by the relevant guidelines. For example, when using a grid sampling system, NSW EPA (2022) provides a table for the minimum number of samples. Similarly, NAGD (2009) also has a table for the required number of samples.

1.11.2 *Aspects of precision*

The number of samples required to detect a difference of size D when the variance of two normally distributed populations is σ^2 is given by Equation (1.8):

$$n \geq \frac{4\sigma^2}{D^2}. \tag{1.8}$$

In practice, σ^2 is not known, so an estimate is required. Some estimates may be available from previous work. An approximation of σ could be the feasible range divided by 5.

From the estimated cost per sample, it is possible to estimate how many samples can be afforded. Often, an upper limit for the sample size can be obtained, which may set the number of samples taken. The cost is a common limitation to the number of samples.

1.12 Obtaining Auxiliary Information

1.12.1 *Site history*

The history of the site can be valuable in determining what contaminants are likely to be found, which areas of the site are more likely to be contaminated and, in the case of mobile contaminants, what sampling depths need to be considered. Information about a fire may indicate the need to assess PFAS (per- and polyfluoroalkyl substances) residues.

The auxiliary variables obtained from site history may be in the form of maps or certain characteristics, which can be used for classification or as a quantitative measure (e.g., height). Such data can be useful not only for the design of the sampling plan but also in statistical analysis.

1.12.2 *Use of aerial imagery*

Aerial imagery has become much cheaper to obtain. A drone can be used to create a map of the area using a range of wavelengths. Those data can then be used for stratification of the site, as a sorting criterion for rank set estimation or as a variable in a regression estimator. The aerial information would be useful as long as it is correlated to a measure of interest.

1.12.3 *Outflow monitoring*

Monitoring outflow from an area can be a useful guide in indicating potential contamination. For example, a low pH may be an indicator of a source of acid mine drainage. The presence of chlorinated solvents in groundwater may be an indicator of previous soil contamination and a potential area of high concentration in the soil.

1.13 Aspects of Costs

Sampling costs consist of multiple components. Costs for sampling depend on access to the study area, sample collection within the study area, transportation of the samples, laboratory analysis, statistical analysis and report writing and presentation. There is a cost per sample so any reduction in sample numbers will reduce of the overall cost of the programme.

Laboratory costs can sometimes be reduced by compositing samples; for example, if a composite of four primary samples is less than a quarter of the trigger value, all primary samples are less than the trigger value. However, when the results of the composite analysis are less clear, a re-analysis of the primary samples is required. Composite sampling will be useful only when most of the samples are well below the trigger level.

Useful gains can sometimes be made through compositing (Correll, 2001), where a simulation revealed that useful information could be obtained by using a composite of four primary samples.

Typically, the estimate of the mean is more important than that of the standard error. Simple random sampling can provide estimates of both the mean and the standard error. There are useful approximations to the variance of a systematic sample (e.g., Fewster, 2011), and most researchers would be willing to sacrifice some precision in the estimate of the variance to gain the field simplicity of a systematic sample, together with an estimate of the maximum distance from any point in a field to a sampling point. Systematic sampling is typically more cost-effective than random sampling.

Some field measurements are cost-effective to measure—for example, soil colour (particularly if it is related to the Munsell colour chart) and field measurements of soil pH. Other more sophisticated measures such as GPR may also be useful. These auxiliary measures are useful provided that they correlate to the measures of interest. A variation of the GPR could potentially be useful in locating drums of contaminants buried at the site.

The auxiliary field measurements may also be useful in defining sampling strata. Stratification is useful if there are marked differences between the strata. Ideally, some strata can be considered uncontaminated, and these areas do not require sampling. Some areas may be considered free from certain COPCs (for example, PAHs) and the corresponding samples would not require analysis for these analytes.

A sampling strategy must then be devised that would give the highest precision obtainable for conclusions drawn both at a single sampling point and across the whole study area. There is a possibility to optimise the balance between precision and cost in a stratified model if the cost of taking a measurement depends on where that measurement is to be taken. For example, in a catchment where certain sampling points are more expensive to visit (or more expensive to manage), it may be wise to consider an optimal design for stratified sampling. We recommend consulting a statistician for those calculations.

The other aspect of cost involves making an erroneous decision based on the analysis of the samples. Such costs are more difficult to appreciate since they refer to long-term consequences of decisions. For example, if the sampling of a catchment is part of a more comprehensive investigation in which political decisions would play an important role, it would be necessary to include such costs in the budgeting of the sampling design.

1.14 Conclusions

- Sampling is typically used to obtain data that are to be used in making decisions. That largely determines the population and sampling process. Typically, the SAP will be constrained by budgetary conditions. That process will require defining the population to be sampled and the sampling pattern that will give appropriate estimates of the characteristics of the population.
- The aim of the sampling must be clearly defined. One component is a listing of COPCs, and another is the areas (strata) that are more likely to be contaminated.
- There should be a well-defined chain of custody for the samples from the field to the laboratory. Using pre-labelled sample containers reduces the risk of field errors. The results from the laboratory must be not only precise but also accurate.
- Many areas of the world have guidelines on sampling contaminated sites. It is recommended that those guidelines be followed.
- Preliminary investigations provide useful information about COPCs and areas of the sites that are more likely to be contaminated.
- The budget for the programme should be estimated, and budget limitations may determine the maximum number of samples that should be taken.
- The sample size ideally should be adjusted to avoid subsampling in the laboratory, perhaps by choosing a corer of an appropriate diameter.
- Systematic sampling offers a cost-effective method that can be readily documented and replicated. There are means of obtaining a useful estimate of its standard error, so it is not a major drawback.
- Composite samples may be useful in lowering laboratory costs. For surface samples, several cores could be combined into a composite sample to achieve better coverage of the sampling location using an appropriate sample weight. Where the background level is well below the trigger values, composite sampling should be considered.

References

ANZECC/ARMCANZ. (2000). *Australian and New Zealand Guidelines for Fresh and Marine Water Quality: Volume 1 — The Guidelines*. Australian and New Zealand Environment and Conservation Council and the Agriculture and Resource Management Council of Australia and New Zealand.

ANZECC/NHMRC. (1992). *Australian and New Zealand Guidelines for the Assessment and Management of Contaminated Sites*. Australian and New Zealand Environment and Conservation Council, and National Health and Medical Research Council.

AS44821. (2005). Guide to the Investigation and Sampling of Sites with Potentially Contaminated Soil Part 1: Non-volatile and Semi-volatile Compounds Standards Australia GPO Box 476, Sydney, NSW 2001, Australia.

Baeza A, Corbacho JA, Guillén J, Salas A, Mora JC. (2011). Analysis of the different source terms of natural radionuclides in a river affected by NORM (Naturally Occurring Radioactive Materials) activities. *Chemosphere*, 83(7):933–40.

Boswell, M. T. and Patil, G. P. (1987). A perspective of composite sampling. *Communications in Statistics — Theory and Methods*, 16(10), 3069–3093.

Canadian Council of Ministers of the Environment. (2013). *Guidance Manual on Sampling, Analysis, and Data Management for Contaminated Sites. Volume I: Main Report*. The National Contaminated Sites Remediation Program. Available at: https://publications.gc.ca/site/eng/9.700166/publication.html.

Cartwright, B., Merry, R. H., and Tiller, K. G. (1977). Heavy metal contamination of soils around a lead smelter at Port Pirie, South Australia. *Australian Journal of Soil Research*, 15(1), 69–81.

Cetintav, B., Esemen, M., and Gurler, S. (2019). RS sampling: A pioneering package for ranked set sampling. *The R Journal*, 11(01), 416–426.

Cochran, W. G. (1977). *Sampling Techniques*. 3rd ed. New York: Wiley.

Correll, R. L. (2001). The use of composite sampling in contaminated sites—a case study. *Environmental and Ecological Statistics*, 8(3), 185–200.

Cressie, N., Frey, J., Harch, B., and Smith, M. (2006). Spatial prediction on a river network. *Journal of Agricultural, Biological and Environmental Statistics*, 11(2), 127–150.

De Gruijter, J. J., Brus, D. J., Gruijter, J. J., Bierkens, M. F. P., and Knotters, M. (2006). *Sampling for Natural Resource Monitoring*. Berlin: Springer-Verlag.

EU. (2014). *Progress in the Management of Contaminated Sites in Europe*. Available at: https://op.europa.eu/en/publication-detail/-/publication/b217d1ca-df14-4866-8cab-addd22fb2184/language-en#:~:text=This%20report%20presents%20the%20current%20state.

Fernandez, L., Lao, W., Maruya, K., White, C., and Burgess, R. (2012). Passive sampling to measure baseline dissolved persistent organic pollutant concentrations in the water column of the Palos Verdes shelf superfund site. *Environmental Science and Technology* (John Wiley & Sons, Ltd., Indianapolis, IN), 46(21), 11937–11947. Available at: https://cfpub.epa.gov/si/si_public_record_report.cfm?Lab=NHEERL&dirEntryId=248732&simpleSearch=1&searchAll=passive+sampling.

Fewster, R. M. (2011). Variance estimation for systematic designs in spatial surveys. *Biometrics*, 67, 1518–1531.

Habib, A., Khan, R., and Phoungthong, K. (2022). Evaluation of environmental radioactivity in soils around a coal burning power plant and a coal mining area in Barapukuria, Bangladesh. Radiological risks assessment. *Chemical Geology*, 600(5), 120865.

Health Canada. (2021). *Federal Contaminated Site Risk Assessment in Canada.* Available at: https://www.publications.gc.ca/collections/collection_2021/sc-hc/H129-107-2021-eng.pdf.

Kader, M., Lamb, D., Correll, R., Megharaj, M., and Naidu, R. (2015). Porewater chemistry explains zinc phytotoxicity in soil. *Ecotoxicology and Environmental Safety*, 122, 252–259.

Kasprzak, P. (2021). *Feasibility Study of Ranked Set Sampling for Typical Modern Agronomy Trials under Different Sampling Designs: Benefits and Challenges of Ranked Set Sampling*. Thesis for Masters of Philosophy (Biometry), University of Adelaide.

Lohr, S. L. (2010). *Sampling: Design and Analysis*. 2nd ed. Boston: Brooks/Cole.

Malaysia DEM. (2024). *Contaminated Land Management and Control Guideline*, No. 1. Available at: https://www.doe.gov.my/.

McIntyre, G.A. (1952). A Method for Unbiased Selective Sampling Using Ranked Sets. *Australian Journal of Agricultural Research*, 3, 385–390.

Merry, R. H., Tiller, K. G., and Alston, A. M. (1983). Accumulation of copper lead and arsenic in some Australian orchard soils. *Australian Journal of Soil Research*, 21, 549–561.

NAGD. (2009). *National Assessment Guidelines for Dredging*. Canberra: Commonwealth of Australia.

National Environment Protection Council (NEPC). (2013). *National Environment Protection (Assessment of Site Contamination) Amendment Measure 2013 (No. 1), Schedule A and Schedules B(1)–B(9)*. Canberra: National Environment Protection Council.

NSW Environment Protection Authority. (2022). *Contaminated Land Guidelines. Sampling Design Part 1 — Application*. NSW: Environment Protection Authority.

Ozturk, O. and Kravchuk, O. (2022). Judgment post-stratified assessment combining ranking information from multiple sources, with a field phenotyping example. *Journal of Agricultural, Biological and Environmental Statistics*, 50(3), 911–927.

Ozturk, O., Rogers, S., Kravchuk, O., and Kasprzak, P. (2021). *Ranked Set Sampling: Easing the Application of Ranked Set Sampling in Practice. R Package Version 0.0.1.* Available at: https://biometryhub.github.io/RankedSetSampling.

Pezzullo, J. C. (2009). *Exact Binomial and Poisson Confidence Intervals.* Available at: https://statpages.info/confint.html.

ProUCL. (2022). *Statistical Software for Environmental Applications for Data Sets with and without Nondetect Observations.* EPA/600/R-07/041. Washington. Prepared by Singh, A., Armbya, N., and Singh, A. K.

Singh, R. and Mangat, N. S. (1996). Regression method of estimation. In: *Elements of Survey Sampling. Kluwer Texts in the Mathematical Sciences*, Vol. 15. Dordrecht: Springer. Available at: https://doi.org/10.1007/978-94-017-1404-4_8.

USEPA. (1972). *Regulations, Guidance, and Additional Ocean Dumping Information. Prevention of Marine Pollution by Dumping of Wastes and Other Matter (London Convention).* Washington, DC: USEPA.

WA Guidelines. (2021). *Guideline Assessment and Management of Contaminated Sites*. Available at: https://www.wa.gov.au/system/files/2023-05/guideline-assessment-and-management-of-contaminated-sites.pdf.

Wahlin, K. and Grimvall, A. (2008). Uncertainty in water quality data and its implications for trend detection: Lessons from Swedish environmental data. *Environmental Science and Policy*, 11(2), 115–124.

WQG. (2000). *An Introduction to the Australian and New Zealand Guidelines for Fresh and Marine Water Quality.* Canberra, Australia: Australian and New Zealand Environment and Conservation Council.

https://doi.org/10.1142/9781800617841_0002

Chapter 2

Strategies for Risk-Based Sampling of Environmental Contamination

R. Brewer[*,§], J. Song[†,¶], and M. Heskett[‡,‖]

[*]*Hawai'i Department of Health, 2385 Waimano Home Road, Suite #100, Pearl City, Hawai'i 96782, USA*

[†]*Institute of Soil Science, Chinese Academy of Sciences, No.298, Chuangyou Road, Jiangning District, 211135, Nanjing, China*

[‡]*Element Environmental, 98-030 Hekaha St #9, Aiea, Hawai'i 96701, USA*

[§]*roger.brewer@doh.hawaii.gov*

[¶]*jingsong@issas.ac.cn*

[‖]*mheskett@e2hi.com*

Abstract

Hastily published guidance for the investigation of environmental contamination in the 1970s and 1980s adopted "grab" sample collection methods already in use for testing of relatively uniform media such as industrial wastewater. Under such scenarios, the spatial and, in the case of air and water emissions, temporal variability of the concentration of a targeted contaminant could reasonably be assumed to be very low, provided that the manufacturing process itself did not change.

Such simplistic scenarios, however, do not apply to the investigation of variable discharges and emissions of contaminants into water and air or to the investigation of contaminated soil, sediment and other particulate matter. Reliable characterisation of such media and assessment of potential environmental risk require a more well-thought-out, systematic approach that includes nine key elements: (1) development of a preliminary conceptual site model; (2) specification of contaminants of potential concern and development of investigation questions; (3) identification of data needs; (4) designation of decision unit areas and volumes of media for sample collection that directly relate to the investigation questions being asked; (5) development of decision statements that specify steps to be taken based on the data collected prior to actual sample collection; (6) preparation and implementation of sampling and analysis plan; (7) evaluation of data quality and usability; (8) identification of potential environmental risks; and (9) refinement of the conceptual site model and recommendations for additional actions based on the previously defined decision statements. Adhering to this nine-step process and the risk-based sampling strategies introduced in this chapter will improve the defensibility and reliability of sample data and reduce the time and effort required to characterise, assess and remediate environmental contamination.

2.1 Introduction

2.1.1 *Background*

Investigations of the contamination of soil, water, air, food and other media are carried out to assess and address potential risks to human health and the environment. This was highlighted in the 1990s by the development of "Risk-Based Corrective Action" guidance and the comparison of sample data to pre-approved, risk-based screening levels (e.g., USEPA, 1991a, 1994, 1996a, 1996b; HIDOH, 1995; ASTM, 1995; Brewer *et al*., 2022). This reduced the time required to identify potential environmental concerns from months and even years to only a few minutes. A more detailed evaluation of identified concerns could be conducted in a site-specific risk assessment, as needed.

Delays in project completion, cost overruns and the discovery of contamination at sites that had been previously declared "clean" persisted, however. "Laboratory error" was initially suspected to be the cause of decision errors. Closer inspection revealed the majority of the errors to be not in the laboratory but in the field. The sample collection methods implemented were appropriate to meet the desired objectives.

The reason behind such errors is now clear—sampling methods as well as methods for processing and testing of samples published in regulatory guidance in the 1980s and 1990s failed to recognise and address the oftentimes complex and random heterogeneity of contaminants in environmental media, particularly at the scale of a typical laboratory aliquot or subsample. As a result, sample data generated by the laboratory were not reliably representative of the sample submitted, and the sample submitted was not reliably representative of the area where it was collected. Decisions made based on individual sample results could not be supported by proper quality assurance standards. Confidence in such decisions could therefore not be assessed—either qualitatively or quantitatively.

"Risk-based" sampling methods described in this chapter are intended to address the cause of decision errors (Brewer *et al.*, 2022). A systematic planning process is presented, which ties methods used to collect, process and test samples directly to methods used to assess environmental risk. The sampling methods described may require more planning and effort at the beginning of a project but will help ensure that decisions regarding assessment of risk and implementation of remedial actions are made with a high degree of confidence and that projects are completed on time and on budget.

The process and methods described will likely seem common sensical to new professionals in the environmental industry. The change in mindset and questions regarding the reliability of "traditional" types of sample data will likely be challenging for seasoned practitioners long accustomed to sampling methods that they were initially trained in, as it was for the authors of this chapter. This is often due to initial confusion over the difference between a "discrete area" and a "discrete sample." Designation of "discrete" areas and volumes of media for characterisation is a key part of risk-based sampling strategies. These are referred to in this chapter as "decision units (DUs)." A "discrete"—or, more appropriately, "grab"—sample, in contrast, is a single, random specimen of the DU that might or might not be representative of the immediately surrounding media. Field and laboratory workers have long questioned the seemingly random variability in contaminant concentrations reported for co-located grab samples and laboratory "duplicates." Risk assessors similarly have no way to assess potential error in estimation of a mean contaminant concentration based on a single set of grab sample data. Statistical tests employed to calculate a mean based on a set of grab sample data only assess potential error with respect to the test itself, rather than the error associated with the collection of representative samples in the field.

These groups will find the concept of DUs pertinent and methods for the collection of single, "risk-based" samples of particulate media such as soil and sediment to be a logical extension of methods long used for the collection of risk-based samples of air and water. The scientific basis of the investigation and sampling methods described can seem complex. Implementation in the field and laboratory is relatively straightforward, however, and is ultimately guided by three simple questions:

1. "What are the investigation question(s)?"
2. "What is the decision unit?"
3. "How can a sample that is representative of the decision unit as a whole be collected?"

The incorporation of more defensible, risk-based sampling methods in characterisation will result in greater satisfaction in the design and implementation of an environmental investigation and increase confidence that environmental concerns have been reliably identified and addressed. The information provided in this chapter is intended for introductory purposes only and is not sufficient to serve as a standalone guidance document. Refer to CRCCARE (2024) and associated references for more in-depth discussions of specific topics.

2.1.2 *Origin of early environmental sampling guidance*

The surge in environmental regulations in economically developed countries such as the United States (US) in the 1970s and 1980s led to the rushed publication of regulatory guidance for the investigation and cleanup of contaminated industrial lands. Much of the developed world soon followed, often replicating these US approaches. An in-depth review of the background of the early sampling guidance is provided by Brewer *et al.* (2017a, 2017b). Much of this guidance, particularly that for testing of particulate matter such as soil and sediment, adopted existing methods for "grab" sample testing of contaminant streams, such as industrial wastewater, or testing of commodities, such as fuels or dielectric fluids.

Concentrations of hazardous chemicals in these media could reasonably be assumed to be uniform at any scale provided that the manufacturing process itself did not change and the process was consistent over time. It seemed logical to authors of early sampling guidance that the concentration of these chemicals in contaminated environmental media would be similarly uniform, even at the scale of a laboratory aliquot.

Such assumptions are clearly stated in early guidance that called for the collection and testing of individual "grab" samples collected from single points within a targeted area to identify the extent and magnitude of contamination and assess environmental risks (emphasis and notations added):

> The implicit assumption (in reliance on discrete/grab soil sample data) that *residual contamination is equally likely to be present anywhere within the sampling area* is reasonable. (USEPA, 1985; refer also to USEPA (1986))
>
> The (discrete/grab soil sampling) method makes use of prior information to divide the target population into subgroups (i.e., "Decision Units") that are *internally homogeneous*. (Gilbert, 1987)
>
> *Any sample* located within the contaminated zone will identify the contamination. (USEPA, 1987)
>
> When there is little distance between points it is expected that *there will be little variability between points*. (USEPA, 1989a)

Concurrent development of guidance for the quantitative assessment of risk emphasised the need to assess risk based on the *mean concentration* of the contaminant in the media which the receptor, for example young children, might be regularly exposed to over many years (USEPA, 1987, 1988, 1989a, 1989b, 1989c, 1989d, 1991b, 1992, 2014). As stated in the USEPA's *Preparation of Soil Sampling Protocols: Sampling Techniques and Strategies* guidance (notations added):

> The concentration term in the (chronic risk) intake equation is an estimate of the arithmetic average concentration for a contaminant (in a defined, exposure area) based on a set of site sampling results. (USEPA, 1992)

In theory, data for a small number of grab samples or even a single sample collected within a targeted area would be adequate to assess risk since contaminants could be assumed to be uniformly distributed within the media, even at the scale of a laboratory aliquot (e.g., 1–10 g).

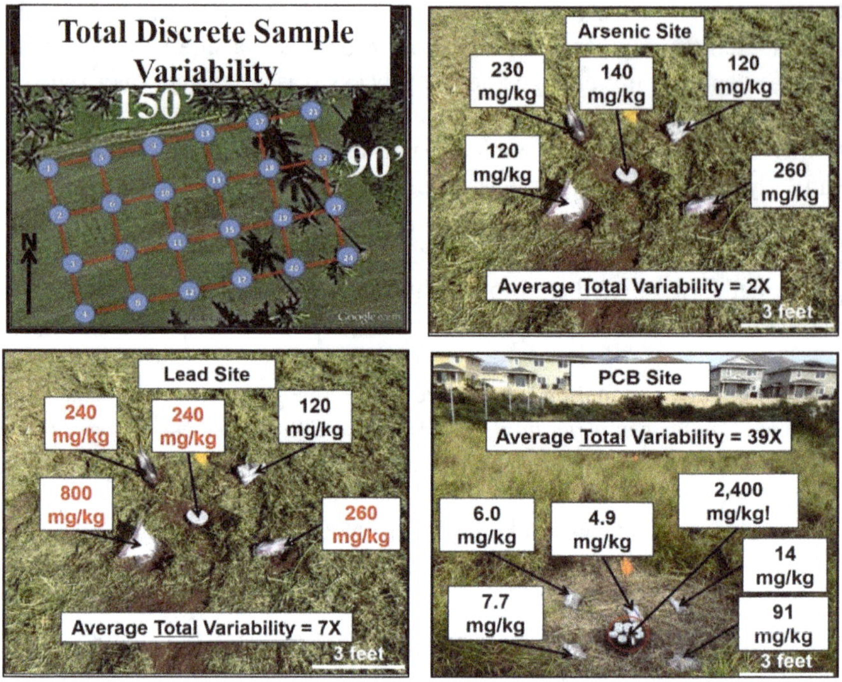

Figure 2.1. Random variability of contaminant concentrations in co-located samples of soil (Brewer *et al.*, 2017a). Noted concentrations of lead varied both above and below target screening levels for 23 out of 24 points tested.

This greatly simplified as well as expedited initial sampling efforts to characterise and remediate contaminated lands.

Such assumptions quickly proved to be premature, however. Oftentimes, high variability between data for co-located grab samples as well as variability between multiple laboratory grab subsamples and estimates of contaminant means called into question early assumptions of contamination uniformity (Figure 2.1; Brewer *et al.*, 2017a, 2017b). Risk assessors struggled to make sense of grab sample data provided to them by field workers. Guidance for estimation of mean contaminant concentrations within a targeted area went so far as to recommend omission of high-concentration sample data (Gilbert, 1987; USEPA, 2013):

> Outliers distort … statistics of interest. Therefore, it is desirable to … not to compute … statistics by accommodating a few low probability outliers. (USEPA, 2013)

Such recommendations often conflicted with observations in the field of obvious "hot spots" of heavy contamination that clearly warranted action. Early warnings of inadequate sample collection methods and resulting uncertainty in estimated mean contaminant concentrations for use in risk assessment were sounded by the United States Environmental Protection Agency (USEPA) staff and others soon after publication of the initial sampling guidance:

> All data not known to be in error should be considered valid. ... High concentrations are of particular concern for their potential health and environmental impact. (USEPA, 1989a)
>
> The so-called grab sample is not really a sample but a specimen of the material that might or might not be representative of the sampling unit. Great care must be exercised when interpreting the meaning of these samples. (USEPA, 1992)
>
> The inherent heterogeneity of soils presents a particular challenge to field personnel who are responsible for sampling hazardous waste sites. The work of George Matheron (1930–2000), father of geostatistics, and Pierre Gy (1924–2015), sampling expert, can provide useful insights for environmental scientists who are faced with sampling a complex matrix for trace contaminants. (USEPA, 1999)
>
> A common error has been to reject "outliers" that cannot be made to fit (a statistical) model. ... The tendency ... has been to make the data fit a preconceived model instead of searching for a (more appropriate way to collect samples). (Pitard, 2009)
>
> It has been clear for some time that the major sources of error in soil sampling for chemical contamination come not from laboratories but from field sampling and subsampling. This situation is—and should be—of concern to environmental forensic scientists. Legal arguments and determinations are based on the prevailing standards of science and practice and often rely on relevant requirements, policies, and guidance from regulatory agencies. Perhaps as a result of deferring to regulatory agencies many of these legal proceedings have focused primarily on the potential for laboratory error rather than on the potential for sampling error. (Hadley and Petrisor, 2013)

The significant error in both site characterisation and assessment of risk inherent in methods used to collect samples in the field was recognised early in the preparation of USEPA and other guidance documents.

Causes behind the failure to quickly acknowledge and address these problems in the early 1990s are multifaceted; they included the following: (1) the unrecognised need for different sampling methods for the characterisation of "finite element" versus "infinite element" media (discussed in Section 2.4.2); (2) the lack of alternative, widely available, field-tested guidance specific to the latter; and (3) a reluctance of regulatory agencies to call into question data being used in court to force responsible parties to address environmental contamination. In the field, the imposed use of what were ultimately unreliable sample data often led to time and cost overruns, "failed" remediation and abandonment of projects that had no clear endpoint in sight.

2.1.3 *Gy's theory of sampling and the commodities industry*

Serious consideration of alternative and more scientifically defensible site characterisation methods in the environmental industry began in the early 2000s, with a more in-depth review of data collection methods long used by the equally sampling intensive commodities industry. Such methods, collectively referred to as *Pierre Gy's theory of sampling* (TOS), for particulate media (Pitard, 2019) had been identified early on by staff within regulatory agencies, but they were unable to get it incorporated into agency-wide policy:

> Particulate sampling theory is new to most environmental investigators even though the techniques used to apply the theory to soil sampling are familiar. (USEPA, 1992)
>
> We searched for a non-conventional statistical sampling theory that … considers the nature of particulate materials. … In 1989, we hit 'pay dirt,' … the Pierre Gy sampling theory for particulate materials. (However) we were only moderately successful at transferring this technology to the environmental community. (USEPA, 2003)

Similar problems with sample data reliability continually plagued the commodities industries up through the middle of the 20th century. The lack of reliable sampling methods was especially acute for the mining industry but also affected the agriculture, pharmaceutical and, later, the recycling industries. Sample data error was more obvious in these industries due in part to routine, third-party confirmation of data quality. The true concentration and, ultimately, mass of a targeted commodity are determined by the extraction of the commodity from crushed ore or previously used material in the mining and recycling industries. Underestimation of the mass of the commodity in the ore results in an economic loss for the mining company. Overestimation of the mass results in cheating the customer and the rejection of future shipments of crushed ore for processing. The efficacy of a medication taken by a patient is closely monitored by the overseeing doctor. Failure to adequately determine nutrient needs in an agricultural field will result in reduced crop yields. Failure to accurately estimate the mass of a commodity in recycled material, for example the mass of rare earth elements in crushed catalytic converters, can lead to problems similar to those faced by mining companies: selling goods which either over- or underestimate the actual items of interest and ultimately going out of business.

Economic risk as well as the risk of legal liability drove the development of more robust and reliable sampling methods in the commodities industries. An intense review of sample databases and the compilation of factors that could lead to data error during collection, processing and testing of samples of particulate matter were initiated in the early 20th century. The more science-based and defensible sampling methods that evolved came to be referred to as Gy's TOS, after the French chemist and sampling statistician Pierre Gy, who worked for the mining industry. Gy's research was eventually compiled and published by his colleague, Francis Pitard (1989, 2009, 2019). An in-depth discussion of Gy's TOS is presented in Section 2.4.7. The concepts and practices of TOS soon spread to the agriculture, pharmaceutical and recycling industries and related industries that require equally rigorous testing of particulate matter. The delay in expansion to the environmental industry was in part due to the lack of English translations of original TOS research and guidance until the late 1980s—too late for consideration in early environmental guidance being published in the US.

2.1.4 *Incorporation of TOS in the environmental industry*

Similar economic and legal incentives, as well as a desire to more reliably assess environmental risk, have belatedly forced the environmental industry to improve sampling methodologies that were published by regulatory agencies at the beginning of the "Superfund" era of the 1980s. This chapter summarises the application of TOS-related sampling methods to the testing of environmental media that include soil, sediment, surface water, groundwater, indoor and outdoor air and subsurface vapours. The methods described partly reflect research and guidance published by the Hawai'i Department of Health (HIDOH) and funded through grants from the USEPA (HIDOH, 2023). Similar guidance is being prepared through cooperative research carried out between HIDOH and the Institute of Soil Science of the Chinese Academy of Sciences (CNSAC, 2023; CRCCARE, 2024). Site investigation and sampling methods described in this chapter reflect a summary of the more detailed guidance published by these entities.

Information presented in this chapter also draws on sampling concepts outlined in the documents *GOODSamples: Guidance on Obtaining Defensible Samples* (AAFCO, 2015) and *Good Test Portions: Guidance on Obtaining Defensible Test Portions* (AAFCO, 2018). The documents present a step-by-step, systematic approach to ensure that the sample collected is directly representative of the question under investigation and that the laboratory data are representative of the sample collected. The guidance was specifically prepared for the livestock feed industry. The concepts of sampling plan design, the collection and analysis of samples and the interpretation of the resulting sample data apply to all types of sampling challenges.

2.2 Systematic Planning

A well-thought-out, systematic approach to the development of a sampling and analysis plan (SAP) is an important first step to ensure that sample data collected are scientifically defensible and appropriate for final decision-making. Systematic planning applies to all types of environmental investigations, regardless of the media involved and the contaminant in question. An SAP need not be overly complicated or excessively time-consuming or costly to prepare. It is important, however, that the SAP lays out the rationale for the collection of samples and specifies decisions to be made once the data are obtained.

Step 1: Define Investigation Scope and Preliminary Conceptual Site Model

The first step in ensuring that sample data are reflective of investigation objectives is preparation of a well-researched, conceptual site model (CSM). A CSM summarises site conditions based on a review of the past and current land use and available environmental data. This can include site walkthroughs and interviews with past or current employees or with other people familiar with operations at the site. A review of past chemical use and known or suspected releases is used to identify contaminants of potential concern (COPCs) and associated potential environmental concerns. Example potential environmental concerns include:

- direct exposure of human and/or ecological receptors to contaminants in soil, water or air via ingestion, dermal contact and/or inhalation;
- potential leaching of contaminants from soil and adverse impacts to nearby surface water or underlying groundwater;
- potential emission of vapours from contaminated soil or groundwater into overlying buildings and adverse impacts to indoor air;
- potential uptake of contaminants into food crops from contaminated agricultural soil or irrigation water;
- potential adverse impacts to aquatic habitats due to natural or induced discharges of contaminated groundwater into a surface water body.

This resulting information and CSM are used to guide subsequent steps of the systematic planning process.

Step 2: Specify Contaminants of Potential Concern and Investigation Questions

Identified COPCs and potential environmental concerns are used to develop "investigation questions" related to subsequent risk assessment and optimise remedial actions. These questions will assist in the designation of specific areas of the site for sample collection. Example investigation questions directly related to risk include the following:

- Does the mean concentration of Contaminant X in the specified area and volume of soil (water, air, food, etc.) exceed a corresponding Risk-Based Screening Level Y or a target chronic risk level for direct exposure of current or future site users?
- Does the mean concentration of Contaminant X in the specified area and volume of soil exceed a corresponding Risk-Based Screening Level Y for potential leaching and adverse impacts to underlying groundwater?
- Does the mean concentration of Contaminant X in the volume of air flowing through the room during the defined exposure period exceed a corresponding Risk-Based Screening Level Y or a target chronic risk level for direct exposure of current or future building users?

- Does the mean concentration of Contaminant X in the volume of vapours within a high-conductivity fill immediately beneath the slab of the building exceed a corresponding Risk-Based Screening Level Y for potential vapour intrusion risk to indoor air?
- Does the mean concentration of Contaminant X in groundwater to be pumped from the well over the exposure duration of the targeted receptor (e.g., a young child) exceed a corresponding Risk-Based Screening Level Y for direct exposure?

Example questions related to remediation include the following:

- What are the lateral and vertical boundaries of contamination above the levels of potential concern?
- Where is the main mass of the contaminant located?
- What is the mass of the contaminant in the area and volume of soil targeted for *in situ* remediation?

The latter set of remediation questions are particularly important when the remediation goal is based on a targeted reduction in the mass of contaminant present (e.g., 80%) rather than cleanup to meet a purely risk-based goal.

Step 3: Identify Data Needs

A common factor for assessment of risk noted in Step 2 is a consideration of the mean contaminant concentration in the media of concern throughout the assumed exposure duration (e.g., six years for a developing child and the assessment of noncancer risk and 30 years for a combined child and adult exposure and the assessment of cancer risk). Risk is always assessed based on the mean concentration of a contaminant for a designated exposure area and volume of soil as a whole. Sample collection must therefore be designed to directly provide the mean contamination concentration for a targeted DU area and volume of targeted media.

Estimation of the mean concentration of a contaminant for a designated DU is accomplished under TOS by ensuring that a single sample is collected in a statistically representative manner in the field. Additionally, statistical manipulation of sample data is unnecessary. This contrasts with the need for statistical manipulation of grab sample data points that might or might not be collectively representative of the DU area and volume of media as a whole. Uncertainty in the latter led to recommendations for calculation and use of a 95% upper confidence level (UCL) of the mean in early USEPA guidance (USEPA, 1989a). Such methods only denote uncertainty in the statistical test employed to estimate a mean for the data provided, however. The error in the sample data itself is unknowable. Recommendations to use a 95% UCL for decision-making do, however, recognise the intent of early USEPA guidance to base decisions on "populations" of soil—referred to in this chapter as "decision units"—rather than on data for single grab sample points. This intent was unfortunately lost, and practitioners as well as regulatory agency staff began to focus on individual data points.

Data, in addition to standard laboratory analytical methods, might be required in some cases (HIDOH, 2023, 2024). For example, assessment of direct exposure risk to contaminants such as arsenic or lead in soil might require additional testing of samples for bioaccessibility if an initial screening level is exceeded. Non-volatile, soluble contaminants, such as nickel salts and chlorinated herbicides, can pose leaching problems and cause contamination of groundwater. Assessment of this concern might require additional soil leaching tests and/or collection of groundwater data. Volatile and soluble contaminants, such as petroleum hydrocarbons and chlorinated solvents, can pose vapour intrusion risks to the indoor air of existing or future overlying buildings. Subslab and/or deeper subsurface soil vapour data are normally warranted. The additional collection of indoor air data might be considered if subslab vapour data exceeds a risk-based screening for potential vapour intrusion risk. Data that can be used to estimate the specific mass of a COPC in soil will be necessary for the design of *in situ* remedial actions.

Step 4: Designate Decision Units

Addressing data requirements to determine the mean concentration of a COPC in a targeted media requires specification of the area and volume of media in question. This is referred to as a "decision unit." A DU can be thought of as the area and volume of a media that you would send to the laboratory for testing as a single sample, if you could. This is usually not practical, so a representative sample of the DU must be collected instead (discussed in Step 6).

Designation of well-thought-out DUs in consideration of the investigation questions prepared in Step 2 is critical for successful completion of a project on time and on budget. Examples of DU designations for different media under different site scenarios are provided in Section 2.3.

There are three basic categories of DUs: exposure area DUs, source area DUs and boundary DUs. DUs are designated in a manner that allows the project investigation question(s) to be directly answered. In most cases, these questions address the desire to assess risk or optimise remediation. The number, locations, sizes and shapes of DUs are designated based on a combined consideration of the CSM and the investigation question(s).

For example, an assessment of direct-exposure risk for young children playing on a soccer field suspected to be contaminated with lead might involve the designation of the entire area of the soccer field to a depth of six inches as a single exposure area DU. Known or suspected areas of known or suspected heavy contamination that could cause a larger exposure area DU to fail, sometimes referred to as "hot spots," are normally isolated and designated as source area DUs for separate testing to help optimise potential remedial actions (CRCCARE, 2024; see also USEPA, 2014). Source area DUs are often surrounded laterally and/or vertically with strategically placed boundary DUs in order to refine the delineation of heavily contaminated volumes of media that might require remediation.

In another example, the physical volume of air circulating through a room over a 24 h period might be designated as a single DU for testing. In the same manner, the water in a water tank slated for use as tap water or even a risk-based volume of groundwater extracted from a well (e.g., volume of water consumed by a developing young child over six years) might be designated as a single DU for testing and comparison with risk-based screening levels.

The designation of DUs is necessarily site-specific, and these are but a few of many potential schemes. The above discussions focus on characterisation of a media in order to assess potential chronic risk. Assessment of acute exposure or potential health effects within a matter of seconds or days due to exposure to very small DU volumes of heavily contaminated media is, in theory, possible. In practice, the extent of sampling necessary to carry out such an assessment with a reasonable degree of confidence, coupled with the lack of acute toxicity factors for most contaminants, negates the feasibility of this type of assessment (refer to CRCCARE, 2024). If a strong possibility of an acute toxicity risk exists, then remedial actions should be taken to eliminate exposure to the subject media.

Step 5: Develop Decision Statements

The investigation questions developed in Step 2 are used to prepare decision statements for each DU. The decision statements specify actions to be taken once sample data are received and reviewed. Example decision statements include the following:

- If the mean concentration of COPC X in the designated DU volume of soil exceeds Screening Level Y, then the entire volume of the DU will be excavated and disposed of.
- If the mean concentration of total arsenic in the designated DU volume of soil exceeds Screening Level Y, then the sample will also be tested for bioaccessible arsenic. If the concentration of bioaccessible arsenic exceeds Screening Level Z, then the entire volume of the DU will be excavated and disposed of.
- If the mean concentration of COPC X in the designated DU volume of indoor air during normal operation of the building exceeds Screening Level Y, then the floor of the building will be sealed to mitigate intruding subsurface vapours.
- If the mean concentration of COPC X in the designated DU volume of groundwater extracted from the well exceeds Screening Level Y, then the well will not be used as a source of drinking water in the absence of treatment.
- If the mean concentration of COPC X in the designated DU volume of soil unexpectedly exceeds the Screening Level Y by less than 25%, then the DU will be immediately resampled as two smaller DUs to determine if disposal of a portion of the DU can be avoided.

Preparing highly specific decision statements for each DU at the beginning of the project and before samples are collected ensures proper planning up front and expedites the final completion of a project. Decision statements should be agreed upon by all stakeholders in the investigation, including risk assessors, remediation experts and the overseeing regulatory agency. The time allotted for the completion of the project and the budget should be considered. Decision statements should be strictly followed when the sample data are obtained.

DUs are ideally designated at the scale necessary for final decision-making at the start, to the extent practicable for the project in question.

Subdivision of initially designated DUs for resampling after data are obtained and determined to exceed screening levels should be minimised and, in many cases, completely avoided with well-thought-out, advanced planning that considers both assessment of risk and optimisation of remedial actions. This will help provide a clear endpoint for a project and avoid time delays and cost overruns. Follow-up subdivision of a site into smaller DUs might be unavoidable in some cases. The need for remobilisation and retesting of sites will decrease with increasing experience of the site investigation team.

The presence of localised areas of heavy contamination within an exposure area DU is oftentimes unknown, however. Investigation in such cases begins with sizing DUs based on risk/receptor scenarios. If risk is such that remediation is required and is cost-beneficial to do so, then smaller DUs can be used to optimise the remediation of the site. If, however, heavily impacted source areas are suspected or known to be present at the beginning of the investigation, then initial isolation and testing of the areas as separate DUs is normally most cost-beneficial.

In other cases, the responsible party might only be willing to initially fund an investigation that focuses on the use of relatively large exposure area DUs to assess risk. Responsible parties should be made aware of potential project completion delays if this approach is followed and weigh the pros and cons.

Step 6: Develop and Implement a Sampling and Analysis Plan

A SAP is prepared and implemented only after completion of Steps 1–5 of the systematic planning process. Brief overviews of SAP preparation and sample processing and testing are provided in Section 2.4. More detailed information is provided in CRCCARE (2024). Characterisation of a targeted DU area and volume of material is carried out by collecting multiple "increments" of the media in a probabilistic manner and combining them into a single Multi Increment® (MI)-type sample (Section 2.4.7). ("Multi Increment" is the registered trademark of EnviroStat, Inc.) A detailed discussion of the scientific basis of MI sample collection methods is provided in Volume 2 of CRCCARE (2024).

Note that existing grab sample data can be useful for initial identification of COPCs and designation of DUs. MI sample data should always be used for final decision-making, however. The use of grab sample data is discussed in more detail in Section 2.6.

Step 7: Evaluate Data Quality/Usability

Following the collection of sample data, each of the preceding steps is reviewed to evaluate the quality and usability for consistency with the workplan and applicability to the original investigation questions (Section 2.5 and Appendix 2A). This includes reviews of DU designation in the field and final methods used to collect, store, ship, process and test samples. Deviations from the SAP due to accessibility or other issues are documented in the sampling report. Limitations on the resulting data (or a lack of data) are noted and considered in the evaluation. Multiple "replicate" samples are collected and tested from a portion of DUs in the field (e.g., 10%) to test the overall precision of the sample collection, processing and testing methods. Multiple subsamples are collected from a portion of submitted samples and tested at the laboratory to assess the precision of the sample processing and analysis. A review of field and laboratory replicate sample data is carried out to evaluate the precision of the overall sampling method and identify errors that might require retesting or even recollection of samples.

Step 8: Identify Potential Environmental Risks

If sample data are determined to be acceptable, then data are compared with pre-established environmental screening values or used to quantify risk in a more detailed risk assessment. All potential environmental concerns compiled in Steps 1–3 of the planning process must be addressed.

Step 9: Refine CSM and Make Recommendations

The final step of the systematic planning process involves revision of the CSM to reflect the data collected and environmental concerns identified. The revision aims to identify any data gaps and determine the scope of work required to complete the site investigation based on the investigation questions and the corresponding decision statements for each DU established in the initial stages of the project. Recommendations for additional actions are then prepared when the investigation is determined to be complete and acceptable.

2.3 Decision Unit Designation

A DU can be thought of as the total volume of media that would ideally be sent to the laboratory for testing and for which a decision is to be made. The locations, areas and volumes of DUs are designated in coordination with risk assessment and remedial experts. DUs are designated to directly reflect the investigation question being asked (Figure 2.2). DUs to assess risk are referred to as exposure area DUs or, simply, exposure DUs. DUs designated to optimise remediation are referred to as source area and boundary DUs (Figure 2.3). These DUs often reflect the division of a larger, initial exposure area DU into smaller DUs in order to provide a better resolution of the extent and magnitude of suspected contamination (refer to Figure 2.3).

An extensive overview of DU designation and configuration for soil and sediment is provided by CRCCARE (2024). Example designations of DUs for the characterisation of contaminated air, subsurface vapours, surface water and groundwater are provided in HIDOH (2023). The designation of DUs for testing of indoor air and confined bodies of water is relatively well established, as previously noted. The designation of DUs for testing of soil, sediment and other particulate media is a relatively new concept in the environmental industry but is gaining adherents. The concept of designating DUs for testing of media such as subsurface vapours and groundwater is even newer. Doing so, however, is an important and necessary step in the successful completion of a risk-based and reliable site investigation.

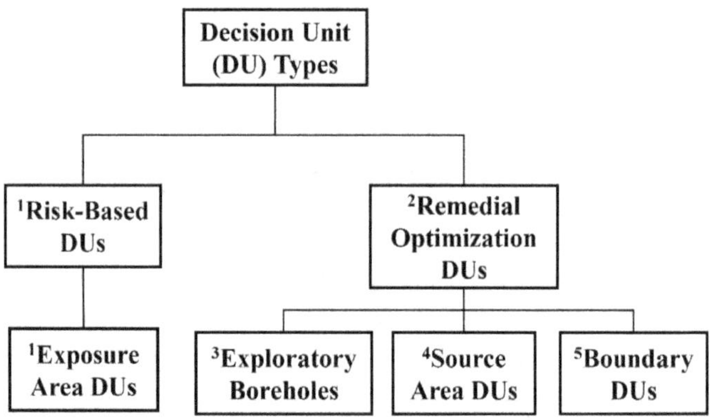

Figure 2.2. Categories of decision units (DUs) used in environmental investigations: (1) exposure area DUs used to directly assess risk to human or ecological receptors; (2) remediation optimisation DUs used to isolate areas of suspected heavy contamination that might require remediation; (3) exploratory borehole DUs based on testing of single cores and used to initially approximate the lateral and/or vertical extent of subsurface contamination; (4) source area DUs designated for anticipated areas of heavy contamination; (5) boundary DUs used to confirm clean perimeters of source area DUs.

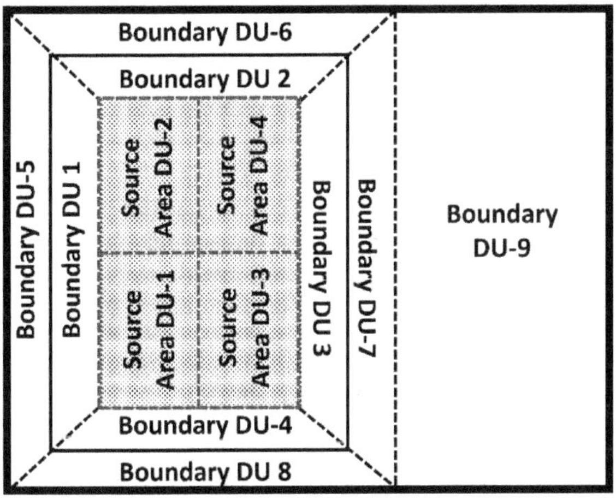

Figure 2.3. Subdivision of a suspected contaminated exposure area DU into source area DUs and boundary DUs for optimisation of remediation. Two rings of boundary DUs are normally established around source area DUs in order to avoid the need for remobilisation should the inner ring fail risk-based screening levels.

2.3.1 *Outdoor and indoor air*

Investigations of ambient air quality might focus on the air moving across a residential area that is downwind of a major highway or industrial area or downwind of a major fire. A potentially relevant investigation question may be the following: "Do the short-term (hours or weeks) or long-term (months or years) mean concentrations of contaminants in the air meet air quality standards appropriate for residential areas?" The volume of air passing through the breathing zone of the subject area over the designated time period represents the DU applicable to the question. The investigation question is answered by collecting and testing a sample representative of the DU volume of air as a whole. Only one primary sample is collected per DU, although replicate samples can be collected to test the overall precision of the sampling methods. This is discussed in more detail in Section 2.4.

In the case of indoor air, the investigation question might be the following: "Does the mean concentration of the targeted contaminant in the volume of air circulating through a subject room (or building) over a specified time period (e.g., single or multiple working hour days) meet the corresponding risk-based screening level?" The entire volume of air flowing through the subject room during the designated time period represents the exposure period DU.

Under certain scenarios, the investigation question could pertain to risks posed by a stagnant volume of air within a room or other enclosed space. Examples include the suspected entrapment of carbon dioxide within low outdoor or indoor areas or the accumulation of methane or other explosive gases within a room or utility vault. In such cases, the fixed volume of air currently within the targeted space represents the DU. If layering of the air and contaminant due to density differences is a concern, then each layer represents a different DU.

Smaller source area DUs of volumes of air representing shorter time intervals can be advantageous for source identification and the optimisation of remedial actions if initial data identify a potential health risk (HIDOH, 2023). Assume, for example, that testing of a risk-based, multi-day DU volume of air identifies a contaminant above the corresponding risk-based screening level. High-frequency, or "continuous," sampling of consecutive, very small source area DU volumes of indoor air over short time periods might, for example, identify specific points in time during the exposure period when high levels of the targeted contaminant are being emitted. This could be related to changes in the building ventilation system (e.g., reduced air exchange at night) or a periodic use of substances that emit the targeted chemicals of concern (e.g., use of a chemical during a specific phase of manufacturing, use of cosmetics in the mornings, or regularly scheduled cleaning). Subsequently, follow-up investigations can be carried out to identify and address the specific source of the contaminant and reduce long-term health risk to an acceptable level (e.g., increased ventilation or use of ventilation hoods).

2.3.2 *Subsurface vapours*

Soil and groundwater contaminated with volatile chemicals in fuels or solvents can pose a vapour intrusion risk to overlying buildings (USEPA, 2012). This risk can be assessed, in part, by characterising vapours immediately beneath the floor of the building. A corresponding investigation question is, "Does the mean concentration of Contaminant X in vapours immediately beneath the building slab and the targeted building area of interest exceed the risk-based screening level for potential vapour intrusion risk and impacts to indoor air?"

A corresponding, risk-based DU volume of vapours beneath the slab of a specified area of a building can be designated for testing based on the volume of vapours that could intrude a known or hypothetical gap in the floor over a specified time period. Such volumes can be predicted based on the indoor–subslab pressure differential, the area of the gap in the floor and the subject duration of exposure. For example, vapour entry rates based on heating and cooling of buildings in different climate zones are predicted to range from approximately 3,000 L/day per 100 m^2 of floor space for tropical climate zones, where windows are often left open or buildings are air-conditioned year-round, to 7,000 L/day per 100 m^2 of floor space for cold climate zones, where buildings are heated for much of the year (Brewer *et al.*, 2014). Vapour intrusion risk is assessed by collecting a continuous sample from a climate-specific volume of vapours purged from a known or hypothetical vapour entry point in the building floor—an approach referred to as "large-volume purge (LVP)" sampling (HIDOH, 2024). Consideration of a five-day vapour entry period to capture and represent heterogeneity within the subslab plume yields the corresponding subslab vapour DU volumes that range from 15,000 to 35,000 L. Vapour intrusion risk is evaluated either by comparing the LVP sample data with the risk-based subslab vapour screening levels appropriate for the climate zone where the building is located or by including the data in a vapour intrusion model specific to the building design and ventilation.

2.3.3 *Surface water*

DUs are designated for surface water bodies in a similar manner as carried out for air. If a lake is used as a source of drinking water, then the entire lake might be designated as a single exposure area DU. In this case, the corresponding investigation question is, "Does the volume of water in the lake as a whole meet drinking water standards?" If the water body is moving, such as a flowing river, then the upstream volume of water anticipated to pass by a designated point, for example, the intake of a drinking water system, over a specified time period might be designated as an exposure DU. These types of DUs are normally used to monitor trends in contaminant levels or other parameters over time. The area and volume of water associated with a specific aquatic habitat could be designated as an exposure area DU as part of an ecological risk assessment. In each case, a single, representative sample is collected from throughout the targeted volume of surface water and used to assess long-term human or ecological health risks.

In other cases, the DU in question might be represented by a specific depth interval of water. For example, the investigation question "Does the mean concentration of bacteria within the designated swimming area meet health safety standards?" or "Does the mean turbidity within a designated distance of a waterway construction area exceed regulatory limits?" might be applied to only the upper few feet of the water body. If a localised area of heavy contamination is suspected, for example a portion of a river directly impacted by an outfall of industrial wastewater, then that area and volume of water can be isolated and tested separately as a source area DU. The time period to be represented by these types of time-integrated samples, including monitoring of industrial discharges of wastewater, is often specified in government regulations.

2.3.4 *Groundwater*

The designation of risk- or remediation-based DUs to help guide the investigation of contaminated groundwater has only recently begun to be developed but is similar in concept to the investigation of more easily accessible surface water (HIDOH, 2023). An example of a risk-based investigation question is the following: "Will the mean concentration of Contaminant X in water pumped from this well over Exposure Time Period Y meet the applicable drinking water standard?" Collection of a sample representative of the entire period of potential exposure will oftentimes not be practicable; however, testing of samples from risk-based volumes of water reflective of shorter but still risk-based time periods normally is. The corresponding DU volume of water for testing could, for example, be designated as the volume of water anticipated to be ingested by a young child over an exposure period of six years. For example, the default drinking water ingestion rate of 0.78 L per day for 350 days a year over a six-year exposure duration equates to an exposure DU volume of 1,638 L (USEPA, 2024). As discussed in Section 2.4, this volume of water could be purged from a current or hypothetical drinking water well, and a subsample representative of the entire DU volume tested for COPCs. The resulting data are used to monitor contaminant trends over time.

If contamination of groundwater is already known or suspected to pose a potential human health or environmental risk, then a remediation-based question might be, "Are localised zones of contamination present that can be targeted for focused treatment?" In this case, subdivision of the targeted interval of groundwater for a more detailed characterisation is appropriate. For example, more permeable units that supply most of the water to a well could be designated as individual DUs for separate sample collection in order to determine if the plume is migrating through a specific interval. Less permeable units that might serve as reservoirs for sorbed contaminants could be similarly isolated for separate sample collection.

Such concepts are fairly common in groundwater investigations. Reference to specific volumes of water produced from a well over a hypothetical time period or specific intervals of the groundwater column as DUs is a relatively new concept, however. As discussed in Section 2.4.6, utilising this concept plays an important role in how samples that are directly representative of the investigation question being asked are collected.

2.3.5 *Soil and other particulate matter*

DUs of soil, sediment, biosolids and other infinite-element particulate media (discussed in Section 2.4.2) are, by default, initially designated for assessment of risk. Smaller, higher-resolution source area DUs are, however, often designated in order to isolate suspected areas of heavy contamination and optimise the design of remedial actions. Designation of source area and associated boundary area DUs in the initial stages of a project will avoid the need to subdivide a large exposure area DUs that fail screening levels and are belatedly determined to be too large and costly to remediate as a whole. DUs should always be designated at a scale appropriate for final decision-making when practicable. Subdivision and retesting of initial DUs add time and cost to a project and can often be avoided by proper planning at the beginning of a project.

For example, an exposure area DU might include all of the upper 10 cm of soil in a 500 m^2 playground for a total of 50 m^3 of soil. The corresponding investigation question might be, "Does the mean concentration of the targeted contaminant in the designated playground DU volume of soil meet the corresponding risk-based screening level?" The question is answered by collecting and testing a representative sample of the DU area and volume of soil as a whole.

If a localised area of significant contamination is known or suspected, an exposure area DU can be subdivided into one or more source area DUs for testing up front in order to optimise anticipated remediation (refer to Figure 2.3). In the case of soil, one or more source DUs are established to isolate heavy contamination. The designation of the DU locations and volumes takes into consideration the optimal resolution of data needed to most cost-effectively remediate the site. Source area DUs are typically surrounded by relatively small boundary DUs in which the targeted volume of soil is anticipated to meet risk-based screening criteria (refer to Figure 2.3).

Particle size is an important part of DU designation for particulate matter. Direct exposure to contaminants in soil is typically assessed based on the concentration of a contaminant in the <2 mm or finer fraction—the fraction documented to be incidentally ingested by young children as well as adults (USEPA, 2011). This requires the removal of larger particles by sieving as part of sample processing. The remaining material represents the DU. Assessment of potential leaching concerns might require inclusion of larger particles. If the target particle size of concern does not exist in the subject area or volume of material, for example a pile of gravel or boulders with no soil-size or smaller particles, then there is no DU, and a sample is not collected.

DUs for characterisation of sediment are designated with similar concepts in mind. The area and volume of sediment encompassing the active benthic zone within an aquatic habitat might be designated as an ecological exposure area DU. Multiple source area DUs might be designated in an active harbour to isolate and remediate areas of suspected heavy contamination. Relatively large DUs might be designated for due diligence characterisation of sand deposits anticipated to be clean and destined to be dredged and reused as fill material or for beach replenishment.

2.4 Sample Collection Methods

2.4.1 *Sampling correctness*

Characterisation of a designated DU volume of soil, sediment, air and water in each of the above examples would, under ideal circumstances, be carried out by collecting the entire volume of media in question and submitting it to a laboratory for testing as a single "sample." This of course will not be practical in most cases, and a sample that is representative of the DU as a whole must instead be collected. Done properly, the sample can defensively be said to be representative of the area and volume of the DU media in question, and the subsample collected for actual analysis at the laboratory can defensively be said to be representative of the sample provided.

Each step of the sampling and subsampling process introduces errors into the final data. Errors associated with testing of subsamples collected from particulate media such as soil can be quite large, as can errors associated with testing of air for aerosols or suspended particulate matter or with testing of water for bacteria. Employing the correct method for collection of samples in the field and collection of subsamples at the laboratory for testing is therefore critical.

2.4.2 *Sampling of finite- versus infinite-element media*

All media can be placed in one of two categories, depending on the investigation question being asked and the nature of the media itself: "finite-element" media and "infinite-element" media (AAFCO, 2015). Finite-element media are those in which a defined group of "elements" will be individually identified and randomly selected for testing. Examples include a community of people being studied for chronic diseases or bags of rice or bottles of water to be sold on the open market. Each person, bag or bottle represents an "element" within a defined DU of elements that is to be characterised. Common investigation objectives, or "data quality objectives" (DQOs), include assessment of the variability of target parameters between individual elements, estimates of mean, minimum or maximum parameter values or other defining or quality control parameters (e.g., age, weight, purity and defects). A sample of the media is prepared by collecting and independently testing individual elements. Each element can be thought of as a "specimen" of the group of media as a whole. This is sometimes referred to as "survey sampling." Individual elements are sometimes referred to as "discrete" or "grab" samples, although the collection of elements as a whole more appropriately represents the "sample." The only variable in the design of the SAP is the number of elements to be selected.

Infinite-element media, in contrast, are defined as media in which individual elements cannot realistically be independently identified and/or randomly selected and tested. Examples include a silo of rice grains, a stockpile of soil, a lake of water and the volume of air that circulates through a room during a specified time period. Each particle in the case of solids or molecule in the case of liquids and air represents a single-element DU volume of media designated for testing. The DQO in most cases is the determination of the mean or "true" value of a particular parameter of interest for the DU volume of media as a whole. Addressing the investigation question requires the collection of a sample that is the corresponding representative of the entire DU.

Most environmental investigations involve the need to collect and test samples of infinite-element media such as soil, water, air and even food. The collection of a representative sample of infinite-element media requires more advanced planning and can be challenging compared to testing of finite-element media. Variables that need to be considered include the number of increments to be collected, the physical method used for collection, the volume or mass of the final sample and the volume or mass ultimately tested by the laboratory. For particulate matter such as soil, the latter may require the consideration of an additional set of parameters such as the size, shape and density of individual grains and even the anticipated concentration of the contaminant in the media.

A precursory overview of risk-based, sample collection methods that can be used to characterise DU media is provided in the following. Refer to the references included in the specific sections for details of the methods described.

2.4.3 *Sampling of indoor and outdoor air*

The collection of indoor and outdoor air samples is common to investigations of potential health effects associated with emissions from industrial manufacturing facilities, combustion engines and the intrusion of subsurface vapours into overlying buildings, among other point and nonpoint sources of volatile chemicals. Assessment of risks associated with contaminants in air typically considers exposure over time periods ranging from a few hours to a few days for potential acute health risks to several months or years for potential chronic health risks. Assessment of the type of risk being assessed requires the collection of a time-integrated sample that is representative of the temporal (distributional) heterogeneity of the contaminant in air over the targeted exposure duration. In practice, this typically involves the collection of a sample during the operational hours of a business (e.g., 8 h) and 24 h or even several days for testing of air inside of a home or in outdoor areas of a residential neighbourhood where continuous exposure to contaminants in the air could occur.

The collection of risk-based, time-integrated samples is well established for investigation of indoor and outdoor air contamination.

Sampling strategies can be grouped into three categories: whole air sampling, sorbent tube sampling, and passive sampling using sorbent materials (HIDOH, 2023). These sampling methods assume that a representative volume of circulating air will pass by and be captured by the collection device. Whole air sampling involves the collection of air from a sample container of fixed volume, such as a Summa canister. The sample is collected by placing the container within the designated DU area (e.g., breathing zone of a room or outside area where exposure might occur) and allowing a continuous flow of air to be drawn into the container during the targeted exposure period. This is achieved by using a flow controller for Summa canisters or a pump to draw a continuous flow of air through a sorbent tube. Passive sampling methods rely on the placement of a sorbent material within the targeted DU. As the name implies, air is allowed to passively pass over the sorbent material throughout the designated exposure period.

At the laboratory, a sample of the air is drawn from the Summa and directly analysed for targeted chemicals. For sorbent tubes, the absorbent material is removed and extracted. The mass of the targeted chemical is then measured and divided by the volume of air drawn through the tube to generate concentration. In the case of passive sorbent samples, the measured mass of the contaminant sorbed is converted into a mean concentration in air based on the total mass sorbed divided by the sampling duration times a compound-specific uptake rate (McAlary *et al.*, 2014; USN, 2015).

Done properly, data based on these sample collection methods can reliably be assumed to be representative of very large volumes of air most pertinent to the assessment of long-term health risks. The assessment of acute risks or risks associated with a fixed DU volume of stagnant air can, in contrast, typically be made by collecting a small-volume grab sample from a random location within the DU. This is most often done with a portable gas detection device specific to the targeted type of vapour or risk.

2.4.4 *Sampling of surface water*

The design and implementation of an investigation of a surface water body adhere to similar concepts utilised for the collection of risk-based samples of air. The method used to collect the sample must ensure that all parts of the designated DU volume of water have an equal chance of inclusion in the sample.

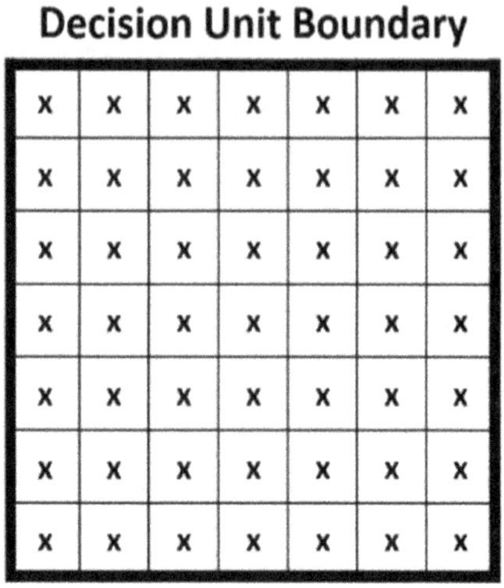

Figure 2.4. Systematic random collection of sample increments.

Non-isokinetic devices are used to collect samples from slow-moving or stagnant water. A sample is prepared by collecting "increments" of water in a systematic random manner throughout the designated DU (Figure 2.4). An increment of water can be collected by slowly pushing down a bottle with two holes in the cap through the targeted depth interval of the DU (HIDOH, 2023). Increments are combined to prepare a single "Multiple Increment," or "Multi Increment," sample. Other types of tools include dippers, Van Dorn samplers and Niskin bottles.

Isokinetic samplers are used to collect representative samples from fast-moving water (HIDOH, 2023). These devices are designed to address changes in flow velocities at different depths as a sample is collected. The sampler(s) are slowly lowered through the targeted water column depth as the water flows past. The water collected from each level of the lowering of the device represents an increment. Increments of water are collected throughout the targeted DU, either as the water moves past a stationary collection point or by physically moving the device(s) to different areas of the DU. Individual increments are again combined to prepare a single Multi Increment sample.

Collection and combination of 30 increments of water will, in most cases, yield a reasonably representative sample. A more precise estimate of the number of increments necessary to capture the distributional heterogeneity of a contaminant within the DU and prepare a representative sample can otherwise be determined by comparing replicate samples collected in the same manner but from different locations within the DU (refer to Section 2.6). A fewer number of increments might be acceptable for the investigation of dissolved-phase contaminants. A greater number of increments might be required to generate reproducible data in the case of particulates or other solid-phase contaminants (e.g., bacteria).

2.4.5 *Sampling of subslab vapours*

The assessment of long-term vapour intrusion risk requires the collection of subslab vapour sample(s) that are representative of the vapours predicted to intrude a specific area of the building slab over a corresponding risk-based period of time (refer to Section 2.3.2). This is most reliably carried out by the collection of LVP vapour samples (HIDOH, 2023; see also McAlary *et al.*, 2010). The volume of subslab vapours to be collected reflects the DU volume of vapours predicted to intrude a specific point in the building floor during the specific exposure period. The collection of a sample is carried out by installing a vapour monitoring point at the designated point in the building slab. For non-explosive vapours, a vacuum pump is used to purge the designated DU volume of vapours from the monitoring point. A continuous sample of the vapour stream is drawn into a Summa canister or through a sorbent tube. Purging is ceased when the target purge volume of vapours is reached. The sampling device is sealed and submitted for testing in the same manner as carried out for samples of indoor or outdoor air. Unlike grab samples, the resulting data are risk-based and directly comparable to vapour intrusion screening levels or used to model potential impacts to indoor air.

Multiple, strategically placed LVP samples can be used to identify and isolate high-concentration areas of vapour plumes under building slabs. A dense grid of small-volume grab samples or passive samples can also be useful for identifying source areas and optimising remediation. Such samples are less reliable for the assessment of vapour intrusion risk, however, due to the small volume of vapours represented. Small numbers of small-volume grab ("discrete") subslab vapour samples are especially unreliable for the estimation of indoor air subslab vapour attenuation factors (Brewer *et al.*, 2014). This is due to both the distributional heterogeneity of vapour plumes under building slabs and the oftentimes lack of knowledge of actual vapour entry points. Errors in the use of such data to assess risk or predict vapour intrusion attenuation factors are random and unknowable.

2.4.6 *Sampling of groundwater*

Characterisation of contaminated groundwater has traditionally been carried out by the collection and testing of individual, small-volume grab samples from monitoring wells. This practice assumes that the concentration of a contaminant in error in the resulting sample data in terms of the investigation question(s) being asked could be small in the case of dissolved-phase contaminants but in practice is largely unknown. Random variability of data among grab samples collected over time could reflect an inherent, random distributional heterogeneity of contaminants in the water. The distributional heterogeneity of contaminant concentrations in groundwater will be controlled by the physiochemical nature and heterogeneity of soil, sediment or bedrock holding the groundwater as well as the nature of the release itself (e.g., pulsed versus steady release over time).

The use of sampling methods directly relatable to the investigation question assigned to the DU will help reduce uncertainty in the data and associated errors in decision-making. LVP sampling methods similar to those employed for testing of vapours under slabs offer a more risk-based approach to the collection of samples if the investigation question pertains to the risk posed by using the groundwater for drinking. A sample could be continuously drawn from a risk-based volume of water (e.g., 1,628 L) purged from a monitoring well installed in groundwater that is considered a source of drinking water. In this sense, the monitoring well is being treated as a hypothetical water supply well.

Sample collection methods more appropriate for optimisation of remedial actions are suitable if contamination of groundwater above levels of potential concern is already known or predicted. In these cases, the collection and testing of samples from one or more targeted intervals of groundwater within the well and representative of the surrounding formations is appropriate. Sample collection devices such as passive diffusion bags, HydraSleeves and Snap Samplers can be used to collect samples across the entire length of the targeted depth intervals of groundwater (HIDOH, 2024). Passive samplers in particular offer time-integrated data that consider flow conditions within smaller-scale zones of the strata under investigation.

Although often recommended to minimise the volume of purge water generated, low-flow sampling methods have the highest potential for error in sample data and decision-making since the water is drawn from only a single point within a well. Such methods risk underestimation of the true concentration of contaminants in targeted intervals of groundwater. The combined use of LVP and grab sample methods representative of targeted intervals rather than single points will provide more reliable data for assessment of risk and optimisation of remedial actions.

2.4.7 *Sampling of soil, sediment and other particulate media*

2.4.7.1 *Gy's theory of sampling*

Methods specifically developed for testing of particulate media such as soil and sediment are described under Pierre Gy's TOS (Pitard, 2005, 2009, 2019; Ramsey and Hewitt, 2005; Minnitt, 2007; Esbensen, 2020). The collection of a representative sample of such material requires consideration of a number of different factors unique to infinite-element, particulate matter. Sampling experts in the mining, recycling, agricultural and pharmaceutical industries have spent decades carrying out statistical analysis of sample data and developing strict protocols for sample collection, processing and testing to ensure that the resulting sample data will be representative of the DU volume of material in question. These industries require highly accurate estimation of the mean concentration and, ultimately, mass of a targeted commodity in a designated DU volume of material. This might include, for example, the mean concentration and mass of gold in a large stockpile of crushed ore that is to be sold to a mineral extraction company. Errors in sample data will ultimately be known through extraction and direct measurement of the commodity from the host material in the case of mining and recycling industries or from decreased yields of crops and decreased efficacy of a drug in the agricultural and pharmaceutical industries.

A brief overview of the key aspects of Gy's TOS is provided in the following sections. The concepts of "compositional heterogeneity" and "distributional heterogeneity" guide the collection of samples in the field and subsamples at the laboratory. The concept of "fundamental error" (FE) guides the minimum mass of the sample and, in particular, the subsample that must be collected. Expanded discussions of these topics are provided in CRCCARE (2024).

2.4.7.2 *Compositional and distributional heterogeneity*

Sample data errors associated with the collection, processing and testing of soil or other particulate media can be controlled by understanding and accounting for factors that cause these errors. This begins with an understanding of "compositional" and "distributional" heterogeneity (Figure 2.5). Compositional heterogeneity refers to variability between individual particles. Heterogeneity is described in terms of the variability of contaminant concentration among individual particles as well as the size, shape and density of the particles themselves. This includes consideration of whether the contaminant is present as a coating on the particle (e.g., arsenic coating grains of iron hydroxide), diffused within the particle (e.g., polychlorinated biphenyls infused in balls of tar) or "liberated" and present as individual nuggets of pure material (e.g., small pieces of lead). Compositional heterogeneity is controlled by ensuring that the sample and, in particular, the subsample collected are of sufficient mass to address the FE.

Distributional heterogeneity refers to the variability of the contaminant within the area of an investigation as a whole, within each individually designated DU volume of material, within the sample collected and even within the subsample collected for actual testing. Large-scale distributional heterogeneity and the need to assess risk and/or optimise remediation for specific areas are addressed through the designation of DUs for separate testing within a project area. Distributional heterogeneity within a DU is addressed by the collection and combination of a sufficient number of increments of the media to prepare a sample.

(a) (b)

Figure 2.5. Example of compositional heterogeneity between individual particles (a) and distributional heterogeneity of particle types within a targeted volume of material (b).

A related misunderstanding in environmental risk assessment involves the concept of "maximum" concentration of a contaminant in soil. Consider the above examples of compositional heterogeneity. If a contaminant is present in soil, then at some scale—the coating on a grain of iron hydroxide or in the form of a pure nugget—the "maximum" concentration of the contaminant will always be one million parts per million, or 100% (Brewer *et al.*, 2017a, 2017b). This is irrelevant with respect to risk assessment, as is the concentration of a contaminant reported for a subsample collected from a grab sample. Risk is always assessed in terms of the *mean* contaminant concentration for the DU area and volume of material designated for testing.

2.4.7.3 *Fundamental error and minimum sample mass*

Gy and other sampling statisticians in the mining industry reviewed decades of data for testing of crushed ore and developed an equation to quantify what he referred to as the "fundamental error" (Pittard, 2005, 2009, 2019; Minnitt, 2007). This equation relates the error induced by compositional heterogeneity to total sample mass. The result is used to determine the minimum mass of a sample required to address compositional heterogeneity. This is used to determine the minimum mass of the subsample to be collected and tested by the laboratory.

Calculation of the FE is most commonly expressed as (Minnitt, 2007)

$$\sigma^2 = \frac{fgcld_N^3}{M_s}, \tag{2.1}$$

where σ^2 is variance or error; f, g, c and l represent the "shape," "granulometric," "mineralogical composition" and "liberation" factors, which quantitatively describe the variability of particle characteristics; d_N is the nominal maximum particle size and M_s is subsample mass. Particle characteristics can often be estimated and combined as a single constant "K" for specific types of ore. Then, the equation becomes simplified as (Minnitt, 2007)

$$\sigma^2 = \frac{Kd_N^3}{M_s}. \tag{2.2}$$

Rearranging Equation (2.2) allows the calculation of the minimum mass of a sample, or, more specifically, the mass of the laboratory subsample required to control the FE at a specified particle size and confidence level:

$$M_s = \frac{Kd_N^3}{\sigma^2}. \tag{2.3}$$

In general, sample collection, processing and testing should be carried out in a manner that will reduce the FE to a maximum of 15% (see USEPA, 2003; Minnitt, 2007). The mass of a subsample required to meet the specified level of the FE is directly proportional to the cube of the particle size. Decreasing the particle size via grinding or crushing leads to a substantial decrease in the FE for a given mass. An alternative is to collect and test a larger subsample mass. Milling of samples is routine in the mining and metal recycling industries, where near 100% extraction of the commodity is targeted.

This solved one aspect of the minimum sample mass required to represent a targeted lot of crushed ore. Gy's research demonstrated that relatively small sample masses could, in theory, be representative of very large volumes of material, provided that the particulate media were crushed to a small enough particle size. Note that in the mining industry, grain size is often measured in centimetres rather than millimetres. This necessitates the collection of very large initial subsamples for testing and grinding of the subsamples to reduce the mass ultimately tested by the laboratory.

2.4.7.4 *Collection of Multi Increment samples in the field*

The FE dictates that a sample mass as small as a few grams could in theory be representative of the mean concentration of a contaminant in a designated DU volume of soil or other particulate material. The only remaining factor might then incorrectly be assumed to be the number of increments to be included in the sample. Field research, however, would quickly indicate that the physical collection of increments in the field was itself a significant source of data error. Addressing this error would require the collection of much larger masses of particulate media, at least 1–3 kg, to obtain a reliably representative sample (CRCCARE, 2024).

A sample of particulate matter is prepared by combining a large number of increments (30–100) from within the target DU (CRCCARE, 2024). Field experience has demonstrated that a default of approximately 50 increments per sample will generate reproducible and reliable data under most field scenarios. As few as 30 increments per sample is often adequate for soil impacted by water-borne contaminants. As many as 75 increments or more per sample are necessary for soil that contains high-concentration "nuggets" of contaminants, such as lead shot or PCB-infused tar or caulking. The collection of more than 100 increments per sample is unlikely to improve sample data quality. Thirty-increment samples are often collected for subsurface soils due to time and expense as well as access limitations. More rigorous sampling can be carried out in the future if the soil is exposed to the surface.

The initial collection of multiple sets of triplicate samples using different numbers of increments (e.g., 30/50/75) and subsequent comparison of data quality can be beneficial for large, complex sites. The results can be used to identify the optimal number of increments for efficient and reliable completion of the investigation. Fewer than 30 increments per sample could be adequate in some scenarios but would need to be demonstrated in comparison to replicate sets of samples consisting of greater numbers of increments. This is unlikely to be cost-beneficial for most environmental investigations.

Each field increment is made up of an equal mass of material (typically 30–50 g) sufficient to produce a final, 1–3 kg sample. Larger bulk masses will necessarily improve the sample representativeness but can pose logistical problems for the laboratory. The increments are combined to form a single, Multi Increment sample for the DU.

A representative sample is most reliably prepared by the collection of increments in a systematic random (grid) manner (see Figure 2.4). Collection points are determined by hypothetically dividing the DU into small, square cells equal in number to the target number of increments. A single increment is collected from the centre point of each cell. The spacing between individual increments, represented by the dimensions of the increment collection cell, is calculated using the following equation:

$$\text{Increment Spacing} = \sqrt{\frac{\text{DU Area}}{\#\text{of Increment Points}}}. \tag{2.4}$$

This simulates the division of the DU into individual increment collection cells with a single, core-shaped increment collected from the centre of each cell across the full thickness of the DU. The spacing is the same regardless of the shape of the DU.

The method used to collect an increment depends on the type and depth of the targeted material (HIDOH, 2023). Simple hand tools such as stainless-steel sampling tubes are effective for shallow, relatively uncompacted soils (Figure 2.6). Independent, replicate samples, normally triplicates (primary sample plus two replicates), are collected from a portion of the DUs (e.g., 10%) and used to test the overall precision of the sample collection method.

Drills, push rigs and even excavators might be required for the collection of sample increments from compact or subsurface soils (Figure 2.7). Each interval of core extracted using a push rig represents a single increment for the targeted DU layer. The entire core increment from each borehole would ideally be combined to prepare a single, Multi Increment sample, as would be done for surface samples. The size of cores collected from push rigs normally requires that a smaller subsample of each increment be collected and combined in the field to prepare a manageable Multi Increment sample (Figure 2.8). The need to subsample DU increments adds error to the representativeness of the sample submitted for analysis. The magnitude of subsampling error is evaluated by testing the precision of data for independent, replicate sets of core increment subsamples from select DU layers (CRCCARE, 2024).

(a) (b)

Figure 2.6. (a) Collection of soil increments from a DU using a stainless steel, open-face sampling tube. (b) Increments are placed in the same container and combined to prepare a single, Multi Increment sample.

(a) (b)

Figure 2.7. Collection of Multi Increment samples from subsurface DU layers: (a) designation of multiple DU layers for the collection of Multi Increment samples; (b) direct-push rigs are often used to collect "core increments" in a systematic random fashion from subsurface DUs, mimicking the approach used to collect surface samples.

(a) (b)

Figure 2.8. Subsampling DU layer increments from borehole cores: (a) DU layers identified in core and subsampled for preparation of a single, Multi Increment sample for the layer; (b) sample to be tested for volatile organic compounds prepared by placing DU layer increment subsample plugs in jar with pre-measured volume of methanol.

Samples to be tested for non-volatile or semi-volatile chemicals can normally be placed in food-grade freezer bags. Increments for samples to be tested for volatile chemicals are ideally placed in an amber bottle with a pre-filled volume of methanol equal to the anticipated volume of soil (refer to Figure 2.8). Alternative methods include rapid freezing of increments and submittal to the laboratory for extraction in methanol. The risk to human health posed by volatile chemicals can alternatively be assessed based on testing of soil vapour samples (HIDOH, 2023).

2.4.7.5 *Processing of Multi Increment samples*

Processing of Multi Increment samples is normally carried out within a fixed laboratory specifically set up to work with these types of samples. Field processing of samples beyond removal of large particles can introduce additional errors in the sample data and should only be carried out when facilities are available to dry and sieve the samples. If required due to limitations of the laboratory, then field processing of samples should be carried out by personnel well trained and experienced in Gy's TOS and methods used to ensure subsamples collected for testing by a laboratory are defensibly representative of the original sample collected.

Multi Increment samples to be tested for non-volatile chemicals are air-dried and sieved to isolate the targeted DU particle size, typically ≤2 mm or smaller (HIDOH, 2023). Additional milling of the sample to reduce the particle size is carried out based on the investigation question being asked and the need to better control the FE. CRCCARE (2024) provides additional discussion of subsample mass based on the maximum particle size of the particulate media and whether or not the sample should be milled.

A subsample of the processed material is then collected and analysed, ideally using a sectoral splitter (Figure 2.9). If a sectoral splitter is not available, then a subsample is collected by moistening the sample (which prevents separation of fines), spreading the material into a thin layer and using a flat-bottomed and square-sided scoop to collect small increments in a systematic random manner (refer to Figure 2.9). This mimics the method used to collect the sample in the field and helps to ensure that the subsample collected is representative of the sample provided.

Grinding a sample to ≤75 microns using a puck mill can, in theory, allow subsamples as small as 1 g to be collected and tested in order to address the FE. However, increment collection error, more specifically termed increment delimitation error or increment extraction error, typically requires the collection of at least 10 g of material, particularly if a manual subsample collection method is used (CRCCARE, 2024). A minimum subsample mass of 10 g is recommended if the sample is not ground but a sectoral splitter is used to collect the subsample. A minimum subsample mass of up to 30 g is required to address the FE if the sample is not ground and a manual subsample collection method is used, particularly if high-density contaminants such as lead are present. Increased testing of laboratory replicate (triplicate) samples should be carried out

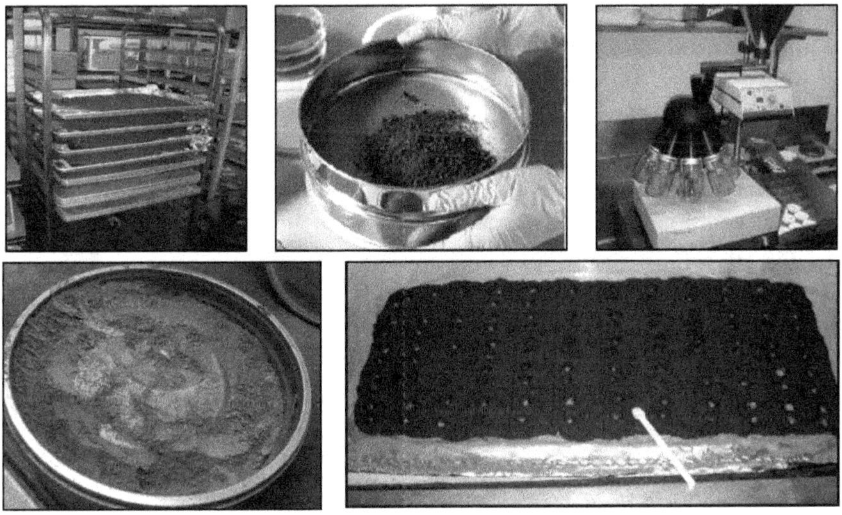

Figure 2.9. Collection of an analytical subsample from an MI sample after air drying and sieving: (top left) samples air dried on baker's trays; (top centre) sample sieved to target particle size; (top right) sectoral splitter used to collect analytical subsample (preferred); (bottom left) sample ground prior to collection of analytical subsample; (bottom right) manual collection of an analytical subsample from an unground and moistened, 2D slab cake.

to better assess the precision of the subsample collection method. Extractions of smaller subsamples can also be combined in some cases in order to generate data representing a larger, total subsample mass. Final methods to process and test samples are documented in the summary report prepared for the investigation.

Milling of ore in order to reduce the FE and better estimate the mean concentration and mass of a commodity present is routine in the mining and metal recycling industries. Milling of soil and other particulate samples for assessment of health risk is more problematic, as this can increase the natural leachable concentration of the contaminant and overestimate the risk to human health and the environment (refer to CRCCARE, 2024). In either case, testing for bioaccessibility should be considered if available for the targeted contaminant. Bioaccessibility tests are normally carried out on the fines fraction of unground samples of soil (e.g., $<100\ \mu m$).

Note that subsample masses recommended in laboratory standard operating procedures (SOPs) typically reflect the minimum mass required for use of the specific analytical method in question rather than the mass required to obtain a reliable subsample of the material. For example, Chapter Three of the USEPA's Solid Waste Compendium 846 for testing of inorganic analytes recommends a minimum subsample (digestion) mass of only 2 g or less for metals (USEPA, 2018). As recognised in earlier USEPA guidance (e.g., USEPA, 2003), this is far too small to address compositional heterogeneity and the FE unless the sample has been milled and the subsample collected using a sector splitter.

The unreliability of the resulting data defeats the effort made to collect a representative sample in the field. Laboratories can normally accommodate testing of larger subsamples but must be notified ahead of time and prior to the collection of samples in the field. If a larger subsample mass cannot be extracted, then the sample should be milled, and a sectoral splitter should be used to collect test portions.

2.5 Evaluation of Data Usability

An overview of guidelines to review the reliability and useability of environmental sample data is provided in Appendix 2A (after CRCCARE, 2024). Each step in the process of sample collection, processing and testing is reviewed. The evaluation begins with a review of the rationale used to designate DUs (Figure 2A.1). Methods used to designate DUs and collect samples in the field are then scrutinised, followed by a review of methods used to process, subsample and test samples at the laboratory. An evaluation of data precision and reproducibility is then carried out based on replicate sample data for DUs, subsurface core increments and laboratory subsamples, as appropriate for the media tested.

The checklist provided in Table 2A.1 can be used as an aid in the evaluation of sample collection in the field and during processing and testing at the laboratory. A review of the analytical errors associated with the method and equipment used to test a sample, referred to as "data validation," is the last step in the data quality evaluation process (USEPA, 2002). Although an important part of data validation, analytical error is rarely the cause of poor sample data reliability. The greatest risk of data error and inappropriate decision-making occurs in the field, not in the laboratory (Hadley and Petrisor, 2013).

The precision of the overall sample collection, processing and testing methods utilised is evaluated based on a comparison of replicate data from samples collected from the same DU (refer to CRCCARE, 2024).

Statistical evaluation of replicate sample data involves a two-step procedure. The first step is to calculate the relative standard deviation (RSD) of the contaminant concentration for the triplicate dataset. The RSD reflects the precision of the total sampling method, including field and laboratory errors. The lower the RSD, the more precise the sampling method used and the more reproducible and reliable the data for DUs where replicate samples were not collected.

The primary source of error in data quality can be assessed by carefully reviewing replicate data for field versus laboratory replicate samples (Table 2A.2). An RSD of <15% is desirable for laboratory subsample replicate data (Section 2.4.7.3), although a higher RSD might be unavoidable for analytical methods with an inherent poor precision. An RSD of ≤35% for field sample replicates and field subsample replicates generally indicates acceptable precision. An RSD of ≤50% is generally acceptable for environmental purposes, given the magnitude of safety margins typically built into risk-based screening levels and risk assessments in general. A review of methods that could be used to improve data quality in future investigations is warranted, however.

Replicate sample RSDs of greater than 50% require a closer scrutiny of the overall sampling method and the potential need to retest samples at the laboratory or even recollect samples in the field. Poor precision for field sample replicates but high precision for laboratory replicates indicates a problem in sample collection in the field. This can normally be addressed by collecting a greater number of increments per sample and a large, final sample mass. Poor precision of laboratory subsample replicates can be addressed by using a sectoral splitter, collecting and testing larger subsample masses and/or milling the sample to a smaller particle size.

Elevated RSDs become more common as the laboratory method detection level is approached and can be unavoidable. An elevated RSD for replicating sample data that are well below applicable screening levels or target risks might also be acceptable. These issues should be discussed with a risk assessor trained and experienced in TOS and DU-MIS investigation methods.

2.6 Related Sampling Guidance

Additional technical documents related to risk-based sampling strategies are referenced in CRCCARE (2024). Decades of research carried out by the mining industry is only now beginning to be discovered and utilised in the environmental industry. Guidance published by the State of Hawai'i (HIDOH, 2024) and modified for use in Australia (CRCCARE, 2024) is the most detailed and field-tested guidance currently available.

The United States-based Interstate Technology and Regulatory Council (ITRC) published related guidance referred to as "Incremental Sampling Methodology," or ISM, in 2010, which was updated in 2020 (ITRC, 2020). A critique of the 2020 update of the guidance was published by HIDOH (2020). The guidance, which focuses on characterisation of soil and sediment, provides a basic overview of the limitations of grab (discrete) sampling methods and implementation of ISM in the field and the laboratory. The fundamentals of Gy's TOS are incorporated into the field and laboratory sections of the document. The recommended default collection of 30-increment samples will be inadequate to generate reliably reproducible replicate sample data under many field scenarios, however. Field experience by the authors and discussions with other field workers suggest that a default of 50 increments per sample will provide more consistently reproducible and representative data and minimise the need for remobilisation and recollection of samples in the field due to failed, replicate samples (refer to Section 2.4.7.4). It is possible that fewer increments might be adequate to characterise soil or other particulate matter in some cases, for example the assessment of natural background concentrations of heavy metals in soil. That would, however, require a detailed field study of the optimal number of increments for the specific site in accordance with Gy's TOS and is unlikely to be cost-beneficial under most project scenarios.

The concept of subdividing DUs into smaller "sampling units" presented in the ITRC guidance for "screening" and independent testing of smaller areas of an initially designated DU is unnecessary. These are more appropriately classified as source area DUs (refer to Figure 2.2). A common mistake is the inappropriate combination and testing of small numbers of increments from different areas of a DU and use of the data to determine if separate areas of a DU are "clean" or "dirty." This fails to recognise inherent, small-scale distributional heterogeneity within subareas of a DU and is likely to result in an underestimate of the true mean concentration of a contaminant for the soil. The default number of sample increments applicable to different contaminant release scenarios applies to all volumes of soil in question, whether the initial DU is to be characterised in the field or the sample is to be subsampled and tested by the laboratory.

The current edition of the ITRC document unfortunately lacks a thorough review and summary of the statistical underpinnings of Gy's TOS and the basis of Multi Increment-type sampling methods. The guidance instead reverts to inferential statistics that are only appropriate for the characterisation of finite element media. Such methods were utilised in the past by risk assessors in an attempt to make sense of environmental grab sample data—the same methods that decades of experience in the mining and other commodity industries demonstrated to be unreliable and not applicable to testing of infinite element particulate matter, such as soil.

The risk assessment section of the guidance mistakenly implies that replicate samples must be collected in every DU and that a 95% UCL of the mean must be used for all decision-making. This confuses past attempts to use a 95% UCL to address variability *between individual points* within a DU and estimate a mean for a single set of grab sample data. Under Gy's TOS, the statistical underpinning of the sample collection method itself is used to capture and represent contaminant heterogeneity within a targeted DU volume of particulate material. The equivalent of requiring the use of a 95% UCL for replicate Multi Increment samples in terms of a traditional grab sampling method would be the calculation of a 95% UCL based on independent replicate sets of grab samples collected from a DU (ignoring the fact that independent sets of grab sample data are unlikely to meet sample collection, processing and testing requirements in general). Such requirements have never been recommended or required in environmental risk assessment guidance.

Consideration of a 95% UCL based on replicate Multi Increment sample data could, however, be warranted in some instances. Examples include cases where there is a minimal safety margin in exposure assumptions or risk-based screening levels being applied and significant, regular exposure of sensitive receptors (e.g., young children) is anticipated. In such cases, statistical manipulation of sample data should only be carried out by risk assessors well trained in Gy's TOS and methods to control sample data error in the field and in the laboratory.

2.7 Discussion

Risk-based sampling strategies such as those introduced in this chapter are sorely needed in the environmental industry in order to reduce the time and effort required to reliably characterise, assess and remediate contaminated land and other environmental media. Environmental regulatory agencies have lagged behind the equally sampling-intensive mining, agriculture, recycling and pharmaceutical industries in part due to the lack of third-party confirmation of sample data quality but also due to the until-recent lack of detailed, field-tested guidance.

The methods described can significantly improve the defensibility and reliability of sample data collected for environmental media and help set a clear endpoint for remedial actions.

No single sampling strategy is applicable to all sites. Sampling strategies can be tailored to specific sites based on the project needs, the investigation questions being asked, the experience of the field workers and the need to control both costs and the time allotted for completion of the project.

2A Appendix: Data Quality Review

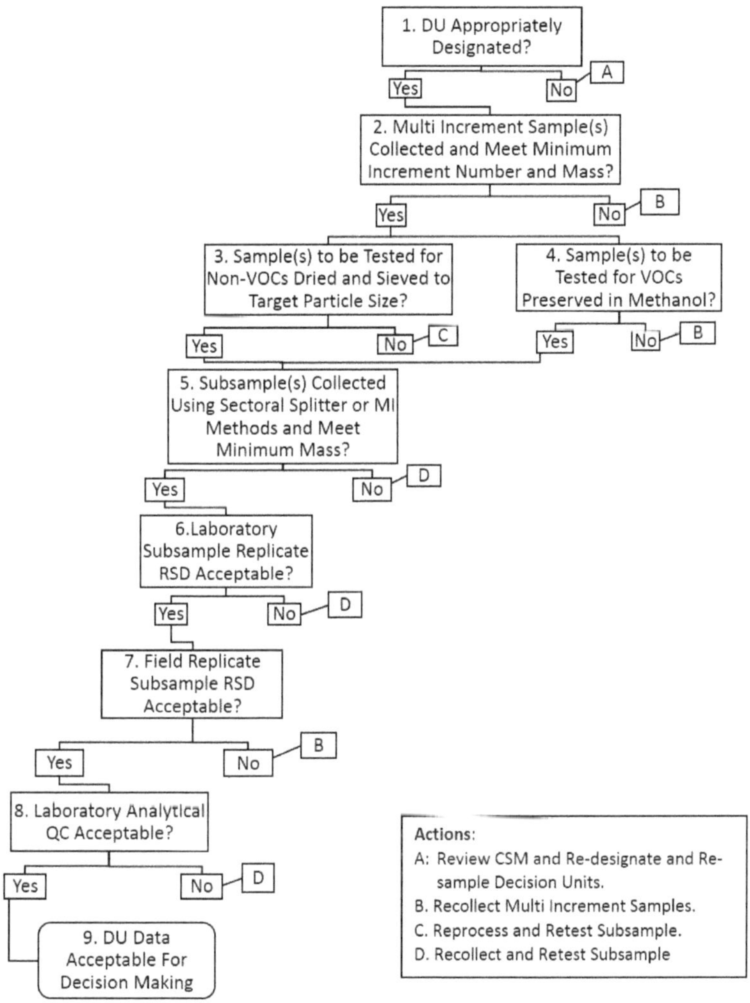

Figure 2A.1. Decision unit (DU) data quality evaluation process.

Table 2A.1. Sample data quality and usability checklist.

Acceptable	Site Investigation Stage
	Conceptual Site Model and Decision Unit (DU) Designation
	• Site history and potential sources and type of contamination well understood?
	• To-scale map depicting location and size of DUs provided?
	• Site investigation questions used to designate DUs for testing clearly stated and based on risk and/or optimisation of anticipated remediation requirements?
	• Questions and decision statements developed for individual DUs presented?
	• Area and total volume of soil associated with each DU noted and acceptable for intended purposes?
	Field Sample Collection
	• Summary of sample collection methods provided, including approximate final mass of each sample and the number of increments included in each sample?
	• Samples prepared by collecting and combining a minimum number of increments appropriate for anticipated types of contaminants?
	• Increments appropriately spaced and collected?
	• Complete, unobstructed access to all portions of the DU soil available for sample collection?
	• Core-shaped increments collected?
	• Samples to be tested for volatile chemicals preserved in methanol in the field or otherwise met requirements for alternative preservation and testing methods?
	• Minimum sample mass of 1–3 kg met (minimum 300 g for samples to be tested for volatile contaminants)?
	• Triplicate Multi Increment samples collected and tested from at least 10% of DUs to test total data precision (minimum 1 set per project)?
	Laboratory Processing and Testing
	• Appropriate methods used to process samples and collect subsamples for analysis employed and documented in a report prepared by laboratory?
	• Samples to be tested for non-volatile chemicals air-dried and sieved to target particle size for each specific DU?
	• Analytical subsample collected using a sectoral splitter or manually collected from at least 30 points?

Table 2A.1. (*Continued*)

Acceptable	Site Investigation Stage
	• Minimum 30 g analytical subsample mass extracted for <2 mm particle size soil?
	• Minimum 10 g analytical subsample mass extracted for <250 μm particle size soil?
	• Triplicate analytical subsamples collected and tested from at least 10% of samples submitted (minimum 1 set)?
	• Sample holding times met?
	• Analytical quality control and quality assessment criteria met (e.g., spikes and blanks)?
	Replicate Sample Collection and Data Precision Evaluation
	• Replicate field sample and laboratory subsample data meet data precision requirements?
	• Source of error for replicate data that exceed an RSD of 35% determined?
	• Excessive field error identified and samples recollected or data adjusted for decision-making purposes?
	• Excessive laboratory subsampling error (<35%) identified and subsamples recollected and tested after grinding and/or larger subsample mass collected?
	Identification and Correction of Field and/or Laboratory Errors
	• Source of sample collection, processing and/or testing methods identified?
	• Need for recollection and/or retesting of samples determined?

Table 2A.2. Recommendations for assessment of data quality based on the relative standard deviation of replicate samples.

Replicate Sample Data Precision	Use of DU Data for Decision-Making
Good (RSD ≤ 35%)	• Data for DUs where replicate samples were not collected can be assumed to be representative without adjustment. • Compare unadjusted MIS data directly with target screening values (use the arithmetic mean of replicate sample data). • Collection of follow-up confirmation samples for DUs, where remedial action is necessary, *is not required* if data for boundary DUs meet target screening levels.

(*Continued*)

Table 2A.2. (*Continued*)

Replicate Sample Data Precision	Use of DU Data for Decision-Making
Moderate **(35% < RSD ≤ 50%)**	• Data for DUs where replicate samples were not collected have lower confidence but are adequate for comparison to screening levels or use in a risk assessment without adjustment. • Review and discuss sampling methods and laboratory processing and analysis methods and summarise potential sources of error in reports for future reference (e.g., inadequate increment collection methods, insufficient number of increments, or inadequate laboratory processing). • Compare unadjusted MIS data directly with target screening values (use the arithmetic mean of replicate sample data). • Collection of more reliable follow-up confirmation samples for DUs, where remedial action is necessary, *is required* even if data for boundary DUs meet target screening levels (e.g., number of increments and total sample mass increased or laboratory processing steps improved).

Poor
(50% < RSD ≤ 100%)
	• Data for DUs where replicate samples were not collected are not reliably representative of the DU mean.
	• Review and discuss field sampling methods and laboratory processing and summarise potential sources of error in reports for future reference.
	• If the majority of the total error is due to subsampling or analysis in the laboratory (less likely), require the laboratory to reprocess and retest the samples, including milling of samples, if necessary, with additional replicate subsamples collected and tested to reassess precision.
	• If replicate sample data precision is still poor, either consider retesting affected DUs using samples with a greater number of increments and total bulk mass or, if determined acceptable by a risk assessor trained in MI sampling methods:
	  ○ for DUs with replicate sample data, compare the highest reported concentration of the contaminant to the screening or cleanup level;
	  ○ for DUs without replicate sample data, adjust the reported contaminant concentration upwards by the RSD calculated for the DU with replicate sample data.

Table 2A.2. *(Continued)*

Replicate Sample Data Precision	Use of DU Data for Decision-Making
	• Additional evidence of data acceptance (or rejection) should be provided for decision-making purposes, including site history and potential for contamination above the level of concern, adequacy of methods used in collecting, processing and analysing samples, closeness of data to screening levels, safety margins built into the screening levels and other information as available and pertinent. • Collection of additional confirmation sampling in DUs, where remedial action is necessary, is required, using samples with a greater number of increments and total bulk mass and the collection of replicate samples.
Very Poor (RSD > 100%)	• Data for all DUs are not reliably representative of the DU mean, including data for DUs where replicate samples were collected. • If the majority of the total error is due to subsampling or (less likely) analysis in the laboratory, require the laboratory to reprocess and retest the samples, including milling of samples, if necessary, with additional replicate subsamples collected and tested to reassess precision. • Review and discuss field sampling methods and laboratory processing and analysis methods and summarise potential sources of error in reports for future reference. • Retesting is not required for DUs where the need for remediation is already clear from the data and other field evidence. • Consider the collection of new samples in DUs using the following approach: (a) If known, designate suspected source areas as separate DUs for individual characterisation; (b) collect a minimum of 75 increments per sample; (c) ensure a minimum 2–3 kg final sample mass; (d) collect replicate samples in all anticipated high-concentration and high-risk DUs. • As an alternative, consult with a risk assessor trained in MI sampling methods regarding the safety level incorporated into the target screening level or cleanup level and the need to resample high exposure risk areas (e.g., all sample data an order of magnitude or more below screening levels).

(Continued)

Table 2A.2. (*Continued*)

Replicate Sample Data Precision	Use of DU Data for Decision-Making
	• Additional evidence of data acceptance (or rejection) should be provided for decision-making purposes, including site history and potential for contamination above the level of concern, adequacy of methods used in collecting, processing and analysing samples, closeness of data to screening levels, safety margins built into the screening levels and other information as available and pertinent. • Collect replicate confirmation samples in all DUs requiring remediation.

References

AAFCO. (2015 October). *GOODSamples: Guidance on Obtaining Defensible Samples*. Association of American Feed Control Officials, Champaign, Illinois. http://www.aafco.org/Portals/0/SiteContent/Publications/GOODSamples.pdf.

AAFCO. (2018 June). *Good Test Portions: Guidance on Obtaining Defensible Test Portions*. Association of American Feed Control Officials, Champaign, Illinois. https://www.aafco.org/Portals/0/SiteContent/Publications/GoodTP_final_web.pdf?v3.

ASTM. (1995). *Standard Guide for Risk-Based Corrective Action Applied at Petroleum Release Sites*. American Society for Testing and Materials, E1739-95.

Brewer, R., Nagashima, J., Rigby, M., Schmidt, M., and O'Neill, H. (2014). Estimation of generic subslab attenuation factors for vapor intrusion investigations. *Groundwater Monitoring & Remediation*, 34, 79–92. DOI: 10.1111/gwmr.12086.

Brewer, R., Peard, J., and Heskett, M. (2017a). A critical review of discrete soil sample reliability, Part 1 – Field study results. *Soil and Sediment Contamination*, 26(1), 1–22.

Brewer, R., Peard, J., and Heskett, M. (2017b). A critical review of discrete soil sample reliability, Part 2 – Implications. *Soil and Sediment Contamination*, 26(1), 23–44.

Brewer, R., Ramsey, C., Heskett, M., and Song, J. (2022). Fake data? The need for theory of sampling concepts in environmental research and investigations. In *Proceedings of the 10th World Conference on Sampling and Blending*, May 31–June 2, 2022, Kristiansand, Norway, pp. 193–206.

CNSAC. (2023 March). *Soil Quality – Decision Unit and Multi Increment Sampling Methods*. China National Standardization Administration Committee, GB/T-42489-2023.

CRCCARE. (2024). *Risk-Based Investigation of Environmental Contamination.* In Brewer, R. and Song, J. (Eds.), *Cooperative Research Center for Contamination Assessment and Remediation of the Environment* (in press).

Esbensen, K. (2020). *Introduction to the Theory and Practice of Sampling.* IM Publications, Charlton Mill, United Kingdom.

Gilbert, R. O. (1987). *Statistical Methods for Environmental Pollution Monitoring.* New York: Van Nostrand Reinhold Company Inc.

Hadley, P. W. and Petrisor, I. G. (2013). Incremental sampling, challenges, and opportunities for environmental forensics. *Environmental Forensics*, 14, 109–120.

HIDOH. (1995 December 13). *Risk-Based Corrective Action and Decision Making at Sites with Contaminated Soil and Groundwater.* Hawai'i Department of Health, Office of Solid and Hazardous Waste.

HIDOH. (2020 November 23). *Critique and Use of ITRC Incremental Sampling Methodology Guidance Document in Hawai'i.* Hawai'i Department of Health, Office of Hazard Evaluation and Emergency Response.

HIDOH. (2024). *Technical Guidance Manual* (and updates). Hawai'i Department of Health, Office of Hazard Evaluation and Emergency Response, Honolulu, Hawaii.

ITRC. (2020 October). *Incremental Sampling Methodology.* Interstate Technology and Regulatory Council.

McAlary, T. A., Nicholson, P. J., Yik, L. K., Bertrand, D. M., and Thrupp, G. (2010). High purge volume sampling — A new paradigm for subslab soil gas monitoring. *Ground Water Monitoring and Remediation*, 30(2), 73–85.

McAlary, T. A., Wang, X., Unger, A., Groenevelt, H., and Górecki, T. (2014). Quantitative passive soil vapor sampling for VOCs– Part 1: Theory. *Environmental Science Processes Impacts*, 16(3), 482–490.

Pitard, F. F. (1989). *Pierre Gy's Sampling Theory and Sampling Practice.* Boca Raton, Florida: CRC Press, Inc.

Pitard, F. F. (2005). Sampling correctness — A comprehensive guideline. In *Sampling and Blending Conference*, Sunshine Coast, Queensland, Australia, May 9–12, 2005.

Pitard, F. F. (2009). Theoretical, practical and economic difficulties in sampling for trace constituents. In *Fourth World Conference on Sampling & Blending*, October 21–23, 2009. The Southern African Institute of Mining and Metallurgy.

Pitard, F. F. (2019). *Pierre Gy's Sampling Theory and Sampling Practice.* 3rd ed. Boca Raton, Florida, USA: CRC Press, Inc.

Ramsey, C. and Hewitt, A. (2005). A methodology for assessing sample representativeness. *Environmental Forensics*, 6, 71–75.

USEPA. (1985 August). *Verification of PCB Spill Cleanup by Sampling and Analysis*. Washington, DC: United States Environmental Protection Agency, Office of Toxic Substances, EPA-560/5-85·026.

USEPA. (1986 May). *Field Manual for Grid Sampling of PCB Spill Sites to Verify Cleanups*. United States Environmental Protection Agency, Office of Toxic Substances, EPA-560/5-86·017.

USPEA. (1987 March). *Data Quality Objectives for Remedial Response Activities*. United States Environmental Protection Agency, Office of Solid Waste and Emergency Response, EPA/540/G 87/003.

USEPA. (1988a April). *Superfund Exposure Assessment Manual*. United States Environmental Protection Agency, Office of Remedial Response, EPA/540/1-881001.

USEPA. (1989a February). *Methods of Evaluating the Attainment of Cleanup Standards, Volume 1: Soils and Solid Media*. United States Environmental Protection Agency, Office of Solid Waste and Emergency Response, EPA 230/02-89-042.

USEPA. (1989b March). *Soil Sampling Quality Assurance User's Guide*. United States Environmental Protection Agency, Environmental Monitoring Systems Laboratory, EPA/600/8-69/046.

USEPA. (1989c December). *Risk Assessment Guidance for Superfund, Volume I, Human Health Evaluation Manual (Part A)*. United States Environmental Protection Agency, Office of Policy, Planning, and Evaluation, EPA/540/1-89/002.

USEPA. (1989d March). *Risk Assessment Guidance for Superfund, Volume II, Environmental Evaluation Manual*. United States Environmental Protection Agency, Office of Policy, Planning, and Evaluation, EPA/540/1-89/001.

USEPA. (1991a December). *Human, Health Evaluation Manual, Part B: Development of Risk-based Preliminary Remediation Goals*. United States Environmental Protection Agency, Office of Solid Waste and Emergency Response, OSWER Directive 9285.7-0lB.

USEPA. (1991b December). *Guidance for Data Usability in Risk Assessment (Part A)*. United States Environmental Protection Agency, Office of Research and Development, EPA/540/R-92/003.

USEPA. (1992 July). *Preparation of Soil Sampling Protocols: Sampling Techniques and Strategies*. United States Environmental Protection Agency, Office of Research and Development, EPA/600/R-92/128.

USEPA. (1994). *Region IX Preliminary Remediation Goals (PRGs) Second Half 1994*. United States Environmental Protection Agency, Region IX, Technical Support Section, San Francisco.

USEPA. (1996a May). *Soil Screening Guidance: Technical Background Document.* United States Environmental Protection Agency, Office of Emergency and Remedial Response, Publication 9355.4-17A.

USEPA. (1996b July). *Soil Screening Guidance User's Guide.* U.S. Environmental Protection Agency, Office of Solid Waste and Emergency Response, Publication 9355.4-23.

USEPA. (1999 March). *Correct Sampling Using the Theories of Pierre Gy.* United States Environmental Protection Agency, Office of Research and Development.

USEPA. (2002 November). *Guidance on Environmental Data Verification and Data Validation.* United States Environmental Protection Agency, Office of Environmental Information, EPA/240/R-02/004.

USEPA. (2003). *Guidance for Obtaining Representative Laboratory Analytical Subsamples from Particulate Laboratory Samples.* United States Environmental Protection Agency, Office of Research and Development. November 2003, EPA/600/R-03/027.

USEPA. (2011 September). *Exposure Factors Handbook.* U.S. Environmental Protection Agency, Office of Research and Development, EPA/600/R-090/052F www.epa.gov.

USEPA. (2012). *Conceptual Model Scenarios for the Vapor Intrusion Pathway.* United States Environmental Protection Agency, Office of Solid Waste and Emergency Response, EPA 530-R-10-003.

USEPA. (2013 September). *ProUCL Version 5.0.00 User Guide.* United States Environmental Protection Agency, Office of Research and Development, EPA/600/R- 07/041.

USEPA. (2014 March). *Hot Spots: Incremental Sampling Methodology (ISM) FAQs.* United States Environmental Protection Agency, Superfund.

USEPA. (2018 December). *SW-846, Chapter 3, Inorganic Analytes.* United States Environmental Protection Agency.

USEPA. (2024 May). *Screening Levels for Chemical Contaminants.* United States Environmental Protection Agency. Prepared by Oak Ridge National Laboratories.

USN. (2015). *Passive Sampling for Vapor Intrusion Assessment.* United States Navy, Naval Facilities Engineering Command, TM-NAVFAC EXWC-EV-1503.

https://doi.org/10.1142/9781800617841_0003

Chapter 3

Soil Sampling and Characterisation of Contaminated Land: Assessing and Understanding Residual Behaviour of Glyphosate

Md Nuruzzaman[*,†], **Saifullah Omar Nasif**[*,†], **W. D. Nipuni. P. Welivitya**[*,†], **Md Abdul Halim**[*,†], **Francisca Munyeza**[*,†], **Alvin Lal**[*,†], **and Ravi Naidu**[*,†,‡]

[*]*Global Centre for Environmental Remediation (GCER), College of Engineering, Science and Environment, the University of Newcastle, Callaghan, NSW 2308, Australia*

[†]*The Cooperative Research Centre for Contamination Assessment and Remediation of the Environment (crcCARE), ATC Building, Callaghan, NSW 2308, Australia*

[‡]*ravi.naidu@crccare.com*

Abstract

Understanding the residual behaviour of pesticides in soil has always been an active field of research for reducing their potential risks to "One Health." Primarily, the goal is achieved through the determination of their residual concentrations and distributions, as well as their interactions with soil, which are influenced by different soil parameters.

Though the concepts of quality control and quality assurance have been established to support laboratory personnel in accomplishing a higher degree of transparency for analytical procedures, including minimising potential sources of error and standardising sample handling procedures, instrumental protocols and data analysis. Ultimately, the rate of non-conformance results is reduced. However, it is essential to ensure appropriate and accurate sampling and sample administration of soil to obtain accurate results. Of the various pesticides, glyphosate has drawn attention from the research community recently as the most widely used herbicide worldwide due to its unique properties. The current chapter aims to briefly review the often-neglected but crucial topics of sample collection, transportation, storage, processing and characterisation methods regarding glyphosate soil residues and put the issue into perspective for the future of glyphosate residue analysis. It also highlights that analysts should validate their sample processing approaches for the analytes/matrices of interest and encourages further studies on sampling and sample mass reduction to produce a test portion. Therefore, this would be helpful to planners, landscapers, herbicide users and stakeholders in the glyphosate application area.

3.1 Introduction

Modern agriculture is governed by pesticide use to control weeds, plant diseases and pests. Globally, over 4.1 million tonnes of pesticides are used each year, and their use has increased by more than 40% over the past two decades (Nasif *et al.*, 2024). A large quantity of applied pesticides is wasted into the soil and consequently dispersed through several processes, including volatilisation, wind erosion, leaching or runoff into the surrounding environment (Nuruzzaman *et al.*, 2016). The indiscriminate use of pesticides raised environmental concerns due to their increasing contamination levels of natural resources (Riedo *et al.*, 2021). For instance, persistent organic pesticides, e.g., DDT and dieldrin, applied to agricultural soils in the past, are a major concern. These chemicals were used directly on soil or applied to crops, contaminating the soil via runoff or spray drift. Similarly, arsenic used in cattle and sheep dip baths has contaminated soils around these facilities. Pesticide contamination may cause residues in stock or crops raised on affected land. Soil residue tests determine if pesticides are present and at what levels, i.e., to determine the nature and extent of contamination, so that the risk of pesticides to livestock or crops can be assessed and managed. The core purpose of managing chemical residues is to minimise risk to the human food chain.

Of the various pesticides, the application of glyphosate ($C_3H_8NO_5P$)-based herbicide formulations is a major concern worldwide due to their intensive and excessive use. Every year, more than 700 thousand metric tonnes of glyphosate are applied globally in both agricultural and non-agricultural settings (Acosta-Cortés *et al.*, 2019). Further, glyphosate persists for a long time (average half-life: 47 days) in soils. It also generates metabolites upon degradation, including aminomethylphosphonic acid (AMPA), which is more persistent in nature. Both chemicals can contaminate the surrounding environment and enter the food chain due to their persistence and the presence of residual concentrations in the ecosystem for a long time. Humans can be exposed to glyphosate directly during their occupations. Figure 3.1 illustrates the ways of human exposure to glyphosate. Thus, humans as well as the whole environment, i.e., "One Health," are impacted by glyphosate (Ferrante *et al.*, 2023).

Recently, the International Agency for Research on Cancer (IARC) has stated that glyphosate is likely carcinogenic to humans. In contrast, the world's leading regulatory bodies have claimed that it is not a carcinogen and poses no risk to human health when used according to the manufacturer-described label (CropLife, 2018). It is also evident that glyphosate in commercial formulations exceeds regulatory limits and is responsible for carcinogenic, hepatorenal, teratogenic and tumorigenic

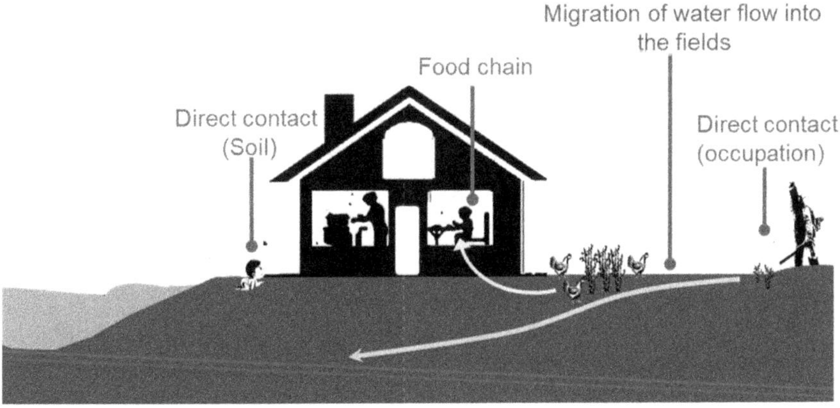

Figure 3.1. Routes of human exposure to glyphosate.

conditions, besides threatening endocrine disruption, metabolic alterations and oxidative stress. These scenarios depend on dose and exposure time (Mesnage *et al.*, 2015). Exposure to glyphosate was even linked to the formation of non-Hodgkin's lymphoma (NHL) in humans (IARC, 2015). So far, it has also become a subject of class-action lawsuits, with thousands of cases filed. Most of these cases claimed to have non-Hodgkin's lymphoma due to glyphosate exposure (Guardian, 2018). Recently, Maggi and coworkers analysed the global environmental contamination by glyphosate and its primary metabolite, AMPA. Worldwide, less than 0.1% of croplands face a high contamination hazard, but approximately 1% of croplands (385,000 km^2) are susceptible to medium-to-high hazards (Maggi *et al.*, 2020). To identify the potential threat of glyphosate and its metabolites, it is required to assess the site contamination.

Generally, soil contamination assessment is performed by collecting representative environmental samples for characterisation and chemical analysis. In this context, proper soil sampling is an important aspect because the assessment's validity is highly influenced by the sampling process, analysis method and data interpretation. However, a risk-based weight-of-evidence (WoE) approach should be considered when deciding whether remediation is required. When the contaminant is sourced from human applications, such as pesticides (including herbicides), the assessment of site contamination is performed for regular monitoring as well. In the case of glyphosate, the maximum residue level (MRL) and regulatory guidance value (RGV) in groundwater and soils can vary from region to region. For example, according to the US EPA's (2003) drinking and groundwater regulation, the Gly MRL is mentioned as 0.7 mg/L, and as per the Australian drinking water guidelines, the MRL of Gly is 1 mg/L (NHMRC, NRMMC, 2011). In Canada, the drinking water MRL of Gly is 0.28 mg/L, with no residues expected in groundwater (Canada, 2023). Similarly, RGVs of Gly in soils are significantly different across countries (Li and Jennings, 2017). Thus, purposeful monitoring of Gly residues and its metabolites in the area and vicinity of regular application is important for a risk-based assessment.

During site contamination assessment, it is also required to ensure that the methods of quantification are valid. Generally, quality control (QC) and quality assurance (QA) concepts are used during sample analysis to achieve a higher degree of transparency of procedures. These concepts minimise potential sources of error and standardise handling of samples, instruments and data during analysis. Ultimately, a decrease in the rate of non-conformance results. So far, several guidelines regarding the residual analysis of pesticides and method validation have been put into effect, such as SANTE (European, 2023), SANCO/825/00 and the Eurachem guideline (The Fitness for Purpose of Analytical Methods) (Magnusson and Örnemark, 2014). An analytical method developed for residual glyphosate and AMPA in soil should also comply with the criteria and validation parameters described in the guidance documents. However, soil sampling in glyphosate-contaminated sites poses distinctive challenges, including matrix interferences, extensive spatial fluctuations, localised heterogeneity within the matrix, variations over time and seasons within the target area and the relatively limited number of measurements compared to the media under evaluation. Consequently, approaches to evaluating site contamination necessitate both physical and numerical information to comprehensively characterise a glyphosate-contaminated site and its environmental context. It is required to ensure representative environmental samples for characterisation and to obtain accurate data about the soil quality of a specific site.

To this end, this chapter underlines soil sampling design and techniques, soil processing and characterisation of soil samples, with an emphasis on the analysis of glyphosate and its primary metabolite, AMPA, as potential contaminants. To get accurate data, progress in the methods of residual concentration analysis has also been demonstrated to ensure the quantification of its amount present in soil. Including this introduction section, the remainder of this chapter will provide a more detailed overview of challenges associated with soil sampling and processing (Section 3.2), characterisation of the soil samples (Section 3.3) and analysis of residual concentration of glyphosate and AMPA (Section 3.4).

3.2 Soil Sampling and Processing

Soil sampling is a fundamental practice in various scientific disciplines, including agriculture, environmental science and geology. This section aims to provide an overview of the key methods, considerations and applications associated with soil sampling, processing and storage, with a special emphasis on glyphosate-contaminated soils. The aim of representative soil sampling is to get accurate data about the soil quality at a specific site; however, its main goal is to fulfil the intended purpose of the sampling (Cline, 1944). In most cases, the purposes of soil sampling include establishing potential threats to public health or welfare or to the environment, identifying and locating sources of contaminants or contamination, defining the extent of contamination and making decisions about treatment or remediation options. Soil sampling requires designing separate sampling plans for monitoring environmental contaminants at different sites which involve several steps and principles.

3.2.1 *Site selection*

In most cases, identification or selection of sites for soil sampling is not required, as the purpose of a specific site investigation is pre-identified, for example accidental chemical spills, explosions, etc., or their intentional use, such as for constructing housing and parks. However, sometimes anthropogenic activities or other reasons may cause soil contamination, where investigation might be required at both the national and local levels to gather soil quality information. In that case, site selection is an important consideration (IAEA, 2004). In the case of glyphosate-contaminated land, the site is specified to be its applied area and surroundings. Glyphosate-based herbicides (GBHs) are applied in both agricultural and non-agricultural sectors (Figure 3.2). If GBHs are applied within non-agricultural settings, such as playgrounds, residential areas, parks, recreational facilities, facilities serving social, health and sports purposes, plots of land used for industrial and commercial purposes and unpaved areas within workplaces or manufacturing plants, regular monitoring is required to avoid the potential threat of direct contact. Further, areas for cultivating varying field crops, such as vegetables and field forage plants, as well as areas used for commercial gardening, vegetable gardens, backyard gardens, small gardens and any other garden areas used for growing food crops, and grasslands or permanent green spaces, are potential targeted sites.

3.2.2 *Conceptual site models*

A systematic plan is required for all site assessment, remediation and validation programmes. It is also required to develop sampling and analysis quality plans. For systematic planning, conceptual site models (CSMs) are used to provide a spatial and temporal overview of contamination at sites and their surroundings. The models highlight the sources of contaminants,

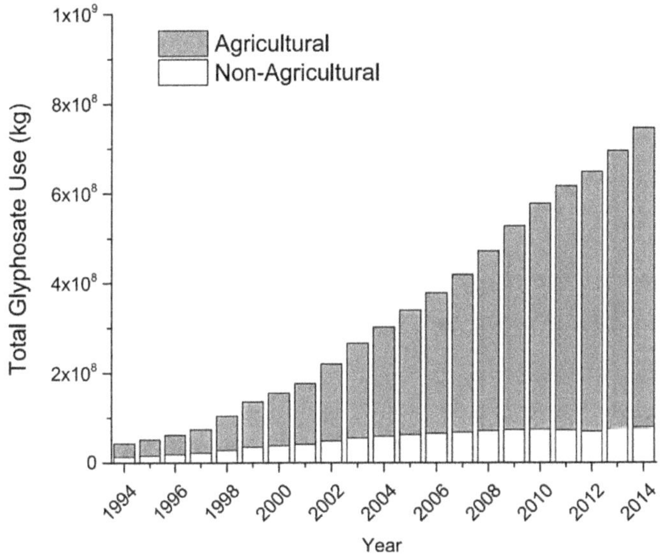

Figure 3.2. Global glyphosate usage for agricultural and non-agricultural purposes.
Source: Adapted from Benbrook (2016).

the receptors and the exposure pathways. The sources of the contaminants of concern often imply the mode of contamination, e.g., accidental spills and their subsurface release or placement of fills, as well as the distribution of the contaminant to affected media such as soil, sediment, groundwater, surface water and air both onsite and off-site. Whereas receptors include both humans and ecological living beings. In a robust CSM, all exposure pathways should be included. It should have the capability to explain the existing information, evidence and data logically from the area under study and be predictive (Byrnes, 2008). If the CSM has poor predictive capabilities, the supporting information and evidence should be reviewed to revise and update the CSM appropriately to identify any data gaps.

CSMs can be created in different formats, including text, tables, graphics and flow diagrams. Additionally, CSMs can take the form of site-specific plans and figures, such as cross-sections, offering a comprehensive presentation of the complex relationship between contaminants and the environmental matrix (Büyüker, 2009). Thus, they provide a structured framework for understanding the dynamics of contamination, ensuring a more informed and targeted approach to site investigations and remediation strategies. In terms of glyphosate contamination, the sources of contamination, the receptors and the exposure pathways could be predicted. The primary source is GBH application to control weeds, whereas secondary sources include diffused concentrations to the surroundings as well as metabolites generated due to degradation.

3.2.3 *Understanding sites' historical context and background information*

Being a key element in understanding soil test results, a site's historical context plays a pivotal role in soil sampling (IAEA, 2004). Soil holds a memory of its past, and previous land management practices can profoundly influence the present status. Historical land-use practices can significantly impact soil composition. Knowing the site's history helps identify past land-use patterns, such as agricultural practices, industrial activities or urban development, which may have left behind residues or contaminants. It is also important to have a clear understanding of past soil amendments and water practices, as they may alter the behaviour of contaminants of potential concern (CoPCs) or soil properties. For instance, the application of fertilisers, lime or organic matter can influence nutrient levels and overall soil fertility, while information on how water has been managed in the past helps in anticipating potential salinity or drainage issues.

More importantly, it is required to identify the potential sources of contamination during the investigation of a site's history. Once these sources are identified, a thorough assessment of CoPCs becomes essential (NSW EPA, 2022). This assessment includes the physicochemical properties of the CoPCs, such as their solubility in water, volatility, miscibility and interactions with various environmental factors. In the context of glyphosate contamination, it involves the development, adoption and widespread use of this herbicide, along with concerns about its environmental impact in both agricultural and non-agricultural settings (Singh *et al.*, 2024). Understanding the historical context of glyphosate-contaminated land involves the intentional application of formulations based on this herbicide over the past 50 years. Soil sampling in glyphosate-contaminated sites poses distinctive challenges, including matrix interferences, extensive spatial fluctuations, localised heterogeneity within the matrix, variations over time and seasons within the target area and the relatively limited number of measurements compared to the media under evaluation. Consequently, approaches to evaluating site contamination necessitate both physical and numerical information to comprehensively characterise a glyphosate-contaminated site and its environmental context.

Glyphosate was first discovered in 1960 by a chemist working for a Swiss company. Monsanto later patented glyphosate in 1970. Then, it introduced Roundup, a glyphosate-based herbicide, to the market in 1974. Its popularity grew rapidly due to its effectiveness in controlling a broad spectrum of weeds (Magin, 2003). The 1980s and 1990s saw the development and commercialisation of genetically modified (GM) crops, particularly Roundup Ready crops. These crops, engineered to be resistant to glyphosate, allowed for the widespread use of the herbicide in agriculture. The introduction of GM crops led to a significant increase in the use of glyphosate, as farmers could use Roundup without damaging their glyphosate-resistant crops (Bain *et al.*, 2017). As glyphosate use became more widespread, concerns emerged regarding its potential impact on the environment, non-target plants and human health. Studies and debates intensified regarding the safety of glyphosate. IARC classified glyphosate as a "probable human carcinogen" in 2015, sparking controversy and regulatory reviews (Benbrook, 2019).

Monsanto faced numerous lawsuits from individuals who claimed that exposure to Roundup caused their cancers. In some cases, substantial legal settlements were reached. Various countries and regions took different regulatory approaches. Some banned or restricted the use of glyphosate, while others maintained its approval with specific guidelines. In 2018, Bayer acquired Monsanto, inheriting the legal and public relations challenges associated with glyphosate. The debate about glyphosate continues, with ongoing research, legal battles and discussions about the environmental and health impacts of its use. However, glyphosate has become one of the most widely used herbicides globally, impacting agricultural practices on a large scale. Whereas concerns include the potential harm to non-target plants, soil health and aquatic ecosystems due to runoff. Given the concerns associated with glyphosate use and its potential effects on "One Health," the significance of soil sampling becomes paramount. Conducting regular soil sampling allows for a comprehensive assessment of glyphosate residues, helping in the monitoring of its impact.

3.2.4 *Site investigation*

Site investigation generally involves the assessment and management of site contamination in sequential stages. This may include a preliminary site investigation, a detailed site investigation, the implementation of the remedial action plan, the validation of the investigation and ongoing monitoring, if necessary. During site investigation, different sampling approaches are applied to achieve the desired goal. In this context, the site investigation of glyphosate contamination may vary depending on the purpose of its application and the area of application. Assessment for contaminated sites consists of a series of preliminary and detailed phases of investigation, which must be completed to confirm the risk from exposure to a contaminated site (Boulding, 2017). Whereas, according to the NSW EPA, a preliminary site investigation (PSI) includes site reconnaissance, design of a CSM, confirmation of potential contamination by sampling soil and analysing its total extractable contaminant (TEC) concentrations and comparison with soil guideline values (SGVs) (NSW EPA, 2022).

The detailed site investigation (DSI) phase usually proceeds when TEC exceeds the SGV (NSW EPA, 2022). The DSI process involves exploring the sources, routes and connections between pollutants and receptors (pollutant linkage). Following this investigation, the CSM is adjusted to accurately illustrate practical exposure scenarios at the contaminated site (NSW EPA, 2022). If the limit of residual glyphosate exceeds their safe limit, the sites must be managed or remediated to minimise the risks, protect receptors of concern or permit proposed changes to land use. In the case of PSI, sampling is typically not mandatory. However, when necessary, judgemental soil sampling methods are utilised, focusing specifically on the target area. Conversely, a DSI involves both judgemental and probabilistic sampling approaches.

3.2.5 *Concerns in soil sampling*

Being a critical aspect of site investigations and risk assessments, soil sampling raises several concerns. One primary concern is the representativeness of the samples collected. The heterogenic nature of soil often makes it challenging to obtain samples that accurately reflect the overall composition and contamination levels, where inadequate sampling may lead to a misunderstanding of potential risks and hinder the development of effective remediation strategies. Another concern is the selection of sampling locations, as improper choices may overlook critical areas of contamination. The depth of sampling is also a crucial consideration, as pollutants may vary across different soil depths. Addressing these concerns is pivotal for ensuring the reliability and relevance of soil sampling data in risk assessment-based decision-making processes. During soil sampling, several aspects need to be considered, including selection parameters, sampling design and approaches.

3.2.5.1 *Selection of area and parameters for sampling*

The sampling areas are selected as those within a site which is to be covered during sampling with the help of analytical and geophysical screening techniques to determine the extent of contamination. While the parameters are targeted during analysis, the samples are normally identified with the problem itself, e.g., a chemical spill or an explosion. However, the parameters are also identified based on the review of the background information for possible present or future release of environmental contaminants. In addition, regular monitoring with respect to their future threats or potential uses could be prioritised in parameter setup (IAEA, 2004). In the case of glyphosate, the selection of sampling areas should prioritise locations where glyphosate application has historically been high or areas susceptible to runoff, such as agricultural fields or those near water bodies. Additionally, parameters for sampling should encompass various soil depths to capture the potential vertical migration of glyphosate. Since glyphosate may bind to soil particles, understanding the soil's composition and texture is crucial for a better understanding of soil test results (Padilla and Selim, 2020). The selection of appropriate parameters should also account for factors such as pH, organic matter content and soil moisture, as these can influence glyphosate's behaviour and mobility in the soil.

3.2.5.2 *Sampling design*

Sampling design in soil sampling refers to the systematic and strategic plan for collecting soil samples from a specific area to accurately represent the overall characteristics of that area. Developing a robust sampling design is important to establish representative soil test results. Specific data quality objectives (DQOs) can also be achieved through sampling design, which guides the collection of reliable and meaningful data in environmental assessments (Bhave and Sadhwani, 2022). A comprehensive and systematic soil sampling design is crucial for accurately assessing soil contamination, such as residual glyphosate, because it enables the collection of representative samples, considers spatial and temporal variability and ensures that the data obtained are reliable for making informed decisions regarding the presence, distribution and potential risks associated with glyphosate or its metabolite in the environment. Sampling designs can be broadly categorised into two main types: probabilistic sampling design (PSD) and judgemental sampling design (JSD).

Conversely, several sampling approaches are generally practised in the assessment of site contamination, such as judgemental sampling, systematic sampling, random sampling and stratified sampling. The strategies of soil sampling are varied and influenced by various factors during the assessment of site contamination. These include land use, topography, climate and the specific objectives of the study. Thus, a combination of strategies is often used, for example, targeted sampling for known features and/or specific media and systematic sampling to ensure adequate site coverage and data suitable for statistical inference. These factors also determine the patterns of sampling points, such as grid-based sampling and zone sampling. This review explores how researchers tailor sampling designs to account for spatial variability and ensure representative results. The applicability of different sampling strategies is described in the following.

3.2.5.2.1 Probabilistic sampling design (PSD)

PSD typically involves random selection, ensuring that each unit in the population being studied has an equal probability of being selected (Tillé and Wilhelm, 2017). This approach yields unbiased and independent data if applied properly. PSD offers several advantages. Firstly, they enable drawing meaningful conclusions through statistical analysis; thus, it is also called statistical sampling design. Secondly, PSD allows for the assessment and quantification of the uncertainty linked to estimates. Additionally, this type of design contributes to producing consistent and replicable results, staying within predetermined uncertainty thresholds. Moreover, PSD provides a framework to incorporate criteria for decision errors into the interpretation process, often communicated through confidence statements. However, optimal design using probabilistic sampling requires an accurate CSM and a clear definition of the population to be sampled (NSW EPA, 2022).

3.2.5.2.2 Judgemental sampling design (JSD)

JSD is a predetermined or targeted approach where decisions on where and when to collect samples are made in advance. This method relies on well-documented site histories and/or clear and distinct site features. Judgemental sampling can be particularly efficient for identifying areas of worst-case impacts and is valuable in situations where site history is insufficient or where the features of concern are obscured or not easily discernible. Generally, judgemental sampling locations are selected based on the probable distribution of contaminants at the site along with known or suspected areas of contamination (Figure 3.3(a)). For example, in a non-agricultural setting, glyphosate is applied to areas where complete eradication of weeds is required, such as fence lines, concrete surfaces, or even garden beds. The application of glyphosate-based herbicides to these areas is mostly seasonal or cyclic depending on the weather conditions. This sampling method makes efficient use of historical site information and field observations; hence, this method is statistically biased. Using this sampling method, an uneven distribution of sampling locations is generated. As this method is mostly location-specific, it may also require additional sampling locations to ensure site coverage (Figure 3.3(b)). Overall, various factors, including land use, topography, climate and the specific objectives of the study, influence soil sampling strategies. A combination of strategies is often used, for example judgemental sampling for known features and/or specific media and systematic sampling to ensure adequate site coverage and data suitable for statistical inference.

3.2.5.2.3 Systematic sampling

Systematic sampling entails selecting sampling points at regular intervals across a specific area, ensuring equal probability of each point being chosen (Malherbe, 2002). This approach helps provide an unbiased and representative assessment of the entire sampling site. In the case of glyphosate-based herbicides applied to a crop field, systematic sampling ensures even coverage of the entire area, as points are selected at regular intervals. This method does not generate clusters of sampling locations,

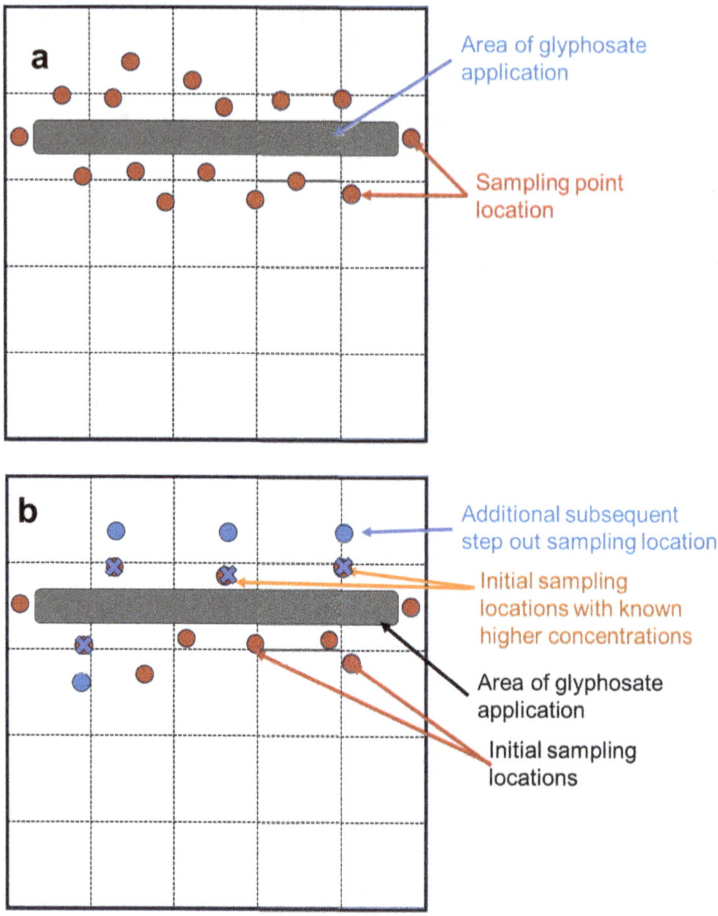

Figure 3.3. Judgemental sampling design.

which can be advantageous when characterising the extent of contamination across the entire site or decision area. The key to maintaining statistical unbiasedness in systematic sampling is the random determination of the coordinates for the first sampling location (NSW EPA, 2022). This initial random selection sets the starting point for the regular intervals, and from there, subsequent points are chosen systematically. As a result, the sampling points collectively provide a comprehensive representation of the entire field, helping to assess the distribution and concentration of glyphosate contamination without introducing bias. By employing systematic sampling in glyphosate-contaminated soil studies, researchers can confidently analyse the data obtained from various points across the field, making informed decisions and recommendations for mitigation or management strategies based on a statistically robust assessment of contamination levels.

3.2.5.2.4 Random sampling

In random sampling, the selection of sampling points is carried out in a non-arbitrary manner using a legitimate uniform pseudo-random number generator, which is often implemented through a computer program to determine the coordinates of sampling locations. This randomisation process ensures unbiased data collection, as every point within the sampling area has an equal probability of being chosen as a sampling point.

According to NEPC (2013, B2), simple random sampling is particularly advantageous when the area of interest is relatively homogeneous, lacking major patterns or hotspots. It finds utility in scenarios such as specific decision areas where no prior information is available, such as in a PSI (NEPC, 2013). However, when employed in assessing site contamination, it is essential to acknowledge and document the limitations of random sampling. While random sampling is statistically unbiased, there is a risk of sample points clustering together by chance. This clustering problem makes random sampling less effective in detecting contamination hotspots and providing a comprehensive spatial distribution overview. In practical terms, random sampling alone has limited applicability in site contamination assessment, unless it is complemented by systematic grid sampling (NSW EPA, 2022). Figure 3.4 shows an example of a systematic sampling pattern as opposed to a randomised offset, where sampling locations (red dots) are placed at random locations within each grid across the investigation area (Figure 3.5).

3.2.5.2.5 Stratified sampling

Stratified sampling is a valuable approach in investigating large and diverse sites with varied uses, features and complex contaminant distributions. This method involves dividing the site into distinct, non-overlapping sub-areas based on geological, geographical, contamination nature, former site usage or other pertinent factors (Cohen, 2011). Each sub-area is then treated as an independent decision unit with tailored sampling patterns and densities. Stratified sampling demands a solid understanding of the

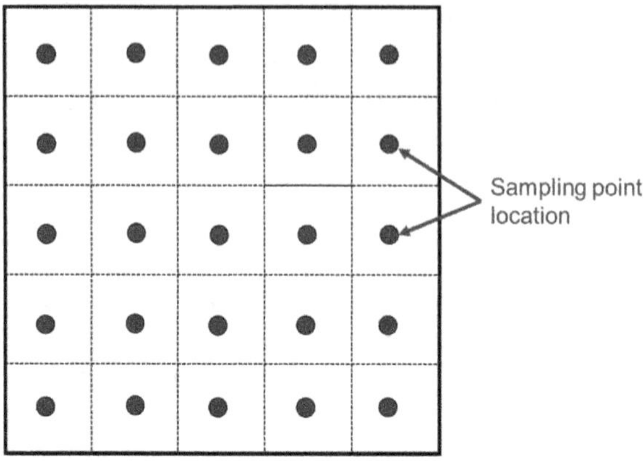

Figure 3.4. Systematic sampling pattern.

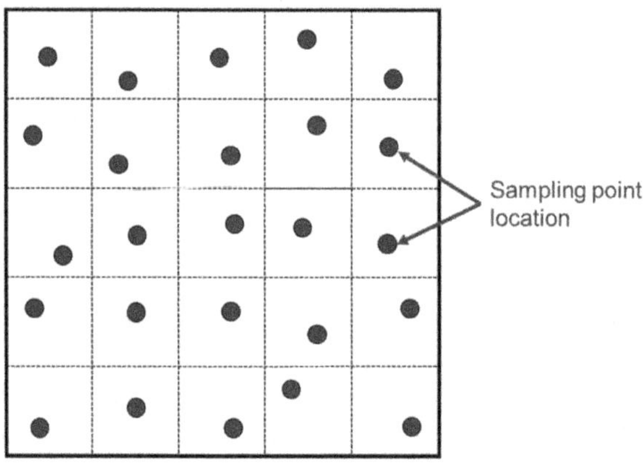

Figure 3.5. Random sampling pattern.

site, as different sub-areas may require diverse sampling strategies. For instance, one area might necessitate a targeted sampling approach, while a neighbouring area may benefit from a systematic sampling. Reliable prior knowledge of the site is crucial for effective implementation (NSW EPA, 2022).

The advantages of employing a stratified sampling pattern include the potential for achieving greater precision in estimates of mean and variance, particularly when the measurement of interest correlates strongly with the variable defining the strata (NEPC, 2013). Additionally, this method allows for the calculation of reliable estimates for subgroups of

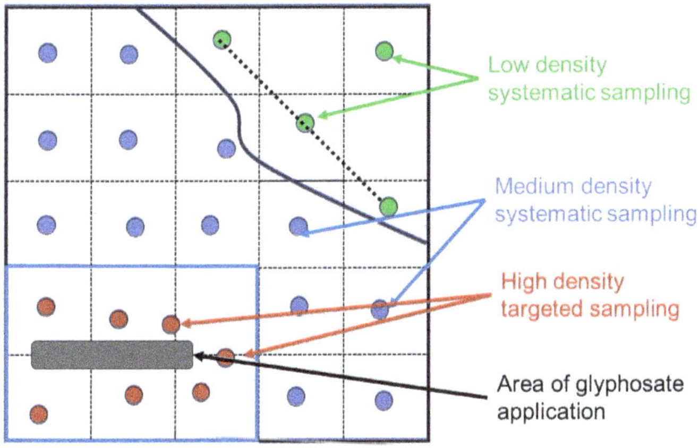

Figure 3.6. Stratified sampling pattern, with three separate investigation areas, sampling strategies and sampling locations due to different characteristics of the site.

special interest. To illustrate, Figure 3.6 depicts a stratified sampling pattern across three distinct investigation areas, each with varying characteristics. In Area 1, a high-density judgemental sampling strategy is employed, with numerous sampling locations clustered around an underground storage tank. Area 2 utilises a medium-density systematic sampling strategy to assess fill material from an unknown source, employing regular intervals in a grid. In contrast, Area 3 employs a low-density systematic sampling strategy to evaluate natural soil without any known contaminants; it uses sampling locations that are farther apart than those in Area 2. The collected data from these areas are treated as separate datasets for analysis.

Despite its merits, stratified sampling has some challenges, including those described by Gilbert (1987), such as the need for reliable prior knowledge and the potential for more complex statistical analyses due to varied sampling patterns. However, the customisable nature of stratified sampling makes it a powerful tool for investigation, offering precision in estimates and the ability to address specific subgroups of interest.

3.2.5.3 *Sampling depth*

The depth at which soil samples are collected is a critical consideration. Agricultural studies often focus on the root zone, while environmental assessments may require samples from deeper horizons to analyse potential contaminant migration. The review explores the implications of sampling depth on data interpretation.

3.2.6 *Sample collection and transport*

3.2.6.1 *Soil sampling tools and techniques/equipment*

Various tools and techniques are employed in soil sampling. Tools such as soil augers and shovels are considered traditional methods. In comparison, with technological progress, various mechanical tools have already been developed to fulfil sampling objectives (Figure 3.7). For instance, when determining contaminant concentrations at each soil horizon interface, the use of a hand auger may not be suitable, as it can disrupt and mix soil horizons, making it challenging to precisely identify the horizon interface (IAEA, 2004). Thus, the depth of sampling is a crucial factor in selecting the appropriate sampling equipment (Maher *et al.*, 1994). A trowel may be suitable for unconsolidated surface soils; however, it may not be the best choice for sampling at greater depths, such as 12 inches, due to changes in soil consistency. Moreover, it is imperative that all sampling devices be of sufficient quality to avoid contributing contamination to the samples (Trick *et al.*, 2008). For instance, painted

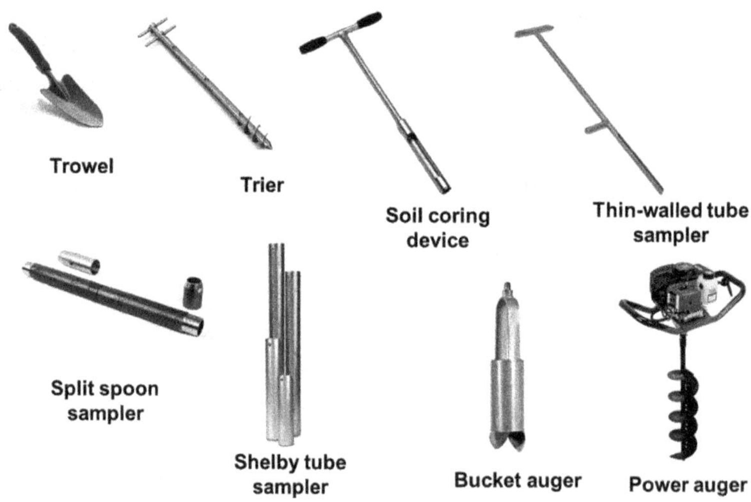

Figure 3.7. Various tools employed in soil sampling.

surfaces on equipment may chip off into the sample, emphasising the importance of using high-quality materials. Additionally, the sampling equipment should either be easily decontaminated or cost-effective if it is considered expendable. Ease of use is another key consideration when selecting sampling equipment.

In contrast, more advanced technologies, such as GPS-guided probes and remote sensing, have emerged. The advantages and disadvantages of each method are discussed within the context of precision, efficiency and cost-effectiveness (Table 3.1). It is essential to carefully weigh these

Table 3.1. Applicability of various soil sampling equipment.

Equipment Name	Soil Type and Depth	Advantages	Disadvantages
Trier	Soft surface soil	• Inexpensive • Easy to use • Easy to decontaminate	• Difficult to use in stony, sandy or dry soil
Scoop or trowel	Soft surface soil	• Inexpensive • Easy to use • Easy to decontaminate	• Trowels with painted surface should be avoided
Tulip bulb planter	Soft soil (0–6 inches)	• Easy to use • Easy to decontaminate • Uniform diameter and sample volume • Preserves soil core	• Limited depth capability • Difficult for hard soil
Soil coring device	Soft soil (0–24 inches)	• Easy to use • Preserves soil core	• Limited depth capability • Difficult to decontaminate
Thin-wall tube sampler	Soft soil (0–10 feet)	• Easy to use • Preserves soil core • Compatible to use with bucket auger • Easy to decontaminate	• Difficult to remove soil cores from sampler

(*Continued*)

Table 3.1. (*Continued*)

Equipment Name	Soil Type and Depth	Advantages	Disadvantages
Split spoon sampler	0 inches–bedrock soil	• Useful for hard soil specially	—
Shelby tube sampler	Soft soil (0 inches–bedrock)	• Excellent depth range • Preserves soil core • Can be shipped the soil from sampling site to lab undisturbed	• Not sustainable for rocky soils
Bucket auger	Soft soil (3 inches–10 feet)	• Excellent depth range • Easy to use • Uniform diameter and sample volume	• May disrupt and mix soil horizon greater than 6 inches in thickness
Hand-operated power auger	6 inches–15 feet soil	• Good depth range	• Destroys soil core • Difficult to decontaminate • Need gasoline powered engine

Sources: Adapted and modified from NJDEP (1988) and USEPA (1991).

factors when deciding on the most suitable soil sampling approach for a given situation.

3.2.6.2 *Sampling*

3.2.6.2.1 Physical observation and climatic conditions

Prior to commencing the sampling process, a thorough physical examination of the sampling location or specific points is important. This initial step involves recording the climate conditions, with particular attention given to factors including soil temperature and moisture levels (IAEA, 2004). This preliminary assessment serves as a crucial point for accurate and representative soil sampling. When dealing with composite samples, a strategic approach should be employed to prevent the accumulation of soils exhibiting different colours or textures. Maintaining the integrity of individual soil characteristics ensures the reliability of the composite sample, thereby enhancing the precision of subsequent analyses. Additionally, it is advisable to assess soil compaction levels at the sampling site (NSW EPA, 2022). Conducting tests to gauge soil compaction aids in the selection of appropriate tools for the sampling process. Understanding the compaction status not only streamlines the sampling procedure but also contributes to the overall efficacy of soil analysis.

3.2.6.2.2 Sample collection

The soil collection process is highly important, as it significantly impacts the representativeness of analytical results. Achieving a substantial level of representativeness involves considering factors such as the number of samples and the volume of each sample. While a greater number of samples and larger sample volumes enhance representativeness, practical constraints, such as sampling budgets and project schedules, often impose limitations.

The required number of samples varies based on the chosen sampling approach. In grid sampling, for instance, a single sample is typically collected at each grid node, irrespective of grid size. Upon identifying contaminated grid nodes, adjacent grid cells can be more thoroughly sampled to delineate contamination areas. For grid cells measuring up to 100 × 100 feet, it is recommended to collect four aliquots from each cell, evenly positioned along the sides and equidistant from each other. Additionally, an extra aliquot from the centre of each cell can be collected, totalling five aliquots per cell. For grid sizes exceeding 100 × 100 feet, nine aliquots, similarly distributed, are recommended. Depending on budget considerations, grid cell aliquots can be analysed individually or combined into one or more samples per cell (NSW EPA, 2022).

Both sample depth and area are crucial factors in determining the appropriate sample volume. Depending on the analytes under investigation, samples may be collected at the surface (0–3 inch or 7.5 cm), at an extended surface (0–6 inch or 0–15 cm) and/or at one-foot (or 30 cm) depth intervals. The properties of contaminants also play a role in determining the sampling depth. An example is when the water solubility of contaminants such as dioxins and PCBs is poor, often encountered within the first six inches of soil. Whereas water-soluble contaminants, such as

Table 3.2. Minimum number of sampling points for a square grid, based on site area.

Size of Site (ha)	Minimum Number of Sampling Locations	Grid Size (m)	Diameter of the Hotspot That Can Be Detected with 95% Confidence (m)
0.05	8	8	9.3
0.1	8	11	13.2
0.2	8	16	18.7
0.3	9	18	21.5
0.4	11	19	22.5
0.5	13	20	23.1
0.6	15	20	23.6
0.7	17	20	23.9
0.8	19	21	24.2
0.9	20	21	25.0
1.0	21	22	25.7
1.5	25	24	28.9
2.0	30	26	30.5
2.5	35	27	31.5
3.0	40	27	32.3
3.5	45	28	32.9
4.0	50	28	33.4
4.5	52	29	34.7
5.0	55	30	35.6

Source: Adapted from NSW EPA (2022).

metals, acids, ketones and alcohols, are frequently encountered at deeper depths in most soils, except for clay. If the contaminants are present in a solution, such as PCPs in diesel fuel and pesticides in solvents, they can penetrate to great depths (e.g., down to bedrock), depending on soil type. When considering the area for surface samples, generally, soils are collected over a square foot surface area per sample. A square cardboard template measuring 12 inch × 12 inch, or a round template with a 12-inch diameter can be used to mark sampling areas. This thoughtful consideration of sample depth and area contributes to the overall effectiveness and reliability of the soil sampling process. NSW EPA has so far provided a chart of sampling numbers based on the sampling area (Table 3.2).

3.2.6.3 *Handling, containment and transport of samples/soil sample handling and dispatch*

3.2.6.3.1 Removal of extraneous material

Proper handling of soil samples is crucial for accurate characterisation of the site, especially in the context of potential glyphosate contamination. Initially, irrelevant or unimportant materials for characterising the sample are identified and removed to prevent errors in subsequent sampling and analytical procedures (US EPA, 2002). Extraneous materials in soil samples, such as glass fragments, twigs or leaves, are typically excluded to ensure the purity and representativeness of the sample (IAEA, 2004). However, not all non-soil materials are necessarily extraneous. In cases where contamination is associated with larger lumps, such as those containing glyphosate residues, excluding them from the sample may lead to an underestimation of the overall contamination levels. For example, if the larger lumps are a significant source of glyphosate, excluding them during sample preparation can result in a substantial reduction in the concentration of the contaminant. Hence, a judicious approach is essential in ensuring that the sampling and preparation process captures the full spectrum of materials relevant to glyphosate contamination for an accurate and comprehensive analysis.

3.2.6.3.2 Sample sieving

Sieving is the process of physically sorting a sample to obtain uniform particle sizes, using sieves with predetermined sizes. Contaminants may exist in various particle sizes; therefore, sieving is required for the separation and analysis of soil particles to understand their distribution within the soil matrix. Additionally, sieving allows for effective subsampling, which is crucial for capturing localised concentrations of contaminants, e.g., glyphosate, and promotes uniformity in subsequent analytical procedures, thereby enhancing the reliability of measurements. Furthermore, it aids in the removal of coarse fragments, such as large stones and plant debris, which may also contain traces of the contaminant, ensuring a focused analysis on soil components directly relevant to targeted contamination. Thus, sieving is an integral part of soil sampling. However, samples may need to be oven-dried or air-dried prior to sieving. Discarding non-soil or non-sieved materials, as well as the sieving process itself, can

Figure 3.8. Sieves with various mesh sizes to perform soil sieving.

result in physical and chemical losses (IAEA, 2004). Notably, sieving may not be advisable when volatile compounds are a concern. To address this, it is recommended to analyse the discarded materials, or a fraction thereof, to assess their contribution to the contamination of the investigated site. Various types of sieves can be used to perform soil sieving (Figure 3.8). For residual analysis of glyphosate and AMPA, air-dried soil samples passed through a sieve with a 2 mm mesh size are used. Whereas for analysing the physicochemical properties, the soil samples may require grinding and passing through a sieve with a smaller mesh size. For example, a 20-micron mesh screen size is recommended for XRF analysis of soils.

3.2.6.3.3 Sample homogenisation

Soil homogenisation is a significant step in the sampling process, as it ensures the uniform distribution of contaminants and properties throughout the collected soil sample, contributing to the accuracy and reliability of subsequent analyses. Soils are inherently heterogeneous, with variations in texture, moisture content and contaminant distribution. Homogenisation involves thoroughly mixing the collected soil, breaking down clumps and aggregates and creating a representative sample. In the context of contaminants such as glyphosate, uneven distribution can lead to inaccurate assessments of contamination levels. Homogenisation minimises the risk of sampling bias, providing a more comprehensive and accurate representation of the site's overall characteristics (Lacorte *et al.*, 2012). Overall, soil homogenisation is essential for obtaining representative samples that reflect the true nature of the soil, enabling reliable evaluations of contamination levels and supporting informed decision-making in environmental assessments. The homogenisation process can be carried out manually by utilising either a stainless-steel spoon or scoop, along with a stainless-steel bucket, or a disposable scoop and pan (NSW EPA, 2022). The sample should be quartered and split; this process should be repeated at least five times to visually ensure homogeneity. A mechanical stirrer (ASTM standard D422-63) can also be used to ensure proper and robust homogenisation.

3.2.6.3.4 Sample splitting

Dividing collected and field-prepared samples into two or more equivalent portions becomes necessary when different segments of the same sample require individual analysis. This process not only facilitates separate analyses but also helps to evaluate sample variability and assess analytical and extraction errors. To ensure uniformity before splitting, adhere to the homogenisation techniques detailed earlier. Specifically, regarding glyphosate-contaminated soil samples, sample splitting is important to accurately represent the sample's composition. After homogenising the sample, the next step is to fill two sample collection jars concurrently, carefully alternating spoonful's (or scoops) of the homogenised glyphosate-containing material. This simultaneous filling procedure ensures that both containers receive a representative portion of the entire sample. For a more integrated approach combining homogenisation and splitting, consider quartering the sample or employing a specialised riffle sample splitter (ASTM Standard C702-87) (Pitard, 2019). This mechanical device allows for a precise division of the glyphosate-containing sample into equivalent parts (Figure 3.9).

3.2.6.3.5 Sample compositing

Compositing involves physically mixing and homogenising multiple individual soil samples to obtain a new sample (Figure 3.10). This process results in an averaged concentration of contaminants across various sampling points, thereby reducing both the number of required lab analyses

Figure 3.9. A riffle sample splitter.

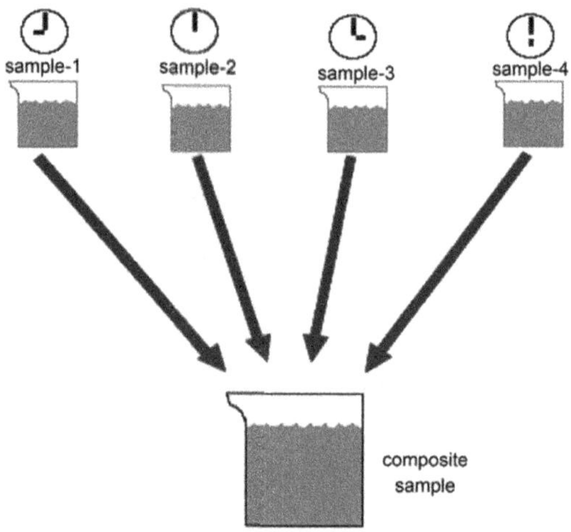

Figure 3.10. Schematic representation of sample compositing.

and the overall sample variability. While compositing can be a beneficial technique, caution must be exercised, particularly in cases involving volatile compounds. It is generally not recommended where the presence of such compounds is a concern.

The selection of aliquots for compositing and the determination of the compositing factor within the sampling plan are critical steps. The compositing factor, indicating the number of aliquots or increments combined into one sample (e.g., 3 to 1 or 10 to 1), must be carefully chosen (IAEA, 2004). Regulatory authorities may have restrictions on compositing, and the selection of the compositing factor should consider detection limits or uncertainties required for parameters of interest. This factor is determined by comparing these limits with the designated action level for a specific parameter.

3.2.6.3.6 Sample preservation

Sample preservation is a crucial aspect of laboratory procedures to ensure the integrity of collected samples during storage and transportation. For example, the choice of sample containers plays a pivotal role, considering factors such as compatibility with the sampled material, resistance to breakage and the required volume. In soil sampling, for instance, wide-mouth glass containers with Teflon-lined lids are often preferred due to their inert nature, which minimises the risk of sample contamination.

Preserving the quality of samples is also significant. Proper preservation techniques involve maintaining specific environmental conditions, particularly for samples to be analysed for organic constituents. Keeping low-to-medium concentration soil samples for organic analysis at a cool temperature of 4°C helps inhibit microbial activity and chemical reactions that may alter the sample composition over time (NSW EPA, 2022).

3.2.6.3.7 Transportation, quality assurance and chain of custody

Sample transportation is a critical link in the chain of custody, ensuring the integrity of samples from the sampling site to the laboratory. To maintain the stability of samples during transportation, a careful protocol is required. In certain scenarios, direct shipment to the laboratory using the sampling equipment is feasible. For instance, the ends of a Shelby tube can be sealed, taped and sent to the laboratory for analysis. This approach minimises potential handling errors and preserves the sample's original state. Additionally, for volatile organic analysis (VOA), using acetate sleeves during soil core collection aids in maintaining sample integrity during transportation (IAEA, 2004).

Upon arriving at the transportation authority, the custodian plays a pivotal role in assessing the condition of the samples. This process includes separating the samples for physicochemical and residue analyses, with clear documentation of the procedures followed on the identification form.

The adherence to quality assurance practices, including meticulous documentation, secure packaging and adherence to specific protocols, establishes a robust chain of custody. This chain ensures that the samples' history and condition are well-documented, providing a foundation for reliable and accurate analytical results in the laboratory.

3.2.6.4 *Decontamination of sampling equipment*

Effective decontamination of sampling equipment is important to ensure the reliability and accuracy of soil sampling and analysis. The process is particularly crucial when personnel, samples or equipment need to leave the contaminated area of a site. Various decontamination methods, categorised as abrasive and non-abrasive, are employed to physically remove or inactivate contaminants (NDEP, 1999). Abrasive methods involve rubbing and wearing away the contaminated surface layer. Techniques include mechanical cleaning using metal or nylon brushes, air blasting for large equipment and wet blast cleaning with suspended fine abrasives. While effective, these methods can generate waste, aerate contaminants and lack control over material removal. Non-abrasive methods, on the other hand, force contaminants off surfaces using pressure while minimising the removal of equipment surface. High-pressure and ultrahigh-pressure water systems use pressurised jets to remove contaminants. Disinfection and sterilisation methods inactivate infectious agents with disinfectants applicable for this purpose. Rinsing methods remove contaminants through dilution and solubilisation (NDEP, 1999).

For equipment cleaning in field sampling, a systematic procedure is recommended. This involves solvent rinses, non-phosphate detergent solution wash, tap water rinse, distilled/deionised water rinse, nitric acid rinse (for trace organics), additional distilled/deionised water rinse, solvent rinse for organic analysis, complete air drying and a final rinse with distilled/deionised water. The selection of decontamination solvents is crucial and depends on the contaminants present at the site. For instance, acetone, hexane or water are typical solvents for organic contaminant removal, while an acid rinse is necessary if metals are present. The decontamination procedure can be modified based on the specific contaminants found at the site, ensuring a tailored and effective decontamination process for sampling equipment. This approach safeguards against cross-contamination and maintains the integrity of samples, thereby contributing to the reliability of environmental analyses.

3.3 Characterisation of Soil Samples

Generally, glyphosate is a small molecule with three different functional groups: carboxylate, amine and phosphonate. Hence, it has high binding affinity for soil through the adsorption process. Various physicochemical properties of soils, including organic matter content, clay content, cation exchange capacity (CEC), metal oxides, phosphorus content and pH, influence the adsorption of glyphosate (Gómez Ortiz *et al.*, 2017). It is also important to understand the behaviour of glyphosate and its physicochemical properties in different representative soils. Thus, the soil samples should be properly characterised. Before characterisation, all samples are processed through sieving and drying. Generally, the soil is passed through a 2 mm sieve. The sieved soils are then either air-dried (at 40°C) or oven-dried (at 105°C) depending on the soil characterisation requirements.

3.3.1 *Soil texture analysis*

Glyphosate is mainly absorbed onto variably charged surfaces (Borggaard and Gimsing, 2008). In soils, clay minerals are the main constituents that provide a variable charge surface on the soil. Further, the soil texture also depends on the clay mineral content. Soils with higher clay content may retain more glyphosate, reducing its mobility. Nevertheless, depending on the clay type, the retention of glyphosate varies (Miles and Moye, 1988; Suwardji and Sudantha, 2021). Therefore, in residual analysis of glyphosate, knowledge about the soil type or soil texture is important. Generally, soil texture is measured through soil particle analysis using the hydrometer method. Some other methods, such as jar testing and micropipette methods, are also employed to measure soil texture.

3.3.2 *Soil pH*

Soil pH is an important parameter to understand the sorption behaviour of any compound, including contaminants and pollutants, in soil. The pH of the soil solution plays the most important role in adsorption, as glyphosate is a polyprotic acid with varying degrees of dissociation. (Borggaard and Gimsing, 2008) (Figure 3.11). As such, the system pH controls the ionic or charge structure of glyphosate molecules as well as the pH-dependent active sites in soil. In most cases, the adsorption of anions is coupled with the release of OH^- ions; therefore, adsorption is favoured at lower pH

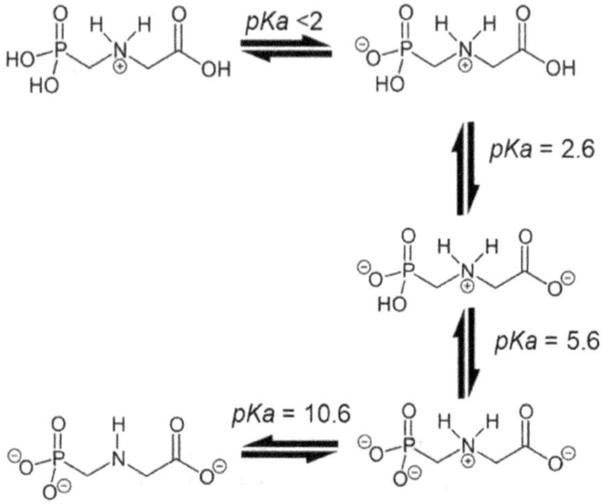

Figure 3.11. Chemical speciation of glyphosate.

values. At higher pH, the sorption of glyphosate declines due to the repulsion forces generated by the high negative charge in the soil as well as the glyphosate ion itself.

3.3.3 *Electrical conductivity*

Soil electrical conductivity (EC) measures the ability of soil pore water to carry electrical current. EC is an electrolytic process that takes place principally through water-filled pores. Cations (Ca^{2+}, Mg^{2+}, K^+, Na^+ and NH_4^+) and anions (SO_4^{2-}, Cl^-, NO_3^- and HCO_3^-) from salts dissolved in soil water carry electrical charges and conduct electrical current. Generally, EC of soil samples is measured using a pH meter, with units of $mScm^{-1}$. During EC measurement a soil/water (1:5, w/v) suspension is generally prepared by shaking using an end-to-end shaker (120 rpm) for 2 h.

3.3.4 *Measurement of moisture content*

The soil moisture content indicates the amount of water present in the soil. It is commonly expressed as the amount of water (in millimetres of water depth) present in a depth of 1 m of soil. The optimal range of soil moisture content for crops depends on the specific plant species, but the range for most crops is between 20% and 60%. The gravimetric method, or drying method, is a direct technique for determining moisture content, involving drying a soil sample and measuring the weight difference between its wet and dry states. This method is precise and reliable, so it is widely recommended in experiments. Moisture content influences shelf life because increased water in a product raises its susceptibility to microbes, which can cause rotting and damage the food. Soil water-holding capacity plays an important role in glyphosate dynamics. For instance, Romano-Armada *et al.* (2019) found a strong correlation between the content of glyphosate in the soil and the soil's available water content, suggesting that glyphosate persists longer in soils with high water retention capacity. Also, for the residual analysis of glyphosate, when obtaining the analytical portion for extraction, the moisture content should be considered because the moisture occupies considerable weight during sample weighing.

3.3.5 *Dissolved organic carbon*

Dissolved organic carbon (DOC) is generally defined as organic matter that can pass through a filter removing materials ranging between 0.70 and 0.22 μm in size (Halewood *et al.*, 2022). The DOC improves soil aeration (oxygen in the soil) and water drainage and retention and reduces the risk of erosion and nutrient leaching. Soil organic carbon is also important for chemical composition and biological productivity, including the fertility and nutrient-holding capacity of a field. One way in which the microbial community and plant growth are tied together is through the presence of carbon and nitrogen in soils. Carbon stimulates the growth of soil microbial communities. It plays a very significant role in the sorption-desorption process of contaminants. To analyse DOC, a soil/water suspension (1:5, w/v) was prepared for each soil sample by shaking it using an end-to-end shaker (at 120 rpm) for at least 2 h. Then, the suspensions were centrifuged to obtain the supernatant at 4,500 rpm. Finally, the supernatant was filtered through a 0.45 μm syringe filter (cellulose acetate) prior to measurement with a TOC analyser.

3.3.6 *Total organic carbon*

Total organic carbon (TOC) influences many soil characteristics, including colour, nutrient-holding capacity (cation and anion exchange capacity) and nutrient turnover and stability, which in turn influence water retention, aeration and workability. Estimated total organic carbon (TOC, expressed in terms of C) is used to assess the amount of organic matter in soils. Total organic carbon forms are derived from the decomposition of plants and animals. They are either capable of decay or products of decay. They contain organic compounds whose molecules contain carbon, oxygen, nitrogen and hydrogen. Whereas the inorganic forms of C present in soils are primarily carbonates and bicarbonates. TOC is directly related to organic matter, which exerts a profound influence on microbial activities in soil. Since glyphosate degradation largely depends on microbes, analysis of TOC in soil samples is important. So far, various techniques have been utilised to analyse TOC, including the wet oxidation method and the combustion method. In wet oxidation, oxidisable matter in the soil is oxidised using a 1 N $K_2Cr_2O_7$ solution. The reaction is assisted by the heat generated when two volumes of H_2SO_4 are mixed with one volume of dichromate. The remaining dichromate is titrated with ferrous sulphate. The titre is inversely related to the amount of C present in the soil sample. Whereas, in the case of the combustion method, dried and pulverised samples are treated with a sufficient amount of HCl for 24 h and kept under the fume hood to remove the inorganic carbon present in the form of carbonates or bicarbonates on clay minerals. Then, the acid-treated soils are mixed with a strong oxidant (COM-CAT™ Accelerator) and placed in a combustion chamber. Then, the TOC is measured automatically by the infrared detector of the CNS analyser by calculating the CO_2 emitted during combustion.

3.3.7 *Soil total elemental analysis*

Soil and sediments are integral parts of Earth and play a major role in food security. The elements in soil significantly influence soil quality and are closely related to human and environmental health. Besides the nutritional elements (e.g., Al, Ba, Ca, Fe, K, Mg, Mn, P and Zn) necessary for plant growth, other trace elements in soil (Cd, Co, Cr, Pb, Mo, Ni, V, etc.), which are considered heavy metals, can also be absorbed by plants through their roots, affecting food safety and posing a potential risk to human health. Nowadays, glyphosate has been reported to enhance trace metal availability in soil, resulting in soil pollution. Routine soil testing to monitor the elemental content in agricultural soils is crucially important for managing such pollution. Among the various analytical techniques, atomic absorption (AA) spectroscopy, inductively coupled plasma optical emission spectroscopy (ICP-OES) and ICP mass spectrometry (ICP-MS) (NexION®) are commonly applied for elemental analysis in soil. Each technique has its advantages and limitations; the most suitable one can be selected based on the number of elements to analyse and their concentrations, sample throughput requirements and the application purpose. ICP-OES is widely used in agricultural laboratories for soil analysis due to its simultaneous multi-element measurement capability, high sensitivity and low detection limits (parts per billion, ppb). This technique has been described in many standards for elemental analysis of wastes and soils, such as US EPA Method 200.7, EN 13346:2000 and China Standard HJ 781-2016. Additionally, US EPA methods 3050B, 3051A and 3052 are the commonly used standards for sample preparation. Depending on the elements of interest and the purpose of analysis, a total digestion or extraction method can be used.

On the other hand, ICP-MS is a popular technique used to analyse samples with low regulatory limits. Due to the differences in metal element detection, the lower detection limit for ICP-MS can extend to parts per trillion (ppt). Although both ICP-OES and ICP-MS can be used for high-matrix samples, sample dilution is often necessary for use with ICP-MS. In addition, if a sample contains analytes of greatly differing concentrations, ICP-MS offers a wider dynamic linear range, so the sample may not need to be diluted to detect these elements at the same time.

3.3.8 *Extractable soil oxides and cations*

Glyphosate sorption in soils has been positively correlated with the presence of amorphous Fe and Al oxides. These oxides in soil are the preferred sorption sites for glyphosate molecules, as they have high affinity, which may be attributed to the relatively high contents of amorphous Fe and Al oxides. The reason behind this is that glyphosate is retained by Fe oxides through both phosphonic acid and carboxylic acid moieties. The high CEC of the soil probably allows GPS complex formation with surface exchangeable polyvalent cations (probably Ca^{2+} in this case) (de Jonge *et al.*, 1999; Dollinger *et al.*, 2015), although ligand exchange at the edge sites of layer silicates may also be of significance. Glyphosate and phosphate may compete more strongly for sorption sites in acidic soils with high Fe/Al oxide contents than in calcareous soils. In addition to this, soil anions lower the zero-point charge of sorption sites such as Fe/Al oxides, potentially increasing the net negative charge on the oxide surfaces and thereby increasing the electrostatic repulsion between glyphosate and soil oxides. Thus, the competitive effect of anions on glyphosate is stronger in soils that are acidic and contain substantial amounts of Fe oxides than in calcareous soils. The soil oxides could be measured by different methods, such as the dithionite-citrate-bicarbonate (DCB) method, oxalic acid method (Rennert *et al.*, 2021) and sodium-citrate method (Reyes and Torrent, 1997). Using the DCB method, extractable Al, Fe and Mn oxides and divalent (Ca^{2+}, Fe^{2+}, Mg^{2+}, Mn^{2+} and Zn^{2+}) and polyvalent cations (combinations of K^+ and Al^{3+} with divalent cations) can be determined.

3.3.9 *Cation exchange capacity and anion exchange capacity*

Cation exchange capacity (CEC) and anion exchange capacity (AEC) are important parameters for understanding the interactions between soil clay particles and pollutants via the ion exchange process. The CEC is measured using the $BaCl_2$ compulsive exchange method (Gillman and Sumpter, 1986). The compulsive exchange method is recommended by the Soil Science Society of America because it is a highly repeatable, precise and direct measure of a soil's CEC. However, this method is time-consuming and generates hazardous waste ($BaCl_2 \cdot 2H_2O$). Thus, this method is unlikely to be well suited for most routine soil testing laboratories, but it is preferred by university laboratories. This method is also used to determine the AEC of soils, where the soils are saturated with barium chloride to saturate soil anion exchange sites with Cl^- ions. Then, a low-ionic-strength molecule such as magnesium sulphate is added to exchange the Cl^- ions with SO_4^{2-}. The amount of Cl^- released was measured to determine the AEC. During AEC measurement, the pH must be held constant to accurately measure the AEC in variable-charge soils (Gillman, 1979). A small difference, driven by variation in soil pH, may misinterpret anion exchange measurements. The differences in pH readings obtained from soil pH measurements taken in 1 M KCl and in deionised water give an indication of the soil charge.

3.3.10 *Soil anions and cations*

Na^+, K^+, Mg^{2+} and Ca^{2+} are the significant cations in alkaline soils, whereas Cl^-, SO_4^{2-}, HCO^{3-} and CO_3^{2-} are the major anions. Mono- and divalent anions are the predominant species present in the soil. However, due to the amphoteric character of Gly and AMPA, both anionic and cationic ions interact with glyphosate and its metabolites, AMPA. Studies have shown that HCO_3^- (Na^+, K^+ and Mg^{2+}) salts and $CaCO_3$ dominate in pH 7.0–8.5, while CO_3^{2-} salts of Na^+ and K^+ dominate above pH 8.5. Again, the amount of dissolved glyphosate increased significantly as pH increased. Therefore, the presence of large concentrations of HCO^{3-} and CO_3^{2-} not only increases pH but also promotes the dissolution of soil glyphosate. The concentration of Ca^{2+} modified this effect; large concentrations of Ca^{2+} increased adsorption and reduced the concentration of glyphosate. For example, inorganic phosphate ions compete with glyphosate sorption in soil. A different report showed that Freundlich coefficients plummeted by 40–65% for glyphosate when phosphate ions were present in the glyphosate solution. Using the ion chromatography technique, the soil anions (F^-, Cl^-, Br^-, NO_2^-, NO_3^-, SO_4^{2-} and PO_4^{3-}) and cation (NH_4^+) were measured. Before measuring, the anions and cations were brought into soil pore water, where a fixed amount of air-dried soils was dispersed into a fixed amount of water and shaken vigorously. Then, the liquid phase (supernatant) was separated by centrifugation, followed by filtration, before analysis.

3.3.11 *Soil micro-morphology and minerology*

With technological progress, the morphology of soil is analysed by using different microscopic techniques, including light microscopy (LM) and electron microscopy. Generally, LM is suited for analysing larger particles (micron-sized), including soil microorganisms such as bacteria. Whereas, to analyse ultrafine particles more robustly, electron microscopy techniques, such as scanning electron microscopy (SEM) and transmission electron microscopy (TEM), are applied. SEM is mostly utilised to investigate the outer surface, whereas the structural skeleton could be analysed using TEM (Figure 3.12). In addition, the elemental composition can be analysed using electron microscopy when equipped with

Figure 3.12. SEM images and an EDX spectrum.

an energy-dispersive X-ray (EDX) spectroscopy analysis system (Figure 3.12). Further, soil mineralogy in solid precipitates is qualitatively identified via X-ray diffraction (XRD) analysis (Kim and Baek, 2019).

3.4 Analysis of Residual Concentration

To determine the residual concentrations of glyphosate and its metabolites effectively, various techniques have been tested, such as spectroscopy techniques, fluorescence-based techniques, chromatographic techniques and colourimetric techniques (Rawat *et al.*, 2023). The applicability of these techniques varies and depends on several factors, including the soil matrix. Sample preparation and the extraction of residual glyphosate and its metabolites are crucial steps for the determination of residual concentrations at any contaminated site. Therefore, it is important to review and understand glyphosate extraction from the matrix (sample preparation) and its quantification through detection.

3.4.1 *Sample preparation and extraction*

After receiving the soil samples, bulk soil should be mixed thoroughly, either with a mixing machine or manually. The mixing processes are done carefully to avoid any possible cross-contamination. The homogenisation and pre-processing have been described earlier in Section 3.2. The sample size required for analysis is determined based on several factors, such as soil organic carbon content, soil type (i.e., sand, silt, clay or loam) and moisture (European, 2023; Wimmer *et al.*, 2022; Todorovic *et al.*, 2013). The analytical portion may range from 1 to 10 g depending on the above factors and the suspected contamination level, which varies based on the sample collection area (e.g., agricultural, non-agricultural or industrial zone). An analytical-grade balance (4–5 digits) is recommended for weighing analytical portions because glyphosate may be present at trace or ultra-trace levels, which means that even 0.01 g of weight might affect the extractable concentration (Anastassiades *et al.*, 2023).

After sample preparation, the glyphosate and its metabolites are extracted from the soil samples for analysis. Several methods are used for this purpose (Koskinen *et al.*, 2016; Rawat *et al.*, 2023; Anastassiades *et al.*, 2023). From simple liquid extraction to highly complex techniques, such as accelerated solvent extraction (ASE), have been utilised in the residual extraction of glyphosate and its metabolites. Both glyphosate and its metabolites are highly polar compounds; therefore, the majority of extraction techniques involve aqueous-based extraction solvents (acidic or basic) or a mixture of water and polar organic solvents, such as methanol. The applicability of the methods differs from each other, as discussed in the following.

3.4.1.1 *Solid-phase extraction*

Solid-phase extraction (SPE) is a solid-liquid extraction technique where dissolved or suspended compounds are separated from other compounds in a mixture or solution (Figure 3.13). This separation process is based on the affinity of the compound towards the stationary phase or soil sorbent. Either a compound will be retained in the stationary phase until selective

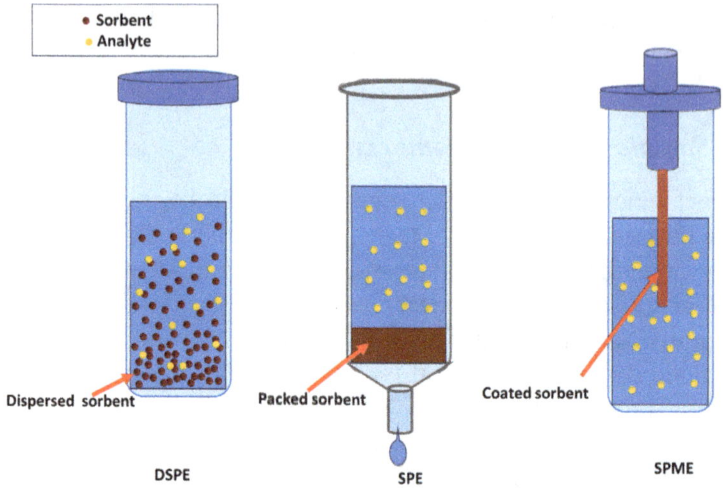

Figure 3.13. Overview of different SPE methods.

elution or interference will retain it, and the compound will be in the eluent. This method is directly used for liquid sample extraction. In glyphosate analysis, SPE has a wide range of applications. In general, SPE could be achieved in three different versions, namely, dispersive solid-phase extraction (DSPE), solid-phase micro extraction (SPME) and packed sorbent solid-phase extraction (SSPE).

3.4.1.2 *Matrix solid-phase dispersion* (*MSPD*)

For solid, semi-solid and viscous samples, extraction of the analyte into homogeneous liquid solvents can be achieved efficiently through matrix solid-phase dispersion (MSPD). In summary, the samples are mixed or blended with sorbent material; then, the mixtures are packed into a column, and the solvent is allowed to pass through it. (Tu and Chen, 2018).

3.4.1.3 *Supercritical fluid extraction*

Supercritical fluid extraction (SCFE) is used for solid matrices, where a supercritical fluid is used as the extraction solvent (Figure 3.14). A supercritical fluid (e.g., carbon dioxide and water) is capable of being gaseous

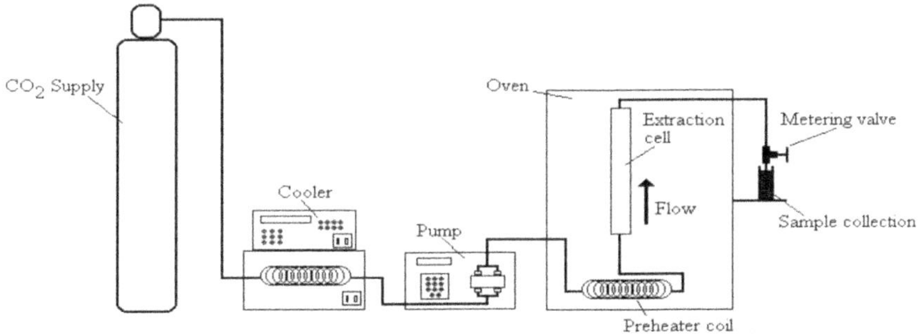

Figure 3.14. Schematic diagram of SCFE.

Source: Adapted from Sapkale *et al.* (2010).

in nature under special conditions (temperature and pressure), penetrate through solid and dissolve materials as a liquid (Figure 3.10).

3.4.1.4 *Accelerated solvent extraction*

Accelerated solvent extraction (ASE) is most appropriate for use in solid or semi-solid sample extraction. Briefly, the extraction is achieved by using an organic solvent under high pressure, where the temperature exceeds the boiling point for a short period, inside a closed sample cartridge or cell. The extract is then purged from the cartridge with the aid of compressed gas (Richter *et al.*, 1996). Thanks to technological advancements, the research community has the luxury of automated ASE machines, in which multiple extractions could be done simultaneously (Figure 3.15).

3.4.1.5 *Dispersive liquid–liquid micro-extraction (DLLME)*

Dispersive liquid–liquid micro-extraction (DLLME) is an organic solvent-based extraction method where a solvent of minimal volume and a high concentration of the analyte are used (Quigley *et al.*, 2016). Prior to proceeding with DLLME, glyphosate in the aqueous extract will be derivatised to a hydrophobic compound; then, an organic solvent will be dispersed into the aqueous extract in a drop-wise manner. The derivative of Gly will be partitioned into the organic solvent droplets. Since the surface area of droplets is high, partitioning will be instantaneous.

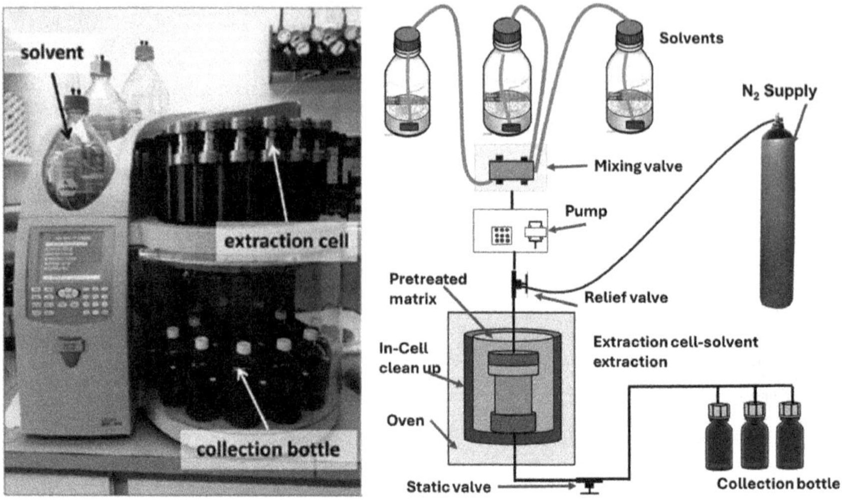

Figure 3.15. ASE instrument: ASE 350 (Thermo Scientific Dionex).
Source: Keržič *et al.* (2019).

Ultimately, the immiscible organic layer could be separated and analysed (Çetin *et al.*, 2017).

3.4.1.6 *Quick polar pesticide extraction (QuPPe)*

Quick polar pesticide extraction (QuPPe) is employed in both food and environmental sample analysis of glyphosate. In this method, water adjustment is the first step; then, the test sample is extracted using acidified methanol. EDTA is added, then centrifuged, filtered and directly sent for detection (Anastassiades *et al.*, 2023).

3.4.1.7 *Quick, easy, cheap, effective, robust and safe method (QuEChERS)*

Quick, easy, cheap, effective, robust and safe (QuEChERS) methods were mainly developed for residual analysis of food. The AOAC Official Method 2007.01 (Acetonitrile Extraction and Partitioning with Magnesium Sulfate) (AOAC, 2007) is the accepted version in the US, while in Europe, EN 15662 Version 2007-10-24 (Acetonitrile Extraction/Partitioning and Clean-up by Dispersive SPE) is accepted (González-Curbelo *et al.*, 2022). However, this method has been modified and used for the analysis of glyphosate residues in soil as well (Mahdavi *et al.*, 2021).

Extraction of glyphosate with only water is possible, but the recovery percentage tends to be lower. In the meantime, the stability of glyphosate hydrolysis ranges from pH 3 to pH 9. Hence, the extraction solvent could be varied from acidic to basic. According to the literature, many different aqueous solutions of acetic acid, formic acid, potassium hydroxide and buffer solutions such as ammonium acetate/ammonium hydroxide, sodium hexametaphosphate and sodium tetraborate are used in extracting glyphosate from solid and semi-solid samples. These acidic and basic extract solutions not only extract glyphosate and its metabolites but also humic substances, metal ions, etc. Therefore, prior to analysis, the extract requires a cleanup.

3.4.2 *Cleanup methods*

In order to mitigate interference from compounds such as lipids, oils, humic substances and metal ions, a cleanup step is followed. The cleanup could be achieved in many ways (Anastassiades *et al.*, 2023; Oulkar *et al.*, 2017; Koskinen *et al.*, 2016).

3.4.2.1 *Liquid–liquid extraction*

In liquid–liquid extraction, the aqueous extracts are shaken or vortexed with organic solvents such as dichloromethane, chloroform and ethyl acetate. Then, with the aid of layer separation, many organic compounds are partitioned into the organic layer. Generally, acetonitrile is mixed and centrifuged for protein precipitation; whereas, along with acetonitrile, C18 sorbent is mixed for fat removal. Cryogenic centrifugation and freeze-out followed by centrifugation are applied for fat and oil removal.

3.4.2.2 *SPE cleanup*

QuEChERS dispersive SPE kits, are the most popular SPE cleanup method uses recently. The dispersive SPE kit contain primary, secondary amine (PSA) sorbent, magnesium sulphate, C18 sorbent and GCB (graphitised carbon black). The appropriate volume from the extract is vigorously mixed with the kit, allowed for a few minutes to settle, then centrifuged and filtered through a suitable filter, resulting in a well-cleaned and clear extract.

3.4.3 *Derivatisation*

Glyphosate is a low-volatility, highly water-soluble small molecule with a zwitterionic structure and no chromophores or conjugated double bonds. Due to these properties, it is challenging to detect glyphosate using spectroscopic detection techniques such as UV–Vis, fluorescence-based methods and Raman scattering (Martin *et al.*, 2023). Since glyphosate is a good metal chelator, the freely available glyphosate present in the extract is reduced, which hinders the detector's response (Mertens *et al.*, 2018). Derivatisation of glyphosate has been shown to hold great potential in addressing these challenges. Various derivatisation agents have been used to analyse glyphosate and its metabolites. When choosing a derivatising agent, the method of detection must be considered. Table 3.3 describes the derivatising methods in relation to detection.

Table 3.3. Summary of different derivatising methods.

Derivatising Agent	Intended Detection Technique
Isobutylchloroformate; isopropylchloroformate	GC coupled with flame photometric detector
Mixture of trifluoroacetic acid, trifluoroacetic anhydride and trimethyl orthoacetate	GC coupled with chemical ionisation mass spectrometry
9-Fluoronomethylchloroformate (FMOC)	LC/HPLC or GC coupled with ionisation mass spectrometry, UV–Vis, fluorescence detectors
4-Methoxybenzenesulfonyl fluoride	LC/HPLC coupled with UV–Vis, fluorescence detector
O-Pthalaldehyde	HPLC coupled with fluorescence detector (λ_{ex} = 340 nm–λ_{em} = 455 nm)
4-Chloro-3,5-dinitrobenzotrifluoride	HPLC coupled with UV detector
2,2-dihdroxynadane-1,3-dione (ninhydrin)	UV–Vis spectrometer; λ_{max} = 570 nm
Carbon disulphite (CS_2)	UV–Vis spectrometer; λ_{max} = 435 nm
4-Chloro-7-nitrobenzofurazan	Fluorescence spectroscopy
Napthelen-2,3-dicarboxaldehyde (NdCA)	Electrokinetic chromatography coupled with laser-induced fluorescence detector
Fluorescein isothiocyanate	Electrophoresis system coupled with laser-induced fluorescence

Source: Gill *et al.* (2017).

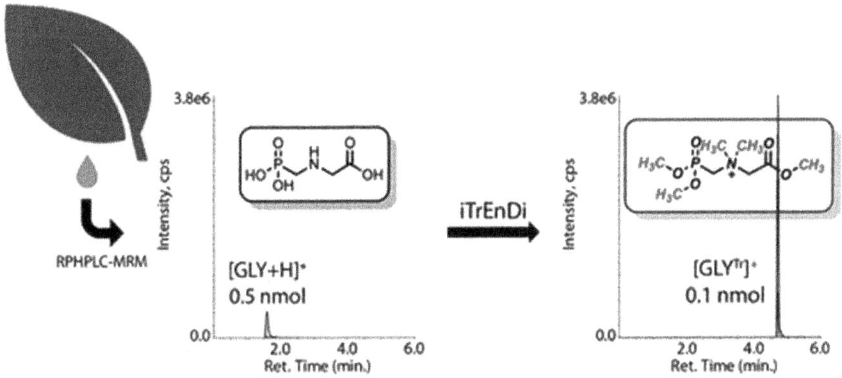

Figure 3.16. Intensity comparison between reverse phase HPLC-MEM of Gly-acid and iTrEnDi derivatised Gly.

Source: Adapted from Rosales *et al.* (2023).

Apart from these derivatising methods, much more advanced and newly introduced methods, such as *in situ* trimethylation enhancement using diazomethane (iTrEnDi), have been reported in Gly analysis with HPLC-MS. When Gly and its major metabolite, AMPA, were derivatised with diazomethane, they produced $[GLY^{Tr}]+$ and $[AMPA^{Tr}]^{+}$ complexes, respectively, which have fixed, permanent positive charges to enhance the detectability of Gly and the intensity of the response in detection (Figure 3.16).

The necessity of derivatisation in the analysis of glyphosate and its metabolites has been emphasised in previous works (Arkan and Molnár-Perl, 2015; Koskinen *et al.*, 2016). Nevertheless, Arkan and Molnár-Perl showed that, out of many derivatising agents, 9-fluoronomethylchloroformate (FMOC) is the most commonly used derivatising agent because derivatisation with FMOC requires only mild conditions, such as room temperature, a pH around 9 and a short derivatising time (20 min to 24 h). An investigation focusing on the effect of derivatising parameters in glyphosate analysis, relevant to [FMOC], [EDTA], [Borate buffer] and the derivatising time (h), revealed that with 1% EDTA, 5% borate, 2.5 mM FMOC-Cl and 4 h of derivatising time could achieve a greater yield of Gly derivative. Additionally, these researchers have developed an improved robust analysis method using LC-MS/MS using this derivatisation method (Martin *et al.*, 2023).

3.4.4 *Separation, detection and quantification*

As discussed in earlier sections, the separation and detection of glyphosate and its metabolites is quite challenging. In qualitative assessment aspects, colourimetric methods such as advanced nano-technological colourimetric detection (Chawla *et al.*, 2018) and enzyme-linked immunosorbent assay (ELISA) methods (Byer *et al.*, 2008) have been studied and validated. For trace-level detection and precise quantification, chromatographic separation coupled with advanced spectroscopic detection is needed. The derivatisation process has also given the opportunity to employ a wide range of analytical techniques for the detection and quantification of glyphosate and its metabolites regardless of the matrix present.

3.4.4.1 *Chromatographic techniques for separation*

Chromatography is an analytical technique used to break mixtures into their components, allowing each part to be analysed separately. Many approaches to detecting glyphosate residues use high-performance liquid chromatography (HPLC), gas chromatography (GC) and ion chromatography (IC). These chromatographic techniques are extensively recognised as methods for detecting glyphosate in food matrices and environmental samples (Kaczyński and Łozowicka, 2015).

The basic principle of chromatography is the distribution of the solute/analyte between the mobile phase (MP) and stationary phase (SP) (Table 3.4). This is named "partitioning" and is an affinity-driven scenario. The solute with a greater affinity for the SP will retain more in the SP and then in the elute, while the solute with minimal affinity for the SP will elute first in the MP. The SP is a modified sorbent material designed to retain a specific compound molecule or a group of compounds with common chemical properties. The SP could exhibit different patterns in sorbent arrangements, such as plates (applied in thin-layer chromatography, TLC) or packed columns. Depending on the MP, the chromatographic technique is classified. When the MP is gaseous, it is called GC, and when it is liquid, it is referred to as liquid chromatography (LC). In GC, prior to sample introduction to the column, volatilisation and ionisation are performed, whereas in LC, these are performed after the sample passes through the column. Ionisation could be electron impact ionisation (EI) or chemical ionisation (CI). In most cases, when using LC to analyse glyphosate, the reverse-phase mode is applied.

Table 3.4. Summary of GC and LC SPs and MPs.

Chromatographic Technique	Common SP Used	Common MP Used
GC (EI or CI)	• Fused silica cross-linked DB capillary column	$He_{(g)}$ $N_{2(g)}$ and $H_{2(g)}$
LC (EI)	• Silica-based C18 columns modified with different materials • Waters XBridge column • *C18, PEEK linked* • Suppressed ion exchange column *Polyvinyl alcohol with quaternary ammonium groups* • Waters Xselect HSS PFP • *Silica-based C18 (pH- 2–8) PFP bonding* • Waters Atlantis T3 Column *silica-based C18 (pH 2–8) T3 bonding* • Gemini® C18 column *organo-silica-based C18 with TMS end capping* • Agilent Zorbax Eclipse C-8 XDB column • *Organo-silane-based dimethyl-N-octyl silane* • Hypercarb column: *porous graphitic carbon*	Methanol (MeOH), acetonitrile (ACN), ammonium acetate, ammonium formate, acidified water/MeOH/ ACN with formic acid or acetic acid, phosphate buffers

Note: MP could be a single solvent, a mixture of solvents in a gradient or an isocratic mixture of two solvents.

In the HPLC technique, various modified methods have been used, which influence the determination power of Gly from a sample, such as HPLC coupled with fluorescence, UV–vis, diode array (DAD) and mass spectrometry (MS/MS) detection (Mallat and Barceló, 1998). GC coupled with mass spectrometry (MS/MS) and flame photometry (DFC) have also been employed for the analysis of glyphosate (Usui *et al.*, 2019). LC-MS/MS and GC-MS/MS have been used for the determination of glyphosate in breast milk (Steinborn *et al.*, 2016). The GC/MS/MS technique involves an extraction step, cleanup on the cation-exchange column and derivatisation of heptafluorobutanol and trifluoroacetic acid anhydride, while for LC/MS/MS, the method involves an anion-exchange column followed by ultrafiltration.

IC is a type of LC in which molecules are retained based on the attraction between charged sites and solute ions bound to the stationary phase. Once Gly becomes an ionic compound (p*K*a1 = 2.27, p*K*a2 = 5.58 and p*K*a3 = 10.25), an anion-exchange column can be used, followed by elution with an alkaline buffer (Valle *et al.*, 2019). The determination of glyphosate from the samples (soybean and corn) has been done by LC-tandem mass spectrometry (LC/MS/MS) using a reversed-phase with weak anion and cation exchange mixed-mode Acclaim Trinity Q1 column. The soybean and corn were shaken with water that contains an addition of acetic acid and ethylenediaminetetraacetic acid disodium salt for 10 min. Interestingly, the sample was analysed using the LC/MS/MS method in only 6 min with no sample concentration or derivatisation step (Chamkasem and Harmon, 2016; Liao *et al.*, 2018).

3.4.4.2 *Electrophoresis for separation*

Capillary electrophoresis (CE) using fused silica beds can acquire good separation of glyphosate and its metabolites (Gotti *et al.*, 2019; Wimmer *et al.*, 2022). In theory, CE separates charged molecules based on the mobility of the molecules under a controlled electric field or a voltage across the silica beds and electrolyte solution. The electrolyte concentration and the intensity of the electric field govern the mobility of charged analytes. The cleanup of the extract and the derivatisation of glyphosate increase the selectivity of electrophoresis separation.

3.4.4.3 *Mass spectrometry detection (MS)*

MS is one of the most powerful analytical techniques for glyphosate determination, and its use is growing in various scientific fields due to its high accuracy, selectivity, sensitivity and high-throughput capability. MS is performed by recording the count of charged masses using a mass counter and plotting the electric signal (m/z) generated against time, resulting in a mass spectrum. MS is a destructive detection technique (Figure 3.17). In this technique, the separated analyte molecule is converted into an ion or charged molecule in a gaseous state. To achieve this, the molecule will either be subjected to a higher voltage or bombarded with concentrated excited electrons at a high temperature within an ionisation chamber. The charged molecule produced is in its unstable excited state, due to which the molecule undergoes further fragmentation, resulting in smaller fragments. These fragments will be passed through a structure built with four metal rods. An electric field is applied to those rods, which generate an electromagnetic field in the space between them. Due to this magnetic field, fragmented charged molecules will acquire different fluctuation patterns corresponding to the m/z ratio and ultimately reach the mass detector with the help of a carrier gas (N_2 or Ar). The detector will generate an electric signal in response to the fragment count, and a mass spectrum is generated. The entire system is housed in a vacuum-sealed unit. This fragmentation pattern became advantageous not only in quantification but also in the identification of compounds.

Once a method is set up for quantitative analysis using GC or LC coupled with MS detection, method validation ensures the efficiency and accuracy of the analysis. Certain validation parameters, such as linear range, R^2, limit of detection (LOD), limit of quantification (LOQ), recovery percentage, relative standard deviation (%RSD) of calibration standards, spiked/QC sample results and uncertainty, should be evaluated cautiously. In general, a method is considered accurate and robust when linear range concentrations reach low levels (μg/L or ng/L) with an $R^2 > 0.99$. LOD and LOQ should also reach lower levels corresponding to the linear range with a low %RSD (<20%) and minimal uncertainty of results (European, 2023; "N° SANTE/11813/2017," 2019). Different analytical methods for glyphosate analysis and some of the critical validation parameters are summarised in Table 3.5.

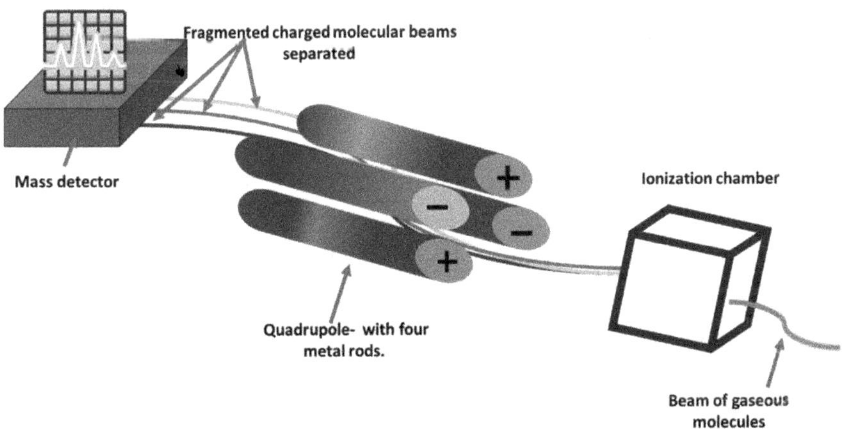

Figure 3.17. Typical schematic diagram of mass spectrometry.

Table 3.5. Analysis methods of glyphosate from environmental samples and critical validation parameters.

Sample Type	Sample Preparation	Derivatisation	Analytical Technique	Linear Range and R^2	LOD and LOQ	Recovery %	RSD %	Reference
Natural samples	QuEChERS	Benzaldehyde or cinnamaldehyde linking at ACN/Tris buffer (pH = 8.14, 10 mM, v/v 2:1)	Colourimetric sensor	–	LOD – 1.78 μM for BNP-Cl and 1.60 μM for CNP-Cl	–	–	Aydin and Keles (2022)
Bell pepper, carrot and strawberry	Extraction with BPS: a. mashed pulp b. runoff wash of chopped sample	–	ElectrochemSENSE	0.01–5 ppm R^2 – 0.98	LOD – 10 ng/L	–	–	Dhamu and Prasad (2020)
Wheat	SPE	FOMC-Cl	FASI sweep-MEKC	0.01–1.5 μg/mL R^2 – 0.999	LOD – 2 ng/L LOQ – 5 ng/L	108.2–114.0	4.3–6.5	Gotti *et al.* (2019)
Soil and sediments	Phosphate addition, alkaline extraction	–	CE-MS	$R^2 \geq 0.9958$	LOD < 20 μg/kg	69–99	<10%	Wimmer *et al.* (2022)
Water	SPE	Heptafluorobutyric anhydride and heptafluorobutanol	GC-FPD	10–1000 ng/mL R^2 – 0.9998	LOQ – 0.37 ng/mL	75.5–86.9	1.3–5.5	Zhang *et al.* (2019)

Water and soil	SPE	Trifluoroethanol and trifluoroacetic anhydride	GC-MS	$R^2 \geq 0.997$	LOQ – 0.01 μg/g	–	–	Bergstrom *et al.* (2011)
Soil	Several SPEs and reconstitutions	Heptafluorobutanol in trifluoroacetic acid	GC-MS/MS	–	LOQ – 10 μg/kg	73–95	–	Mamy *et al.* (2008)
Water	Lyophilisation and reconstitution	FMOC in the presence of EDTA	LC-FLD + MS/MS	0.1–50 μg/L R^2– 0.995	LOD – 0.085 μg/L	–	–	Ramirez *et al.* (2014)
Soil	Magnetic DSPE	–	HPLC-PDA	72–2000 μg/L $R^2 > 0.9902$	LOD – 21–22.5 μg/L LOQ – 72–75 μg/L	86–118	–	Surapong and Burakham (2021)
Water	SPE	FMOC	LC-MS/MS	1–500 μg/L R^2 – 0.995	LOQ – 50 ng/L	89–106	<9	Ibanez *et al.* (2005)
Soil	QuEChERS	NA			LOQ – 0.05 mg/kg	90–92	<7	

3.4.4.4 *Spectroscopy techniques for glyphosate detection*

Following separation, quantitative detection is the critical aspect of the analysis. In Gly analysis, various detection techniques have been studied. The majority of detection techniques in the literature fall under spectroscopic techniques. There are several phenomena included in spectroscopy, such as absorption, emission, nuclear magnetic resonance (NMR) and scattering. These phenomena occur due to the interaction between matter and electromagnetic radiation. And these types of detection are known as non-destructive methods, which obey the Beer–Lambert law. Some examples of non-destructive spectroscopic detectors are as follows:

i. ultraviolet and visible (UV–Vis) detectors,
ii. nuclear magnetic resonance (NMR) detectors,
iii. fluorescence detectors,
iv. flame photometric detectors (FPD),
v. photodiode array (PDA)/diode array detector (DAD),
vi. surface-enhanced Raman scattering (SERS) detectors,
vii. electrochemical detectors (ECD).

Spectroscopy techniques reveal the interaction between matter and emits electromagnetic radiation as a function of its wavelength or frequency (Pasquini, 2003). These techniques use various phenomena such as absorption, emission, SERS, NMR and surface plasmon resonance to detect glyphosate from different food materials (Chang, 2013). Spectroscopic techniques are advantageous in that they are simple, eco-friendly, precise, economical and require a minimal amount of reagents and samples for the determination of glyphosate. However, most spectroscopy techniques are only suitable for water and food matrices and not applicable for soil samples, which require chromophore groups and coloured reagents (da Silva *et al.*, 2011). To overcome this problem, a transformation of the glyphosate's amino acid to dithiocarbamate acid is required. Dithiocarbamate acid is used as an optical colour changer of polyvinyl alcohol that changes the colour from blue to yellow (De Almeida *et al.*, 2015; Tai *et al.*, 2022).

NMR explores the magnetic properties of atomic nuclei and can determine the glyphosate content in the liver, blood and urine in the sample postmortem (Dickson *et al.*, 1998). Surface plasmon resonance is used as an optical biosensor to monitor the interaction between the analyte and the (bio)element immobilised using special types of electromagnetic waves (Homola, 2003; Ding *et al.*, 2021; Gokila *et al.*, 2022).

UV–Vis was applied to water samples, where silver nanoparticles served as a colourimetric probe. The UV–Vis spectrum showed the LSPR band at 400 nm. And in the presence of glyphosate, the red colour shifted relative to the original band at 600 nm. Detection limits of 0.9 and 3.2 mg/L have been reported for UV–Vis extinction and surface-enhanced spectroscopy, respectively (De Góes *et al.*, 2017; Liao *et al.*, 2018). The flow injection spectrophotometry technique was used for the detection of glyphosate and its residues from wheat grains that showed a limit of detection and quantification of 0.20 and 0.70 μg/mL, respectively, via condensation reaction with p-dimethylaminobenzaldehyde (Jan *et al.*, 2018).

A combination of ninhydrin reaction and surface-enhanced Raman scattering spectroscopy was done to detect the glyphosate content in water and food. The direct correlation of glyphosate and the reaction product of glyphosate, ninhydrin (purple colour dye, PD) and the silver nanoparticles was found to be surface enhanced Raman scattering spectroscopy-active. This proposed method resulted in a detection limit of 1.43×10^{-8} mol/L for glyphosate. Electrochemical impedance spectroscopy (EIS) was used to determine the concentration of glyphosate from water and chitosan-molecularly imprinted polymer (CS-MIPS) modified with gold microelectrodes. EIS includes high and low frequencies for the electron transfer-limiting process and the diffusion process, respectively. With a slight increase in the concentration of glyphosate, the graph of EIS increased gradually, and the absorption of glyphosate by non-imprinted chitosan was negligible (Zouaoui *et al.*, 2020).

3.4.4.5 *Fluorescence-based glyphosate detection*

Fluorescence spectroscopy is a non-invasive spectroscopic technique where the fluorescence from the material of interest is measured after excitation with a beam of light (usually in the ultraviolet spectrum). Fluorescence-based glyphosate detection is a highly sensitive, yet simple, method used for the detection of glyphosate. This method uses emitted light, where electrons, upon absorbing the light, gain energy, become excited and release energy when returning to the ground state. The excitation of the component to fluorescence is shown by the exciting wavelength

(Nsibande and Forbes, 2016; Wan *et al.*, 2022). Recently, a fluorescence assay was developed for the detection of glyphosate content based on carbon dots, quantum dots, paper-based and aptamer-based approaches (Wang *et al.*, 2016; Yang *et al.*, 2022; Cheng *et al.*, 2022).

Graphene quantum dots and cysteine-capped silver nanoparticles were used in an analytical method for the detection of glyphosate from pulsed (peas and lupins), and QuEChERS was used for sample treatment. The LOD of glyphosate from pulses was 9 ng/mL (Jiménez-López *et al.*, 2020). A fluorescence-based technique was developed by enhancing the fluorescence of *N*-methylmesoporphyrin IX (NMM) with G-quadruplex (G4) and Cu^{2+} as a catalyst in the presence of Gly (Fang *et al.*, 2023). The LODs in water, rice, maize and wheat samples using the proposed method were 0.065, 0.13, 0.14 and 0.27 μM, respectively, with a recovery range of 88–103%.

3.4.4.6 *Visual or onsite detection of glyphosate by metal nanoparticles as a colourimetric probe*

Recently, scientists and researchers have been looking for novel technologies to improve the conventional analytical methods. Nanotechnology is one of the new technologies that has been developed to improve the detection methods of Gly. Advanced techniques using nanoparticles are among the most attractive methods compared to other analytical approaches such as chromatography, spectroscopy and fluorescence-based detection techniques. Various nanoparticles are used in nanotechnology-based sensing approaches. Visual or colourimetric detection is used for determining the chemical solution based on the colouring reagent and by comparing the colour changes of solutions, which determine the concentration of the analyte. Similarly, when light at a specific wavelength passes through a sample, it results in two possibilities. Firstly, the light can be absorbed by the sample, and secondly, the light can be transmitted through the sample, which is then detected. The colourimetric detection of Gly using metal nanoparticles is less time-consuming and more accurate. Metal nanoparticles that could be used as a colourimetric probe include gold (Au), silver (Ag), calcium (Ca), zinc (Zn), copper (Cu), iron (Fe) and silicon (Si). These metals are responsible for energy levels in the d–d transition, where the electron in a d-orbital is excited by a photon to another d-orbital of higher energy than the light emitted by the absorption and the scattering of the photons. Furthermore, an advantage of localised surface plasmon resonance (LSPR)-based colourimetric detection is its simplicity, where colour changes in the solution can be seen by the naked eye, and hence there is no requirement for sophisticated instruments for the measurements (Aldewachi *et al.*, 2018; Ma *et al.*, 2019).

These nanoparticles possess unique chemical and optical properties, such as biocompatibility, LSPR and resonance light scattering, which are of interest to researchers, scientists and general readers alike (Kumar *et al.*, 2017; Chawla *et al.*, 2018; Tai *et al.*, 2022). Visual or colourimetric detection techniques using nanoparticles are based on the phenomenon of LSPR, which has been shown to detect several residual herbicides in the samples. LSPR is the phenomenon in which light, having a shorter wavelength than the occurrence wavelength, interacts with charged nanoparticles.

Visual detection of glyphosate is categorised into two different mechanisms. The first mechanism is based on the catalytic growth of metal nanoparticles, which is directly influenced by enzymatically hydrolysed analytes, resulting in a colour change. The second mechanism is based on the aggregation of metal nanoparticles. The enzymatically hydrolysed analyte is reacted with the functional group of the metal nanoparticles, which results in colour changes. The detection of Gly from an environmental water sample using cysteamine-stabilised gold nanoparticles (CS-AuNPs; 30 nm) and HAc-NaAc buffer (20 mM at pH 4) was used, and Gly was quantified using UV–Vis with an absorption ratio of A_{650}/A_{524}.

3.4.5 *Quality control and quality assurance*

The determination of analytes, such as glyphosate, at ultra-trace detection levels in samples with increasingly complex matrices is one of the main developmental trends in chemical analytics. In general, this is a complicated task which poses a great challenge to analytical chemists. Therefore, it is required that appropriate attention be paid to the quality assurance (QA) and quality control (QC) of the obtained measurements. Ensuring the reliability of soil sample data requires robust QA and QC measures. The review explores strategies for minimising sample contamination, maintaining sample integrity during transportation and validating laboratory results.

3.5 Conclusion

This chapter comprehensively synthesises the current state of knowledge on soil sampling, emphasising the evolution of techniques, key considerations in sampling design and emerging trends in technology. A deeper understanding of soil sampling is vital for researchers, practitioners and policymakers engaged in sustainable land management and environmental protection. When the CoPCs are glyphosate and its metabolite AMPA, it is challenging to assess them due to their highly polar characteristics. Their interactions with soil components are established in several ways; therefore, it becomes difficult to predict their fate and behaviour. Further, analysis in the laboratory is also challenging. This chapter has outlined possible ways of soil sampling, soil characterisation and the analytical progress in glyphosate analysis.

References

Acosta-Cortés, A. G., Martinez-Ledezma, C., López-Chuken, U. J., Kaushik, G., Nimesh, S., and Villarreal-Chiu, J. F. (2019). Polyphosphate recovery by a native *Bacillus cereus* strain as a direct effect of glyphosate uptake. *The ISME Journal*, 13, 1497–1505.

Aldewachi, H., Chalati, T., Woodroofe, M. N., Bricklebank, N., Sharrack, B., and Gardiner, P. (2018). Gold nanoparticle-based colorimetric biosensors. *Nanoscale*, 10, 18–33.

Anastassiades, M., Wachtler, A.-K., Kolberg, D. I., *et al.* (2023). Quick method for the analysis of highly polar pesticides in food involving extraction with acidified methanol and LC- or IC-MS/MS measurement I. Food of plant origin (QuPPe-PO-Method). *EU Reference Laboratory for Pesticides Requiring Single Residue Methods (EURL-SRM)*. Stuttgart, Germany: CVUA, p. 164.

Arkan, T. and Molnár-Perl, I. (2015). The role of derivatization techniques in the analysis of glyphosate and aminomethyl-phosphonic acid by chromatography. *Microchemical Journal*, 121, 99–106.

Association of Official Agricultural Chemists (AOAC) International. (2007). Pesticide Residues in Foods by Acetonitrile Extraction and Partitioning with Magnesium Sulfate; Gas Chromatography/Mass Spectrometry and Liquid Chromatography/Tandem Mass Spectrometry. AOAC Official Method 2007.01, p. 10. http://www.aoacofficialmethod.org/index.php?main_page=product_info&products_id=2155.

Aydin, Z., Akın, Ş., Çenet, E. N., Keskinateş, M., Akbulut, A., Keleş, H., and Keleş, M. (2022). Two novel enzyme-free colorimetric sensors for the detection of glyphosate in real samples. *Journal of Food Composition and Analysis*, 112, 104674.

Bain, C., Selfa, T., Dandachi, T., and Velardi, S. (2017). 'Superweeds' or 'survivors'? Framing the problem of glyphosate resistant weeds and genetically engineered crops. *Journal of Rural Studies*, 51, 211–221.

Benbrook, C. M. (2016). Trends in glyphosate herbicide use in the United States and globally. *Environmental Sciences Europe*, 28, 1–15.

Benbrook, C. M. (2019). How did the US EPA and IARC reach diametrically opposed conclusions on the genotoxicity of glyphosate-based herbicides? *Environmental Sciences Europe*, 31, 1–16.

Bergstrom, L., Borjesson, E., and Stenstrom, J. (2011). Laboratory and lysimeter studies of glyphosate and aminomethylphosphonic acid in a sand and a clay soil. *Journal of Environmental Quality*, 40, 98–108.

Bhave, P. P. and Sadhwani, K. (2022). Sampling in environmental matrices: A critical review. *Environmental Forensics*, 23, 75–92.

Borggaard, O. K. and Gimsing, A. L. (2008). Fate of glyphosate in soil and the possibility of leaching to ground and surface waters: A review. *Pest Management Science*, 64, 441–456.

Boulding, J. R. (2017). *Description and Sampling of Contaminated Soils: A Field Guide*. United Kingdom: Routledge.

Büyüker, B. (2009). A decision support tool for conceptual site model development at contaminated sites. MS Thesis, Middle East Technical University, Turkey.

Byer, J. D., Struger, J., Klawunn, P., Todd, A., and Sverko, E. (2008). Low-cost monitoring of glyphosate in surface waters using the ELISA method: An evaluation. *Environmental Science & Technology*, 42, 6052–6057.

Byrnes, M. E. (2008). *Field Sampling Methods for Remedial Investigations*. USA: CRC Press.

Çetin, E., Sßahan, S., Ülgen, A., and Sßahin, U. U. (2017). DLLME-spectrophotometric determination of glyphosate residue in legumes. *Food Chemistry*, 230, 567–571.

Chamkasem, N. and Harmon, T. (2016). Direct determination of glyphosate, glufosinate, and AMPA in soybean and corn by liquid chromatography/tandem mass spectrometry. *Analytical and Bioanalytical Chemistry*, 408, 4995–5004.

Chang, R. (2013). *Surface Enhanced Raman Scattering*. Netherlands: Springer Science & Business Media.

Chawla, P., Kaushik, R., Shiva Swaraj, V. J., and Kumar, N. (2018). Organophosphorus pesticides residues in food and their colorimetric detection. *Environmental Nanotechnology, Monitoring & Management*, 10, 292–307.

Cheng, Y., Li, G., Huang, X., Qian, Z., and Peng, C. (2022). Label-free fluorescent turn-on glyphosate sensing based on DNA-templated silver nanoclusters. *Biosensors*, 12, 832.

Cline, M. G. (1944). Principles of soil sampling. *Soil Science*, 58, 275–288.

Cohen, M. P. (2011). 'Stratified sampling'. In Lovric, M. (Ed.), *International Encyclopedia of Statistical Science.* Berlin: Springer, pp. 1547–1550.

Croplife, A. (2018). Glyphosate — The Facts. Available at: https://www.croplife.org.au/media/issues-campaigns/glyphosate/ (Accessed on 01 July 2023).

Da Silva, A. S., Fernandes, F. C. B., Tognolli, J. O., Pezza, L., and Pezza, H. R. (2011). A simple and green analytical method for determination of glyphosate in commercial formulations and water by diffuse reflectance spectroscopy. *Spectrochimica Acta, Part A: Molecular and Biomolecular Spectroscopy*, 79, 1881–1885.

De Almeida, L. K. S., Chigome, S., Torto, N., Frost, C. L., and Pletschke, B. I. (2015). A novel colorimetric sensor strip for the detection of glyphosate in water. *Sensors and Actuators B: Chemical*, 206, 357–363.

De Góes, R. E., Muller, M., and Fabris, J. L. (2017). Spectroscopic detection of glyphosate in water assisted by laser-ablated silver nanoparticles. *Sensors*, 17, 954.

De Jonge, L. W., Jacobsen, O. H., and Moldrup, P. (1999). Soil water repellency: Effects of water content, temperature, and particle size. *Soil Science Society of America Journal*, 63, 437–442.

Dhamu, V. N. and Prasad, S. (2020). ElectrochemSENSE: A platform towards field deployable direct on-produce glyphosate detection. *Biosensors and Bioelectronics*, 170, 112609.

Ding, S., Lyu, Z., Li, S., Ruan, X., Fei, M., Zhou, Y., Niu, X., Zhu, W., Du, D., and Lin, Y. (2021). Molecularly imprinted polypyrrole nanotubes based electrochemical sensor for glyphosate detection. *Biosensors and Bioelectronics*, 191, 113434.

Dollinger, J., Dagès, C., and Voltz, M. (2015). Glyphosate sorption to soils and sediments predicted by pedotransfer functions. *Environmental Chemistry Letters*, 13, 293–307.

European Commission (2023). Guidance Document on Pesticide Analytical Methods for Risk Assessment and Post-approval Control and Monitoring Purposes. SANTE/2020/12830, Rev.2. EFSA.

Fang, H., Zhang, X., Ma, L., Xiao, Y., Yang, H., and Zhou, Y. (2023). Fluorescence detection of glyphosate based on G-quadruplex and porphyrin metalation. *Journal of Food Composition and Analysis*, 116, 105074.

Ferrante, M., Rapisarda, P., Grasso, A., Favara, C., and Oliveri Conti, G. (2023). Glyphosate and environmental toxicity with "One Health" approach, a review. *Environmental Research*, 235, 116678.

Gilbert, R. O. (1987). *Statistical Methods for Environmental Pollution Monitoring*. USA: John Wiley & Sons.

Gill, J. P. K., Sethi, N., and Mohan, A. (2017). Analysis of the glyphosate herbicide in water, soil and food using derivatising agents. *Environmental Chemistry Letters*, 15, 85–100.

Gillman, G. P. (1979). A proposed method for the measurement of exchange properties of highly weathered soils. *Soil Research*, 17, 129–139.

Gillman, G. P. and Sumpter, E. A. (1986). Modification to the compulsive exchange method for measuring exchange characteristics of soils. *Soil Research*, 24, 61–66.

Gokila, N., Muthumalai, K., Haldorai, Y., and Kumar, R. T. R. (2022). Electrochemical non-enzymatic sensor based on Co-H_2ABDC metal organic framework for detection of glyphosate. *Chemical Physics Letters*, 795, 139481.

Gómez Ortiz, A. M., Okada, E., Bedmar, F., and Costa, J. L. (2017). Sorption and desorption of glyphosate in Mollisols and Ultisols soils of Argentina. *Environmental Toxicology and Chemistry*, 36, 2587–2592.

González-Curbelo, M. Á., Varela-Martínez, D. A., and Riaño-Herrera, D. A. (2022). Pesticide-residue analysis in soils by the QuEChERS method: A review. *Molecules*, 27, 4323.

Gotti, R., Fiori, J., Bosi, S., and Dinelli, G. (2019). Field-amplified sample injection and sweeping micellar electrokinetic chromatography in analysis of glyphosate and aminomethylphosphonic acid in wheat. *Journal of Chromatography A*, 1601, 357–364.

The Guardian (2018). Landmark Lawsuit Claims Monsanto Hid Cancer Danger of Weedkiller for Decades, pp. 186–206. Available at: https://www.theguardian.com/business/2018/may/22/monsanto-trial-cancer-weedkiller-roundup-dewayne-johnson (Accessed on 11 September 2023).

Halewood, E., Opalk, K., Custals, L., Carey, M., Hansell, D. A., and Carlson, C. A. (2022). Determination of dissolved organic carbon and total dissolved nitrogen in seawater using high temperature combustion analysis. *Frontiers in Marine Science*, 9, 1061646.

Health Canada. (2023). Guidelines for Canadian Drinking Water Quality. Government of Canada. Available at: https://www.canada.ca/en/health-canada/services/environmental-workplace-health/reports-publications/water-quality/guidelines-canadian-drinking-water-quality-summary-table.html#t2.

Homola, J. (2003). Present and future of surface plasmon resonance biosensors. *Analytical and Bioanalytical Chemistry*, 377, 528–539.

Ibanez, M., Pozo, O. J., Sancho, J. V., Lopez, F. J., and Hernandez, F. (2005). Residue determination of glyphosate, glufosinate and aminomethylphosphonic acid in water and soil samples by liquid chromatography coupled to electrospray tandem mass spectrometry. *Journal of Chromatography A*, 1081, 145–155.

International Agency for Research on Cancer (IARC). (2015). Some organophosphate insecticides and herbicides. In *IARC Monographs on the Evaluation of*

Carcinogenic Risks to Humans, Vol. 112, p. 464. The International Agency for Research on Cancer, Lyon, France.

International Atomic Energy Agency (IAEA). (2004). Soil Sampling for Environmental Contaminants. Available at: https://www.iaea.org/publications/7113/soil-sampling-for-environmental-contaminants (Accessed on 02 November 2023).

Jan, M. R., Shah, J., Muhammad, M., and Ara, B. (2018). Flow injection spectrophotometric determination of glyphosate herbicide in wheat grains via condensation reaction with P-dimethylaminobenzaldehyde. *South African Journal of Chemistry*, 71, 39–45.

Jiménez-López, J., Llorent-Martínez, E. J., Ortega-Barrales, P., and Ruiz-Medina, A. (2020). Graphene quantum dots-silver nanoparticles as a novel sensitive and selective luminescence probe for the detection of glyphosate in food samples. *Talanta*, 207, 120344.

Kaczyński, P. and Łozowicka, B. (2015). Liquid chromatographic determination of glyphosate and aminomethylphosphonic acid residues in rapeseed with MS/MS detection or derivatization/fluorescence detection. *Open Chemistry*, 13, 000010151520150107.

Keržič, E., Vek, V., Poljanšek, I., and Oven, P. (2019). Optimization of accelerated solvent extraction (Ase) of silver fir wood (*Abies alba* Mill.). *Les/Wood*, 68, 69–77.

Kim, E. J. and Baek, K. (2019). Selective recovery of ferrous oxalate and removal of arsenic and other metals from soil-washing wastewater using a reduction reaction. *Journal of Cleaner Production*, 221, 635–643.

Koskinen, W. C., Marek, L. J., and Hall, K. E. (2016). Analysis of glyphosate and aminomethylphosphonic acid in water, plant materials and soil. *Pest Management Science*, 72, 423–432.

Kumar, A. M., Suresh, B., Das, S., Obot, I. B., Adesina, A. Y., and Ramakrishna, S. (2017). Promising bio-composites of polypyrrole and chitosan: Surface protective and in vitro biocompatibility performance on 316L SS implants. *Carbohydrate Polymers*, 173, 121–130.

Lacorte, S., Bono-Blay, F., and Cortina-Puig, M. (2012). Sample homogenization. In Pawliszyn, J. (Ed.), *Comprehensive Sampling and Sample Preparation*. United Kingdom: Oxford Academic Press, pp. 65–84.

Li, Z. and Jennings, A. (2017). Worldwide regulations of standard values of pesticides for human health risk control: A review. *International Journal of Environmental Research and Public Health*, 14, 826.

Liao, Y., Berthion, J.-M., Colet, I., Merlo, M., Nougadère, A., and Hu, R. (2018). Validation and application of analytical method for glyphosate and glufosinate in foods by liquid chromatography-tandem mass spectrometry. *Journal of Chromatography A*, 1549, 31–38.

Ma, X., He, S., Qiu, B., Luo, F., Guo, L., and Lin, Z. (2019). Noble metal nanoparticle-based multicolor immunoassays: An approach toward visual quantification of the analytes with the naked eye. *ACS Sensors*, 4, 782–791.

Maggi, F., La Cecilia, D., Tang, F. H. M., and Mcbratney, A. (2020). The global environmental hazard of glyphosate use. *Science of the Total Environment*, 717, 137167.

Magin, R. W. (2003). Glyphosate: Twenty-eight years and still growing—The discovery, development, and impact of this herbicide on the agrichemical industry. Pesticide formulations and application systems. *ASTM International,* 23, 149–157.

Magnusson, B. and Örnemark, U. (2014). *Eurachem Guide: The Fitness for Purpose of Analytical Methods — A Laboratory Guide to Method Validation and Related Topics*. 2nd ed. Available at: www.eurachem.org.

Mahdavi, V., Heris, M.-E. S., Dastranj, M., Farimani, M. M., Eslami, Z., and Aboul-Enein, H. Y. (2021). Assessment of Pesticide Residues in Soils Using a QuEChERS Extraction Procedure and LC-MS/MS. *Water, Air, & Soil Pollution*, 232(4), 159. https://doi.org/10.1007/s11270-021-05104-4.

Maher, W., Cullen, P., and Norris, R. (1994). Framework for designing sampling programs. *Environmental Monitoring and Assessment*, 30, 139–162.

Malherbe, L. (2002). Designing a contaminated soil sampling strategy for human health risk assessment. *Accreditation and Quality Assurance*, 7, 189–194.

Mallat, E. and Barceló, D. (1998). Analysis and degradation study of glyphosate and of aminomethylphosphonic acid in natural waters by means of polymeric and ion-exchange solid-phase extraction columns followed by ion chromatography-post-column derivatization with fluorescence detection. *Journal of Chromatography A*, 823, 129–136.

Mamy, L., Gabrielle, B., and Barriuso, E. (2008). Measurement and modelling of glyphosate fate compared with that of herbicides replaced as a result of the introduction of glyphosate-resistant oilseed rape. *Pest Management Science*, 64, 262–275.

Martin, P. J., He, K., Blaney, L., and Hobbs, S. R. (2023). Advanced liquid chromatography with tandem mass spectrometry method for quantifying glyphosate, glufosinate, and aminomethylphosphonic acid using pre-column derivatization. *ACS EST Water*, 3, 2407–2414.

Mertens, M., Hoss, S., Neumann, G., Afzal, J., and Reichenbecher, W. (2018). Glyphosate, a chelating agent-relevant for ecological risk assessment? *Environmental Science and Pollution Research*, 25, 5298–5317.

Mesnage, R., Defarge, N., Spiroux De Vendômois, J., and Séralini, G. E. (2015). Potential toxic effects of glyphosate and its commercial formulations below regulatory limits. *Food and Chemical Toxicology*, 84, 133–153.

Miles, C. J. and Moye, H. A. (1988). Extraction of glyphosate herbicide from soil and clay minerals and determination of residues in soils. *Journal of Agricultural and Food Chemistry*, 36, 86–491.

Nasif, S. O., Nuruzzaman, M., and Naidu, R. (2024). Porous silica nanocarriers: Advances in structural orientation and modification to develop sustainable pesticide delivery systems. *ACS Agricultural Science & Technology*, 4, 144–172.

National Environmental Protection Council (NEPC). (2013). Schedule B2, Guidelines on Site Characterization. Available at: https://www.nepc.gov.au/nepms/assessment-site-contamination (Accessed on 23 August 2023).

National Health and Medical Research Council (NHMRC), National Resource Management Ministerial Council (NRMMC). (2011). Australian Drinking Water Guidelines Paper 6 National Water Quality Management Strategy. National Health and Medical Research Council, National Resource Management Ministerial Council, Commonwealth of Australia, Canberra. https://www.nhmrc.gov.au/sites/default/files/documents/reports/aust-drinking-water-guidelines.pdf.

Nevada Division of Environmental Protection (NDEP). (1999). Sampling Equipment Decontamination. Available at: https://ndep.nv.gov/environmental-cleanup/brownfields/quality-assurance-plans (Accessed on 04 September 2023).

New Jersey Department of Environmental Protection (NJDEP). (1988). Field Sampling Procedures Manual. USA: New Jersey. Available at: https://www.nj.gov/dep/srp/guidance/fspm/ (Accessed on 24 December 2023).

New South Wales Environmental Protection Authority (NSW EPA). (2022). Sampling Design Part 1 — Application: Contaminated Land Guidelines. Available at: https://www.epa.nsw.gov.au/search?q=Sampling+Design+Part+1+-+Application%3A+Contaminated+Land+Guidelines (Accessed on 14 August 2023).

Nsibande, S. A. and Forbes, P. B. C. (2016). Fluorescence detection of pesticides using quantum dot materials — A review. *Analytica Chimica Acta*, 945, 9–22.

Nuruzzaman, M., Rahman, M. M., Liu, Y., and Naidu, R. (2016). Nanoencapsulation, nano-guard for pesticides: A new window for safe application. *Journal of Agricultural and Food Chemistry*, 64, 1447–1483.

Oulkar, D. P., Hingmire, S., Goon, A., Jadhav, M., Ugare, B., Thekkumpurath, A. S., and Banerjee, K. (2017). Optimization and validation of a residue analysis method for glyphosate, glufosinate, and their metabolites in plant matrixes by liquid chromatography with tandem mass spectrometry. *Journal of AOAC International*, 100, 631–639.

Padilla, J. T. and Selim, H. M. (2020). Environmental behavior of glyphosate in soils. *Advances in Agronomy*, 159, 1–34.

Pasquini, C. (2003). Near infrared spectroscopy: Fundamentals, practical aspects and analytical applications. *Journal of the Brazilian Chemical Society*, 14, 198–219.

Pitard, F. F. (2019). *Theory of Sampling and Sampling Practice*. United States: CRC Press.

Quigley, A., Cummins, W., and Connolly, D. (2016). Dispersive liquid-liquid microextraction in the analysis of milk and dairy products: A review. *Journal of Chemistry*, 2016, 1–12.

Ramirez, C. E., Bellmund, S., and Gardinali, P. R. (2014). A simple method for routine monitoring of glyphosate and its main metabolite in surface waters using lyophilization and LC-FLD+MS/MS. Case study: Canals with influence on Biscayne National Park. *Science of the Total Environment*, 496, 389–401.

Rawat, D., Bains, A., Chawla, P., *et al*. (2023). Hazardous impacts of glyphosate on human and environment health: Occurrence and detection in food. *Chemosphere*, 329, 138676.

Rennert, T., Dietel, J., Heilek, S., Dohrmann, R., and Mansfeldt, T. (2021). Assessing poorly crystalline and mineral-organic species by extracting Al, Fe, Mn, and Si using (Citrate-) ascorbate and oxalate. *Geoderma*, 397, 115095.

Reyes, I. and Torrent, J. (1997). Citrate-ascorbate as a highly selective extractant for poorly crystalline iron oxides. *Soil Science Society of America Journal*, 61, 1647–1654.

Richter, B. E., Jones, B. A., Ezzell, J. L., Porter, N. L., Avdalovic, N., and Pohl, C. (1996). Accelerated solvent extraction: A technique for sample preparation. *Analytical Chemistry*, 68, 1033–1039.

Riedo, J., Wettstein, F. E., Rösch, A., Herzog, C., Banerjee, S., Büchi, L., Charles, R., Wächter, D., Martin-Laurent, F., Bucheli, T. D., Walder, F., and Van Der Heijden, M. G. A. (2021). Widespread occurrence of pesticides in organically managed agricultural soils- the ghost of a conventional agricultural past? *Environmental Science & Technology*, 55, 2919–2928.

Romano-Armada, N., Amoroso, M. J., and Rajal, V. B. (2019). Construction of a combined soil quality indicator to assess the effect of glyphosate application. *Science of the Total Environment*, 682, 639–649.

Rosales, C. A., Shields, S. W., Aulenback, C. L., *et al*. (2023). Improved chromatography and MS-based detection of glyphosate and aminomethylphosphonic acid using iTrEnDi. *Journal of the American Society for Mass Spectrometry*, 34, 948–957.

Sapkale, G. N., Patil, S. M., Surwase, U. S., and Bhatbhage, P. K. (2010). Supercritical fluid extraction. *International Journal of Chemical Science*, 8, 729–743.

Singh, R., Shukla, A., Kaur, G., Girdhar, M., Malik, T., and Mohan, A. (2024). Systemic analysis of glyphosate impact on environment and human health. *ACS Omega*, 9, 6165–6183.

Steinborn, A., Alder, L., Michalski, B., Zomer, P., *et al*. (2016). Determination of glyphosate levels in breast milk samples from Germany by LC-MS/MS and GC-MS/MS. *Journal of Agricultural and Food Chemistry*, 64, 1414–1421.

Surapong, N. and Burakham, R. (2021). Magnetic molecularly imprinted polymer for the selective enrichment of glyphosate, glufosinate, and aminomethylphosphonic acid prior to high-performance liquid chromatography. *ACS Omega*, 6, 27007–27016.

Suwardji, S. and Sudantha, I. M. (2021). The fate of glyphosate in soil and water: A review. *Jurnal Penelitian Pendidikan IPA*, 7, 389–399.

Tai, S., Qian, Z., Ren, H., Barimah, A. O., Peng, C., and Wei, X. (2022). Highly selective and sensitive colorimetric detection for glyphosate based on B-CD@DNA-Cuncs enzyme mimics. *Analytica Chimica Acta*, 1222, 339992.

Tillé, Y. and Wilhelm, M. (2017). Probability sampling designs: Principles for choice of design and balancing. *Statistical Science*, 32, 176–189.

Todorovic, G. R., Mentler, A., Popp, M., *et al.* (2013). Determination of glyphosate and AMPA in three representative agricultural Austrian soils with a HPLC-MS/MS method. *Soil and Sediment Contamination*, 22, 332–350.

Trick, J. K., Stuart, M., and Reeder, S. (2008). Contaminated groundwater sampling and quality control of water analyses. In De Vivo, B., Belkin, E. H., and Lima, A. (Eds.), *Environmental Geochemistry*. Netherlands: Elsevier, pp. 25–45.

Tu, X. and Chen, W. (2018). A review on the recent progress in matrix solid phase dispersion. *Molecules*, 23, 2767.

United States Environmental Protection Agency (US EPA). (1991). Compendium of ERT Soil Sampling and Surface Geophysics Procedures. Available at: https://www.epa.gov/.

United States Environmental Protection Agency (US EPA). (2002). Soil Screening Guidance: Technical Background Document (TBD). Available at: https://www.epa.gov/.

United States Environmental Protection Agency (US EPA). (2023). Ground Water and Drinking Water. US- EPA. Retrieved January 14 from https://www.epa.gov/ground-water-and-drinking-water/national-primary-drinking-water-regulations#Organic.

Usui, K., Minami, E., Fujita, Y., Kubota, E., *et al.* (2019). Application of probe electrospray ionization-tandem mass spectrometry to ultra-rapid determination of glufosinate and glyphosate in human serum. *Journal of Pharmaceutical and Biomedical Analysis*, 174, 175–181.

Valle, A. L., Mello, F. C. C., Alves-Balvedi, R. P., Rodrigues, L. P., and Goulart, L. R. (2019). Glyphosate detection: Methods, needs and challenges. *Environmental Chemistry Letters*, 17, 291–317.

Wan, C. Q., Pang, Y. H., Feng, Y. W., and Shen, X. F. (2022). A ratio fluorescence sensor based on rhodamine B embedded metal-organic framework for glyphosate detection in agri-food products. *Food Chemistry*, 394, 133446.

Wimmer, B., Neidhardt, H., Schwientek, M., *et al.* (2022). Phosphate addition enhances alkaline extraction of glyphosate from highly sorptive soils and aquatic sediments. *Pest Management Science*, 78, 2550–2559.

Yang, Y., Ghalandari, B., Lin, L., Sang, X., Su, W., Divsalar, A., and Ding, X. (2022). A turn-on fluorescence sensor based on Cu^{2+} modulated DNA-templated silver nanoclusters for glyphosate detection and mechanism investigation. *Food Chemistry*, 367, 130617.

Zhang, W., Feng, Y., Ma, L., An, J., Zhang, H., Cao, M., Zhu, H., Kang, W., and Lian, K. (2019). A method for determining glyphosate and its metabolite aminomethyl phosphonic acid by gas chromatography-flame photometric detection. *Journal of Chromatography A*, 1589, 116–121.

Zouaoui, F., Bourouina-Bacha, S., Bourouina, M., *et al.* (2020). Electrochemical impedance spectroscopy determination of glyphosate using a molecularly imprinted chitosan. *Sensors and Actuators B: Chemical*, 309, 127753.

https://doi.org/10.1142/9781800617841_0004

Chapter 4

Soil Sampling and Analysis for Heavy Metals at Contaminated Sites

S. B. Shah

Department of Chemistry, School of Sciences, College of Engineering & Technical Vocational Education and Training, Fiji National University, Suva, Fiji

sofia.shah@fnu.ac.fj

Abstract

Sampling and assessment of contaminated sites are fundamental to identifying and understanding the extent of contamination and potential risks to human and environmental health. Environmental investigations of contaminated sites usually include site characterisation, sampling design, sample collection and analysis and data interpretation to make informed decisions about site remediation and management. Contaminated sites, including soils, are complex matrices; therefore, precautionary measures need to be taken so that sampling and assessment protocols are followed correctly. Sampling needs to be conducted after a careful consideration of all exposure pathways and the creation of well-thought-out and well-designed conceptual site diagrams capturing them.

Contaminated sites can contain various types of pollutants, with heavy metals being a common contaminant among them. Heavy metals are elements with high atomic weights and densities greater than 5 g/cm^3, and they are regarded as harmful at higher concentrations. The origin of these heavy metals, including lead, cadmium, mercury and arsenic, is attributed to different anthropogenic activities, such as mining, industrial discharges, various land-use types and the use of metal compounds in domestic and agricultural products. Soils contaminated with heavy metals pose an environmental threat in terms of food security, affecting human and ecosystem health and degrading land for agricultural production.

Hence, it is imperative to carry out soil sampling in order to assess the amount of heavy metals present at various contaminated sites and determine the degree of contamination as per environmental regulations governing heavy metal concentrations in soil. Thus, this chapter provides information about heavy metals in soils, looks into risk assessment indices of soils and provides sampling and assessment techniques for analysing heavy metal concentrations in contaminated soils.

4.1 Introduction

Soils are the basis for life on Earth along with water and air. Soils are finite resources with ecological and economic significance. Being the topmost weathering layer of the Earth's crust, soils consist of minerals, humus, water, air and living organisms. Soils filter and purify water, store and break down substances and regulate the natural cycles of water, air and organic and mineral matter. In fact, soils are classified as an indispensable medium for material cycling and energy exchange between plants and the natural environment (Zhou *et al.*, 2023).

Nuralykyzy *et al.* (2021) reported that half of the Earth's surface is occupied by agricultural lands (cultivated lands, managed pastures and permanent crops), and as such, land use has a significant impact on both the chemical and physical properties of soils. Soil health plays a pivotal role in the quality and quantity of agricultural crop production, ensuring food security (Chaudhry *et al.*, 2024).

Soils are regarded as the ultimate sinks for environmental pollution (FAO & UNEP, 2021; Wuana and Okieimen, 2011). Soil pollution refers to the buildup of persistent toxic compounds, chemicals, salts, radioactive materials or disease-causing agents in soils at higher-than-normal concentrations, which tend to have adverse effects on ecosystems (Okrent, 1999; Münzel *et al.*, 2023).

Soil contamination and pollution occur due to a variety of intended, accidental or naturally occurring activities. These include manufacturing, mineral extraction, waste disposal, accidental spillages, leaks from storage tanks, pesticide use, fertiliser application, wastewater irrigation, coal combustion residues, spillage of petrochemicals, discarding of electronic/metal waste on land, urban development and atmospheric deposition (Wuana and Okieimen, 2011). Geological as well as pedological processes in soils release naturally occurring contaminants such as trace elements, radionuclides and asbestos (ISO, 2015; Tian *et al*., 2017), often with natural soil background concentrations, which depend on the mineralogical composition of the soil parent material and paedogenetic processes (Kabata-Pendias and Pendias, 2001; Wilson *et al*., 2008).

Some common soil contaminants include heavy metals (HMs), organic pollutants, asbestos, nutrients and fertilisers, pesticides, contaminated water and leachates, soil acidification, salinisation, radionuclides, petroleum, pesticides and solvents, as well as microplastics and per- and polyfluoroalkyl substances (PFASs). This chapter explores and discusses the sampling protocols and assessment of soils contaminated with HMs.

HMs are inorganic, persistent and biologically toxic pollutants; the HM pollution at contaminated sites usually results from rapid urbanisation, industrial discharges, mining, solid waste and sewage irrigation, as well as increased municipal, agricultural, domestic, medical and technological applications (Tang *et al*., 2010; Li *et al*., 2014; Wang *et al*., 2015; Uchimiya *et al*., 2020; Mehr *et al*., 2021; Xu *et al*., 2019; Rashid *et al*., 2023). HMs and their associated compounds are pervasive in the natural world (Hu and Cheng, 2013).

Both natural and anthropogenic sources contribute to HM contamination in soils and sediments (Cui *et al*., 2022). For example, iron (Fe) is naturally occurring; however, anthropogenic activities have significantly increased its level in many soils.

Soil contamination by HMs results from emissions caused by the rapid expansion of industrial areas, mine tailings, leaded gasoline and paints, fertiliser and pesticide use, animal manures, coal combustion residues, spillage of petrochemicals and atmospheric deposition (Khan *et al*., 2008; Zhang *et al*., 2010).

Soils contaminated with HM are regarded as a global problem, as they pose risks and hazards to human and ecosystem health, contaminate groundwater, reduce food quality and damage infrastructure (McLaughlin *et al*., 2000a; Rehman *et al*., 2023). Being persistent, toxic and non-degradable, the accumulation of HMs in soils might pose a serious threat to food safety and human health (Ma *et al*., 2013; Rehman *et al*., 2023). For example, Nunez *et al*. (2017) reported that long-term exposure to HMs led to mental and behavioural disorders and an increased risk of cancer.

Soil pollution is now emerging as a significant global issue due to increasing industrialisation, urbanisation, agricultural production and inadequate waste management. Particularly, agricultural soils become long-term sinks for HMs (Huang *et al.*, 2015; Wong *et al.*, 2002); thus, they become less productive (Zhou *et al.*, 2023). HM contamination in agricultural soils has become a global issue (Rashid *et al.*, 2023), an obstacle to sustainable agricultural development and a societal concern (Zhou *et al.*, 2023) since HMs have the potential to degrade crop health and production at elevated concentrations (Shahid *et al.*, 2015; Rashid *et al.*, 2023).

Soils contaminated with HMs may pose risks and hazards to humans and the ecosystem through direct ingestion or contact with contaminated soil, the food chain, consumption of contaminated groundwater, reduction in food quality via phytotoxicity, degradation of agricultural land leading to food insecurity and land tenure problems (McLaughlin *et al.*, 2000a, 2000b; Ling *et al.*, 2007). Accumulation and contamination of HMs in agricultural soils have aroused worldwide attention due to their toxicity, non-biodegradability and persistence (Alloway, 2013). Especially, the risk assessment for HM contamination in agricultural ecosystems is crucial for achieving the Sustainable Development Goals (SDGs) for food security (Wu *et al.*, 2021).

Understanding and addressing the sources of soil contamination is also crucial for environmental protection and human health. Regulatory frameworks, site assessments and remediation efforts are essential components of managing and mitigating the impact of these various activities on soil quality.

Environmental investigations of contaminated soil sites usually include site characterisation, sampling design, sample collection and analysis and data interpretation to inform decisions about site remediation and management.

Sampling and assessment of contaminated sites are fundamental to identifying and understanding the extent of contamination and potential risks to human and environmental health. A contaminated soil site refers to an area where the soil has been adversely affected by the presence of hazardous substances or pollutants, thus posing a risk to human health and the environment. Contaminated soils, especially those containing chemicals and toxic substances, could be lethal to humans as well as to the environment.

The focus of this book chapter is to elaborate on HM pollution in soils, soil geochemical risk indices assessing the extent of pollution, soil sampling related to contaminated sites and assessment techniques of analysing HMs in contaminated soils.

4.2 Heavy Metals and Soil Pollution

4.2.1 *Heavy metals in soils*

Soil pollution by HMs refers to the presence of elevated concentrations of metallic elements in the soil, which can have detrimental effects on soil quality, plant and microbial life. HMs are elements with high atomic weights and densities greater than 5 g/cm^3 (Duffus, 2002; Hubbard, 2005; Oves *et al.*, 2012; Li *et al.*, 2019). HMs are persistent, toxic and non-biodegradable substances; they tend to accumulate within organisms and across food chains. These substances can persist for longer periods in the environment, thus bioaccumulating in the food chain and posing risks to humans and wildlife (Upadhyay *et al.*, 2024).

Natural occurrence of HMs in soils is from the weathering of parent material (igneous and sedimentary rocks) and variations in the intensity of paedogenic processes (Bocardi *et al.*, 2020), with levels that are regarded as trace (<1000 mg/kg) (Kabata-Pendias and Pendias, 2001; Alloway, 2013). Furthermore, natural events such as volcanic emissions, ocean salt sprays, wind-borne soil particles, forest fires and rock weathering also contribute to the presence of HMs in soil (Priya *et al.*, 2023). Biogenic sources, such as the decay of organic matter, also release HMs into soil (Yan *et al.*, 2020).

Anthropogenic activities resulting in soil pollution by HMs include waste disposal, atmospheric deposition, waste incineration, urban effluents, mining, fertiliser application, animal manures, sewage sludge, pesticides, coal combustion residues, spillage of petrochemicals, traffic emissions, emissions from rapidly expanding industrial areas, industrial discharges, disposal of high metal wastes and long-term application of wastewater in the agricultural sector (Khan *et al.*, 2008; Zhang *et al.*, 2010; Cachada *et al.*, 2012; Lv *et al.*, 2015; Priya *et al.*, 2023).

For instance, Cheng *et al.* (2014) reported that the total emissions of Cd in China were estimated at about 743.77 metric tonnes for the year 2009, of which the contributions from industrial processes and combustion sources were approximately 56.6% and 43.4%, respectively. Xing *et al.* (2019) determined HM concentrations in farmland soil near a Pb smelting area in Jiyuan City, China, and reported mean concentrations of total Cd, Pb, Cu and Zn to be 2.10, 184, 31.8 and 99.2 mg/kg, respectively.

Table 4.1 highlights some studies which determined the levels of HM contamination in different soil types, with some exceeding the WHO permissible limits of HMs in soils.

Common HM pollutants (Figure 4.1) in the soil environment include arsenic (As), cadmium (Cd), chromium (Cr), mercury (Hg), lead (Pb), copper (Cu), zinc (Zn) and nickel (Ni) (Jorgensen *et al.*, 2005; Li *et al.*, 2019; Zhao *et al.*, 2022; Priya *et al.*, 2023), usually found at contaminated sites. Four of these HMs, namely As, Cd, Hg and Pb, are considered extremely harmful to both plants and humans, according to the Agency for Toxic Substances and Disease Registry (ATSDR) (Mansoor *et al.*, 2020), with Pb and Hg having hazardous properties (Le Roux *et al.*, 2019; Jiang *et al.*, 2006).

Essential elements such as iron (Fe), Cu and Zn have important physiological roles at minute concentrations; however, they become toxic to humans and biota at higher concentrations (Parvez *et al.*, 2023). Consumption of food contaminated with HMs can lead to various health issues including kidney damage, neurological disorders, respiratory problems and even cancer (Xing *et al.*, 2020), with the exposure pathway being soil–plant–human. Table 4.2 highlights some common HMs, their sources and health impacts.

HM contamination has led to several environmental catastrophes, impacting both environmental and public health worldwide. The most notable environmental disasters have been associated with improper handling of industrial waste/discharge and mining activities as well as gas leaks; they are highlighted in Figure 4.2 (briefly summarised in Box 4.1). Common environmental disasters are associated with HMs such as Hg, Cd, Pb and As.

The total contents of HM in soil are the sum of the concentrations of elements derived from minerals in the geological parent material on which the soil has developed (lithogenic source) and inputs from a wide range of possible anthropogenic (contamination) sources (Alloway, 2013).

A simple mass balance of HMs in the soils is as follows (Alloway, 2013; Lombi and Gerzabek, 1998):

$$HM_{total} = (HM_p + HM_a + HM_f + HM_{ag} + HM_{ow} + HM_{ip}) - (HM_{cr} + HM_l), \quad (4.1)$$

where *HM = heavy metal; *p = parent material; *a = atmospheric deposition; *f = fertiliser application; *ag = agricultural sources; *ow = organic waste sources; *ip = inorganic pollutants; *cr = crop removal and *l = losses by leaching and volatilisation.

Table 4.1. Studies that have determined the HM contamination in different soil types.

Study	Type of Soil Analysed	Site/Country	Source of Pollution	Metal Causing Pollution	Mean Levels (mg/kg)	*WHO Permissible Limits (mg/kg)
Xing *et al.* (2019)	Farmland soil near lead smelters	Jiyuan City, China	Lead smelter	Cd	3.29	0.8
				Pb	184	85
				Cu	31.8	36
				Zn	99.2	50
				Ni	45.6	35
Xing *et al.* (2020)	Non-utilized soil areas near a Pb smelting facility	Jiyuan City, China	Lead smelter	Cd	22.3	0.8
				Cu	139	36
				Pb	1870	85
				Zn	262	50
Yu *et al.* (2022)	Soil samples near an abandoned recycled lead site	Zhoukou, Henan Province, China	Smelting of old batteries	Pb	1921.77	85
				Mn	598.21	200
				Zn	162.29	50
				Cr	84.65	100
				Cu	15.16	36
				Cd	1.8	0.8

(*Continued*)

Table 4.1. (*Continued*)

Study	Type of Soil Analysed	Site/Country	Source of Pollution	Metal Causing Pollution	Mean Levels (mg/kg)	*WHO Permissible Limits (mg/kg)
Chen *et al.* (2023)	Soil samples	Henan Yellow River Wetland Reserve, China	Various land-use types	As	9.46	20
				Cd	0.3	0.8
				Cr	63.8	100
				Cu	25.8	36
				Hg	0.068	0.03–2**
				Pb	26.6	85
Gao *et al.* (2021)	Surface soil of agricultural land	Gaomi City, China	Weathering	Cu	17.4	36
				Pb	23.3	85
				Zn	44.3	50
				Ni	21.7	35
				Cr	58.9	100
				Cd	0.1	0.8
				As	8.79	20
				Hg	0.0252	0.03–2**
Kamunda *et al.* (2016)	Gold mining area	Witwatersrand gold-mining basin, Johannesburg, South Africa	Mining	As	79.4	20
				Pb	4.79	85
				Hg	0.09	0.03–2**
				Cd	0.05	0.8

				Cr	278.76	100
				Co	25.56	10*
				Ni	112.06	35
				Cu	42.51	36
				Zn	51.3	50
Ali and Al-Tamimi (2018)	Shewasoor sub-basin	Northeastern Iraq	Erosion processes	As	7.23	20
				Pb	11.59	85
				Cd	0.21	0.8
				Cr	76.53	100
				Co	18	10*
				Cu	26.53	36
				Ni	117.15	35
				Zn	52.26	50

Notes: *FAO/WHO (2011).
**WHO/FAO (2001) and de Vries *et al.* (2003).

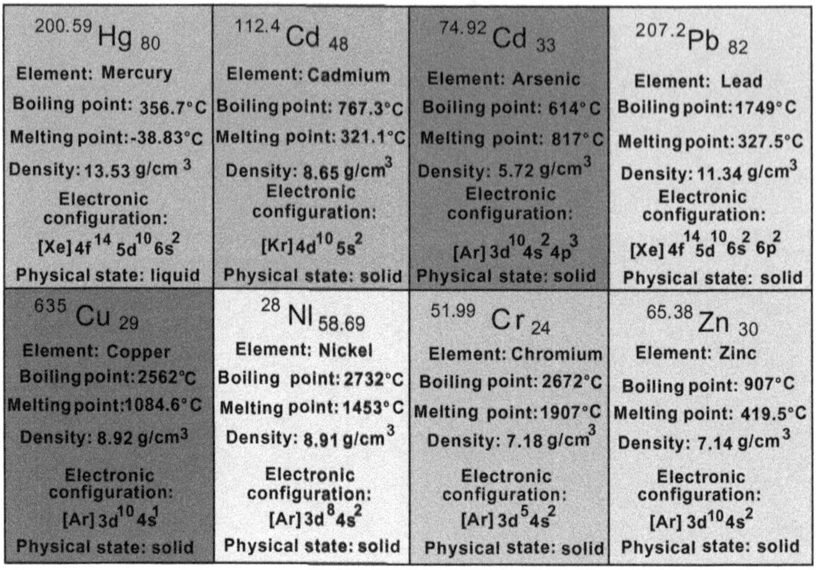

Figure 4.1. Comprehensive overview of common heavy metals' location in the periodic table and their characteristics.

Table 4.2. Sources and health impacts of selected HMs.

Heavy Metal	Sources (Natural and Anthropogenic)	Human Health Impacts	References
Pb	Naturally occurring as a mineral combined with other elements; non-essential elements; lead-based paints and pigments; industrial emissions; mining and metallurgy; leaded gasoline; combustion of fossil fuels; pesticides; manufacturing of lead-acid batteries; municipal sewage	Neurological damage, development issues in children, cardiovascular disease, kidney damage, liver fibrosis, skin irritation, thyroid damage	Wuana and Okieimen (2011) Baldwin and Marshall (1999) Obeng-Gyasi (2019) Lone *et al.* (2008) Adnan *et al.* (2024)
Cd	Geogenic sources; non-essential element; industrial activities; phosphate fertilisers; mining; PVC; pigments and paints; sewage sludge; fossil fuel burning	Kidney damage, bone disorders, carcinogenic effects	Wuana and Okieimen (2011) Yuan *et al.* (2019) Lone *et al.* (2008) Adnan *et al.* (2024)

Table 4.2. (*Continued*)

Heavy Metal	Sources (Natural and Anthropogenic)	Human Health Impacts	References
As	Occurs in a wide variety of minerals; non-essential; mining and metallurgy; agricultural pesticides; industrial processes; combustion of fossil fuels; pharmaceuticals; semiconductors; wood preservatives; animal feed additives	Carcinogenic effects, skin disorders, cardiovascular effects, brain damage, thyroid, liver fibrosis, bladder cancer	Wuana and Okieimen (2011) Bhardwaj *et al.* (2020) Lone *et al.* (2008) Adnan *et al.* (2024)
Hg	Naturally occurring; non-essential; only liquid metal; coal and fossil fuel combustion; industrial discharges; peat and wood burning; mining; florescent bulbs; incineration of municipal wastes; pesticides; fertilisers	Neurological damage, development issues, cardiovascular effects, gastro disease, kidney damage, intestinal disease, liver fibrosis, skin irritation, thyroid	Wuana and Okieimen (2011) Bhardwaj *et al.* (2020) Lone *et al.* (2008) Adnan *et al.* (2024)
Cu	Essential element; occurs in minerals; industrial discharges mining and metallurgy metal processing paints and pigments pesticides; textile industry	Cardiovascular disease, gastro disease, kidney damage, intestinal disease liver fibrosis, pulmonary disease, thyroid	Wuana and Okieimen (2011) Izydorczyk *et al.* (2021) Lone *et al.* (2008) Adnan *et al.* (2024)
Zn	Naturally occurring; industrial discharges; mining and metallurgy; metal waste; fertilisers; electroplating; galvanisation	Brain damage, gastro disease, intestinal disease, skin irritation	Wuana and Okieimen (2011) Lone *et al.* (2008) Adnan *et al.* (2024)

(*Continued*)

Table 4.2. (*Continued*)

Heavy Metal	Sources (Natural and Anthropogenic)	Human Health Impacts	References
Cr	Doesn't occur naturally in elemental forms but found in compounds; industrial discharges; mining and metallurgy; metal processing; industrial dust and fumes; tanning; fertilisers	Brain damage, gastro disease, kidney damage, intestinal disease, liver fibrosis, skin irritation, pulmonary disease	Wuana and Okieimen (2011) Sharma *et al.* (2021) Kerur *et al.* (2021) Lone *et al.* (2008) Adnan *et al.* (2024)
Ni	Essential element in small doses; naturally occurring; industrial discharges mining and metallurgy coal combustion; batteries; landfill food processing industries; waste incineration	Kidney damage, bladder cancer, pulmonary disease	Wuana and Okieimen (2011) Obeng-Gyasi (2019) Lone *et al.* (2008) Adnan *et al.* (2024)

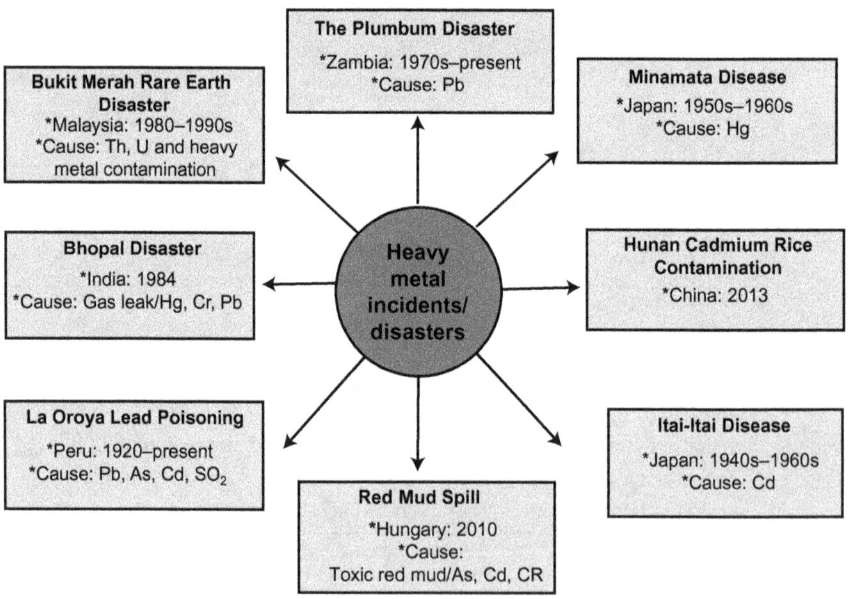

Figure 4.2. Major global HM disasters.

In order to understand the speciation (i.e., the different chemical forms), bioavailability and remediation of HMs in soils, pertinent knowledge about the chemical, environmental and associated health effects of the HMs is necessary (Wuana and Okieimen, 2011). The speciation of a HM contributes to its fate and transport. Within the soil environment, HMs undergo reactions such as adsorption and desorption, mineral precipitation and dissolution, complexation reactions, ion exchange, mobilisation and immobilisation as well as uptake by plants.

Box 4.1: Environmental disasters around the globe.

Major HM Disasters around the Globe

Minamata disease was the result of methylmercury (MHg) poisoning due to the discharge of industrial waste from a chemical plant (Chisso Co. Ltd.) in Minamata Bay. The marine products (which were the staple food for the population) in Minamata Bay displayed high levels of Hg contamination (5.61–35.7 ppm) (Harada, 1995), leading to severe neurological symptoms such as loss of motor control, intellectual disability and even death.

Itai-itai disease was caused by Cd contamination from mining waste released by Mitsu Mining & Smelting Co. Ltd. into rivers, which were used to irrigate rice fields. Chronic Cd poisoning was the result of prolonged oral Cd ingestion (Inaba *et al*., 2005) and led to severe pain in the bones and joints, kidney failure and osteoporosis, especially in women.

The plumbum disaster: The Kabwe mine in Zambia polluted the environment with extremely high levels of toxic Pb, leading to severe poisoning affecting children with symptoms such as cognitive impairment and development delays.

Bukit Merah rare earth disaster: Mitsubishi Chemicals of Japan opened a rare earth refinery in Bukit Merah, leading to thorium (Th), uranium (U) and HM contamination. Though the plant is closed now, the local population is still facing health issues such as leukaemia and birth defects as a result of exposure to radioactive waste and HMs.

Lead poisoning: La Oroya, an area which is highly contaminated by mining waste, dust and acidic fumes in Peru. hosts a polymetallic smelter since 1922, operated by Doe Run Company. Health issues are prevalent, especially among children with Pb poisoning, respiratory problems and developmental delays. Fraser (2009) mentioned that 99.7% of the children residing in the neighbourhood nearest the smelter had high levels of Pb in their blood.

Hunan Cd rice contamination: Another HM disaster was experienced in Hunan, China, as a result of Cd contamination in rice due to industrial pollution from mining and smelting processes. Hunan is a major rice-growing province in China, and it also hosts mining industries and other industrial activities. Contamination of paddy soils with Cd is a result of mining and smelting activities as well as the irrigation of paddy soils with contaminated water (Wang *et al.*, 2019). Consumption of Cd-contaminated rice, a staple food, led to widespread public health concerns such as kidney damage and bone disease.

Bhopal gas tragedy: Leakage of methyl isocyanate from a pesticide plant, Union Carbide, in Bhopal led to the death of thousands of people immediately and long-term health impacts for survivors (Broughton, 2005) with the HMs left in the water and soil exacerbating health issues for a long period.

Red mud disaster, Hungary: There was a catastrophic release of toxic red mud at the Ajkai Timföldgyár Alumina plant in Western Hungary. The sludge, containing HMs such as As, Cd and Cr contaminated local water bodies and soil, leading to health risks. After-effect investigations have shown that soils affected in the red mud were moderately to strongly alkaline, and the levels of trace metals such as Cd, Ni and Cr exceeded threshold concentrations for soil (Winkler *et al.*, 2018).

4.2.2 *Soil contamination assessment methods using soil geochemical risk indices*

Soil assessment or hazard risk assessments need to be carried out so as to determine the nature and degree of contamination, identify hazards connected with contaminants, identify possible targets and routes of exposure, and assess the risks related to current and future use of the site and neighbouring land (Paetz and Wilke, 2005). Hazard risk assessment is an effective scientific tool, enabling decision-makers to manage contaminated sites in a cost-effective manner (Zhao and Kaluarachchi, 2002).

Qualitative, semi-quantitative and quantitative methods based on laboratory and field observations and analysis provide information on the quality of soils. Qualitative assessment of HM contamination in soils can be obtained through pollution indices, geo-accumulation and enrichment factors (Kampa and Castanas, 2008; Rai *et al.*, 2019). Paz-Kagan *et al.* (2014) mentioned that the soil quality index is used as a measure of soil function since it considers multiple soil properties to provide an overall assessment of soil health and productivity.

In order to classify the potential risks of HMs in soils and characterise appropriate solutions for risk control, it becomes imperative to identify the total amount, distribution pattern and contamination levels of the HMs present (Mao *et al.*, 2023). The total amount of soil HM reveals information about the overall pollution status of the contaminated area. Furthermore, soil parameters such as pH, organic matter, texture and acidic sulphate soil (ASS) also indicate the amount of HM present and available for uptake by plants. Soil pollution and assessment indices reveal the degree of contamination at contaminated sites and are used to assess a particular element's contamination in soils. This allows for an estimation of human health risk assessment, which might result from exposure to HM contaminants. Some of these indices are discussed in the following, together with relevant examples from specific studies.

These are single or integrated indices, including the geo-accumulation index (Igeo), enrichment factor (EF), contamination factor (CF), potential ecological risk indices (RI or PERI), single-factor pollution index, Nemerow comprehensive pollution index and target hazard quotient (HQ) index (Hakanson, 1980, Ran *et al.*, 2016; USEPA, 2019; Liu *et al.*, 2020; Wu *et al.*, 2021; Ferreira *et al.*, 2022; Yang *et al.*, 2022; Koner *et al.*, 2023; Napoletano *et al.*, 2023).

4.2.2.1 *Geo-accumulation index*

The geo-accumulation index (Igeo) determines the degree of contamination by comparing the current concentration of a given metal with the natural level contained in the investigated soils (Mueller *et al.*, 2010), as given in Equation (4.2):

$$\text{Igeo} = \frac{\log_2[\text{sample}]}{1.5*[\text{reference}]}, \tag{4.2}$$

where:

*[sample] = concentration of the investigated element in soil; *[reference] = geochemical background concentration; *1.5 = factor which is applied to minimise value fluctuations due to the wide heterogeneity of the Earth's crust; *Igeo index = values range from 0 to >5 and can be grouped into seven classes (Mueller *et al.*, 2010) according to Table 4.3.

Table 4.3. The classification of Igeo index values.

Igeo Index	Soil Contamination Classification
0	Uncontaminated soil
0–1	Uncontaminated to moderately contaminated soil
1–2	Moderately contaminated soil
2–3	Moderately to strongly contaminated soil
3–4	Strongly contaminated soil
4–5	Strongly to extremely contaminated soil
>5	Extremely contaminated soil

In their study of the Yellow River Estuary in China, Wang *et al.* (2022) evaluated the presence of V, Cr, Mn, Co, Fe, Ni, Cu, Zn, Cd and Pb in the surface sediments during summer and autumn sampling and calculated the Igeo index for these elements. The Igeo values for Pb varied from (−3.4) to (−1.4), indicating that Pb did not contaminate the intertidal zone of the river, while the Igeo values (>1) for Cd in autumn showed contamination. Sellami *et al.* (2022) analysed 36 soil samples of urban and peri-urban sites from Setif City, eastern Algeria, and determined Cd, Cr, Cu, Pb and Zn for these samples. The Igeo values for these soil samples showed that there is no anthropogenic contamination by Cr (<0), while Zn showed moderate pollution. Usese *et al.* (2017) investigated As contamination in sediment samples collected from a tropical open lagoon in southwest Nigeria and found that Igeo varied between −5.10 and −1.86, indicating that the lagoon is uncontaminated.

4.2.2.2 *Degree of contamination* (C_d) *and contamination factor* (*CF*)

The values of CF are proposed to define the factor of contamination, which is calculated using the formula shown in Equation (4.3) (Muller, 1969):

$$\mathrm{CF} = \frac{c_{\mathrm{metal}}}{c_{\mathrm{background\ value}}}, \tag{4.3}$$

where:

*CF = contamination factor; *c_{metal} = concentration of metal; *$c_{\mathrm{background\ value}}$ = reference value for each element.

The degree of contamination (C_d), which is a complex index, is an evaluation tool for contamination in soils and sediments (Ferreira *et al.*, 2022). It is described as the sum of all CF (Muller, 1969), as shown in Equation (4.4):

$$C_d = \sum_{i=1}^{n} \mathrm{CF}_i, \tag{4.4}$$

where:

*CF_i = contamination factor for each element (i) investigated.

Table 4.4 describes the classification for CF and the degree of contamination (C_d).

The CFs of eight HMs in estuarine sediments from the Iguacu and Meriti Rivers, Rio de Janeiro, Brazil, were studied by Monte *et al.* (2021). Both rivers showed high contamination for Zn (CF > 6), while Cd showed moderate (CF < 3) contamination. However, Pb showed high contamination (CF > 6) for the Meriti River and considerable contamination for the Iguacu River (CF < 6). Likewise, the Meriti River showed moderate contamination for Ni (CF between 2.65 and 3.03), while the Iguacu River showed an absence of contamination (CF < 1) (Monte *et al.*, 2021). Usese *et al.* (2017), on the other hand, investigated As contamination in sediment samples collected from a tropical open lagoon in southwest Nigeria and found low contamination (CF between 0.04 and 0.34). Swarnalatha *et al.* (2013) evaluated the contamination of sediment samples obtained from a shallow lake in southern India for the degree of contamination. The C_d for station 4 was 3.56 ($C_d < 8$), indicating a low degree of contamination, while that of station 5 was 16.18 ($8 \leq C_d \leq 16$), indicating a moderate degree of contamination.

Table 4.4. The classification for CF and C_d.

Contamination Factor (CF)	Degree of Contamination (C_d)	Classification
CF < 1	$C_d < 8$	Low
1 ≤ CF < 3	$8 \leq C_d < 16$	Moderate
3 ≤ CF < 6	$16 \leq C_d < 32$	Important/considerable
CF ≥ 6	$C_d \geq 32$	Very high

Source: Hakanson (1980).

4.2.2.3 *Enrichment factor*

The EF estimates the amount of a given HM derived from human activities by comparing the element concentration in the sample with an uncontaminated background after normalisation with iron (Fe) (Okedeyi *et al.*, 2014). Iron, being the most significant part of soil, is chosen as a reference element (Al-Dahar *et al.*, 2023). The EF can be calculated using Equation (4.5):

$$\mathrm{EF} = \frac{\left(\frac{C_X}{C_{Fe}}\right)\text{sample}}{\left(\frac{C_X}{C_{Fe}}\right)\text{reference}}, \quad (4.5)$$

where:

*(C_x/C_{Fe})sample refers to the concentration of the investigated metal and Fe in the sample; *(C_x/C_{Fe})reference refers to the concentration of the investigated metal and Fe in the background.

Based on the EF, there can be five contamination categories, with an increase in the EF indicating an increase in the anthropogenic origins' contributions, as described in Table 4.5.

For instance, Moukhchan *et al.* (2013) determined the EF for metals Cu, Mn, Ni and tin (Sn), which were obtained from marine sediments from Palma de Mallorca Bay in Spain, and found them to be 3, 5, 3 and 2, respectively, indicating moderate contamination from anthropogenic sources. The average EF for Pb among stations 1–7 was found to be 39, indicating substantial contamination (Moukhchan *et al.*, 2013). Through their study, Arisekar *et al.* (2022) found that the EF of various metals (Cd, As, Pb, Co, Zn, Ni, Cr and Mn) from sediment samples obtained from the

Table 4.5. Summary of the contamination category with respect to the enrichment factor.

Enrichment Factor	Contamination Category
<2	Minimal enrichment
2–5	Moderate enrichment
5–20	Significant enrichment
20–40	High enrichment
>40	Extremely high enrichment

Thamirabarani River located in the Western Ghats of South India was lower than 2, indicating natural sources of elements in the river.

4.2.2.4 *Pollution load index*

The pollution load index (PLI) considers all investigated HMs (Cd, Cu, Fe, Mn, Ni, Pb and Zn) and evaluates their overall impact on soils using the formula (Equation (4.6)) proposed by Tomlinson *et al.* (1980):

$$\text{PLI} = n\sqrt{\text{CF1} \times \text{CF2} \times \text{CF3} \times \cdots \times \text{CFn}}, \tag{4.6}$$

where:

*n = number of metals studied; *CF = contamination factor as calculated in Equation (4.3).

The PLI provides simple but comparative means for assessing a site's quality (Al-Dahar *et al.*, 2023). The PLI values assume >1 or <1 values, as indicated in Table 4.6.

Jahan and Strezov (2018) evaluated the PLI of six seaports, with one port (Yamba) classified as not polluted (PLI = 0.01), Port Jackson pollution at baseline level (PLI = 1.02) and the remaining ports heavily polluted (PLI > 1). Swarnalatha *et al.* (2013) evaluated the contamination of sediment samples obtained from a shallow lake in southern India for PLI. The PLI for stations 4 and 5 were found to be 0.52 and 2.25, respectively, with station 4 being classified as having no contamination and station 5 as being polluted.

4.2.2.5 *Potential ecological risk index*

The RI or PERI provides information about the impact of HMs on environmental and ecological processes (Hakanson, 1980; Ahamad *et al.*, 2020) by comparing the contents of the target HMs with their background

Table 4.6. The significance of the PLI values.

PLI	Indication
<1	No contamination
=1	Baseline levels of pollutants are present
>1	Deterioration of site quality

and standard values (Mao *et al.*, 2023). The potential ecological risk for a given contaminant can be calculated according to Equations (4.7) and (4.8) by Hakanson (1980):

$$\mathrm{RI} = \sum E_r^i, \tag{4.7}$$

$$E_r^i = T_r^i \times \mathrm{CF}, \tag{4.8}$$

where:

*CF = contamination factor; *T_r^i = toxic response factor of a given HM. The T_r^i for Fe, Mn and Zn is 1; Cr is 2; Cu, Ni and Pb is 5 and Cd is 30 (Pejman *et al.*, 2015; Ahamad *et al.*, 2020; Napoletano *et al.*, 2021). *E_r^i = potential ecological risk factor; *i = HM of interest.

The RI and risk factor E_r^i are classified (Ahamad *et al.*, 2020) as shown in Table 4.7.

Together with CF, Monte *et al.* (2021) also assessed the risk factor and the PERI of the estuarine sediments. The risk factor, E_r^i, for Ni ranged from 2.80 to 29.64, indicating a low ecological risk from this metal for both Iguacu and Meriti rivers. However, for Cd, low ecological risk was obtained for the Iguacu River (16.73–48.93) and moderate to considerable ecological risk for the Meriti River (67.76–95.30) (Monte *et al.*, 2021). A similar pattern was obtained for Pb for both rivers, i.e., the Iguacu River had low ecological risk (<40) and the Meriti River had moderate ecological risk (50.20–76.17). The assessment of PERI showed that the Iguacu River had a low ecological risk (PERI < 150), while the Meriti River presented a moderate ecological risk (150 < PERI < 300).

Table 4.7. Classification of the RI (PERI) and E_r^i.

Potential Ecological Risk Index (RI or PERI)	**Potential Ecological Risk Factor E_r^i**	**Classification**
<150	<40	Low risk
$150 \leq \mathrm{RI} < 300$	$40 \leq E_r^i < 80$	Moderate risk
$300 \leq \mathrm{RI} < 600$	$80 \leq E_r^i < 160$	Considerable risk
$\mathrm{RI} \geq 600$	$160 \leq E_r^i < 320$	Very high risk for RI; high risk for E_r^i
–	$320 \leq E_r^i$	Very high risk

Almeida *et al.* (2020) analysed sediment samples obtained in rainy as well as dry periods from the Joanes River in Bahia State, Brazil, and evaluated it to have low ecological risk (PERI < 150).

4.2.2.6 *Single-factor and Nemerow comprehensive pollution index*

The single-factor pollution index assesses the pollution level of a single HM (Hakanson, 1980), while the Nemerow comprehensive pollution index assesses the overall pollution caused by the simultaneous presence of multiple HMs, incorporating the mean and the maximum value of a single-factor pollution index (Yang *et al.*, 2013), thus evaluating the environment quality. Since different HM pollutions have different impacts on the environment, the weight coefficients of different HMs need to be considered.

The single-factor pollution index is determined using Equation (4.9):

$$\text{Pi} = \text{Ci}/\text{Si}, \tag{4.9}$$

where:

*Pi = single-factor pollution index of HM *i* (a larger Pi value is an indication that the HM pollution of soils is more serious); *Ci = measured value of HM *i*; *Si = second-level standard of the Environmental Quality Standard for Soils of HM *i*.

The second-level standards of the Environmental Quality Standards for Soils are for soils with pH > 7.5, with threshold values for Cd, Cr, As, Pb, Cu and Zn being 0.6, 250, 25, 350, 100 and 300 mg/kg, respectively (SEPA, 1995).

The Nemerow comprehensive pollution index can be determined according to Equation (4.10):

$$\text{NPI} = \sqrt{\left(\left(\frac{\text{Ci}}{\text{Si}}\right)_{\text{max}^2} + \left(\frac{\text{Ci}}{\text{Si}}\right)_{\text{ave}^2}\right) \Big/ 2}, \tag{4.10}$$

where:

$*(\text{Ci}/\text{Si})_{\text{max}}$ = the maximum value in the single-factor pollution index; $*(\text{Ci}/\text{Si})_{\text{ave}}$ = the average value of Ci/Si among the tested metals for each soil sample.

Table 4.8. The pollution grade for Pi and NPI indices.

Grade	Single-Factor Index (Pi)	Pollution Grade	Nemerow Pollution Index (NPI)	Pollution Grade
1	Pi ≤ 1	No pollution	NPI ≤ 0.7	Clean
2	1 < Pi ≤ 2	Low pollution	0.7 < NPI ≤ 1	Warn limit
3	2 < Pi ≤ 3	Moderate pollution	1 < NPI ≤ 2	Slight pollution
4	Pi > 3	High pollution	2 < NPI ≤ 3	Moderate pollution
5			NPI ≥ 3	Heavy pollution

Table 4.8 shows the grade standard for soil HM pollution for the single-factor (Pi) and Nemerow comprehensive pollution index for each sampling site (Liu *et al.*, 2018).

In their study, Chen *et al.* (2023) evaluated the single-factor pollution index as well as the Nemerow comprehensive pollution index on the river basin soil of the Henan Yellow River Wetland Reserve, China, for Pb, Hg, Cd, As, Cu and Cr. The Pi and NPI obtained were less than 1, thus giving an indication that there was no risk of pollution.

4.2.2.7 *Hazard quotient (HQ) and hazard index (HI)*

The exposure assessment of HMs can be determined by measuring the average daily intake (ADI) through ingestion, inhalation and dermal contact by humans. The ADI exposure via these three ways is shown using the following equations (Kamunda *et al.*, 2016):

a. Ingestion through soils

$$ADI_{ing} = \frac{C \times IR \times EF \times ED \times CF}{BW \times AT} \tag{4.11}$$

where:

*ADI_{ing} = average daily intake of HM ingested from soil in mg/kg/day; *C = concentration of HM in mg/kg for soil; *IR = ingestion rate in mg/day; *EF = exposure frequency in days/year; *ED = exposure duration in years; *BW = body weight of the exposed individual in kg; *AT = time period over which the dose is averaged in days; *CF = conversion factor in kg/mg.

b. Inhalation through soil particulates

$$ADI_{inhal} = \frac{C_s \times IR_{air} \times EF \times ED}{BW \times AT \times PEF}, \qquad (4.12)$$

where:

*ADI_{inhal} = average daily intake of HM inhaled from soil in mg/kg/day; *C_s = concentration of HM in soil in mg/kg; *IR_{air} = inhalation rate in m^3/day; *PEF = particulate emission factor in m^3/kg; *EF, ED, BW and AT as described above.

c. Dermal contact with soil

$$ADI_{derm} = \frac{C_s \times SA \times FE \times AF \times ABS \times EF \times ED \times CF}{BW \times AT}, \qquad (4.13)$$

where:

*ADI_{derm} = exposure dose via dermal contact in mg/kg/day; *C_s = concentration of HM in soil in mg/kg; *SA = exposed skin area in cm^2; *FE = fraction of the dermal exposure ratio in soil; *AF = soil adherence factor in mg/cm^2; *ABS = fraction of the applied dose absorbed across the skin.

The hazard quotient (HQ) characterises non-carcinogenic hazards in order to estimate the potential health risks of pollutants, as shown in Equation (4.10). A HQ ≤ 1 indicates no risk, while a HQ > 1 indicates that risks do exist (USEPA, 1989):

$$HQ = \frac{ADI}{RfD}, \qquad (4.14)$$

where:

*ADI = average daily dose (mg/kg/day); *RfD = reference dose (mg/kg/day) of a specific HM.

The hazard index (HI) is the sum of all the HQs for *n* number of HMs and their non-carcinogenic effects on the population (USEPA, 1989). If the HI value is < 1, then the exposed population is unlikely to experience adverse health effects, whereas if the HI value is > 1, then there is a concern for potential non-carcinogenic effects (USEPA, 1989). Equation (4.11) describes the HI:

$$\mathrm{HI} = \sum_{k=1}^{n} \mathrm{HQ}_k = \sum_{k=1}^{n} \mathrm{ADI}_k / \mathrm{R}f\mathrm{D}_k, \tag{4.15}$$

where:

$^*\mathrm{HQ}_k$, ADI_k and $\mathrm{R}f\mathrm{D}_k$ are values of HM k.

Conducting soil HM pollution assessments helps in assessing the effects of single- or multi-element contamination in the environment. The soil geochemical indices provide the overall degree of contamination and the environmental pollution load, especially at contaminated sites. Hence, it becomes imperative to carry out soil sampling in order to assess the levels of HMs present at various contaminated sites and determine the degree of contamination as per environmental regulations governing HM concentrations in soils.

4.3 Soil Sampling

Soil sampling refers to the process of collecting small representative portions of soil from a given area to analyse and assess its physical, chemical, biological and radiological properties in order to understand its characteristics. Soil sampling is usually employed in fields such as agricultural, chemical, biological, physical and environmental sciences and geology, as well as in construction areas. There are a variety of parameters which can be determined from the sampled soils, such as the levels of essential nutrients (e.g., nitrogen, phosphorus, potassium and micronutrients), pH, moisture content and contaminants, as well as texture and structure assessment.

Analysis for HMs at contaminated sites is a crucial process for assessing the extent of pollution and determining potential environmental and health risks. A contaminated site is defined as an area where soil pollution has been identified, which poses a prospective risk to humans, ecological systems or other receptors (Li *et al.*, 2017).

Soil sampling is a valuable tool for making informed decisions in agriculture, land management, environmental protection and construction projects. It helps optimise resource use, enhance crop yield and address environmental concerns. Soils can be sampled for the determination of the general soil quality, the characterisation of soils for soil mapping, supporting regulatory actions and as part of hazard or risk assessment. An effective soil sampling strategy is crucial for obtaining representative and accurate data about soil properties. A generalised overview based on the large amount of literature is given for the soil sampling process.

4.3.1 *Soil sampling process*

For any soil sampling process to be carried out, a strategic plan needs to be formulated before the actual sampling is carried out. The background information about the contaminated site has to be studied, and the sampling objectives need to be figured out. Accessibility to the sampling site should be considered; the specific sampling sites have to be identified, as well as the number of samples to be collected from each sampling site. The physiochemical parameters (depending on the contaminated site) need to be recorded. The sample transport to the laboratory, pretreatment, storage and cost of analysis and instrumentation need to be worked out.

Whenever a volume of soil is required for analysis/characterisation, it is generally impossible to examine the whole volume of soil; therefore, it becomes necessary to take samples for analysis and assessments. The soil samples which are collected need to be as fully representative as possible, with necessary precautions being taken to ensure that samples do not undergo any changes in the interval between sampling and examination (Paetz and Wilke, 2005).

Some important factors that need to be considered when formulating a soil sampling process include selecting the sampling technique, sampling strategy, sampling method, sample pretreatment and storage. These are elaborated as follows:

a. *Selecting the sampling technique*: The sampling technique is dependent on the sampling objectives, which are the major determining factors of the sampling programme.
b. *Sampling strategy*: The sampling strategy focuses on locations, the number of samples analysed and the cost of investigation. Initially, a preliminary site investigation is carried out. For soils contaminated with HMs, sampling requires a careful and targeted approach to accurately assess the extent and severity of contamination. An effective soil sampling strategy should emphasise the following:

    - *Site description*: This involves a thorough description of the site identifying all potential sources of heavy metal contamination. There is a need to figure out the historical land use, industrial/agricultural activities and the nature of contaminants involved in the contamination.

- *Description of the purpose of sampling*: The objectives of the sampling need to be defined clearly, i.e., the reason the sampling is taking place plus the type of heavy metal contaminant to be assessed and the effects of the contaminant on human and environment health.
- *Outline of contaminated areas*: The boundaries of the contaminated sites need to be set up to assist in sampling.
- *Sampling approach*: A sampling approach needs to be implemented such that the entire contaminated area is covered.
- *Depth profiling*: This has to be performed, as soil samples need to be collected at different depths to gauge the depth up to which the contaminant has reached.
- *Marking hot spots*: The hot-spot areas need to be prioritised for sampling purposes.
- *Sampling tools*: The sampling tools to be used should be non-contaminating and corrosion-resistant, and they need to be properly cleaned before use.
- *Personal protective equipment (PPE)*: Appropriate PPE need to be worn by laboratory personnel to minimise the risk of exposure to HMs. Sampling PPE such as gloves and respiratory protection are a must.
- *Decontamination procedures*: A strict decontamination procedure needs to be implemented.
- *Field documentation*: Detailed information at each sampling location, depth, visual observations and any specific conditions needs to be documented.
- *Preservation and storage*: Samples need to be preserved appropriately to prevent changes in metal speciation. Samples need to be stored in clean containers or polyethylene bags.
- *Lab analysis*: Samples need to be analysed for HMs according to laboratory or accredited procedures or the Environment Protection Agency (EPA) analytical methods. Instruments such as atomic absorption spectrometry (AAS), inductively coupled plasma mass spectrometry (ICPMS) and inductively coupled plasma optical emission spectroscopy (ICPOES) could be used for the analysis. Appropriate reference standards and QA/QC should be followed during these analyses.

- *Data interpretation*: The collected data need to be interpreted within the context of relevant regulatory standards and guidelines. The risks posed by the HM concentrations need to be assessed, and necessary remediation and monitoring measures should be determined.

4.3.2 *Soil sampling methods*

Soil sampling can be done in order to assess the biological, physical, chemical or radiological parameters alone or in association with each other. The type of sampling required is dependent on the type of parameters sought. For instance, for a chemical investigation of soils, precautionary measures need to be taken into consideration during sampling so that there is little to no disturbance/alteration to the soil sample. In the case of sampling soils contaminated with HMs, representative and reliable sampling should be performed to ensure that the samples accurately reflect the extent and distribution of HM contamination. Commonly used sampling strategies include the following:

- simple random sampling,
- stratified random sampling,
- systematic (or grid) sampling.

Simple random sampling is the simplest method of sampling, where the selection of samples is left totally to the luck of the selection without regard for the variation in a soil population. Simple random sampling accounts for variability in contamination throughout the sampling area, thus preventing bias and providing a statistically representative sample. In systematic sampling, the sampling sites are determined at random, thus providing more accurate results than simple random sampling (Tan, 2005). Systematic sampling ensures a representative distribution of samples from a sampling location. Stratified random sampling is usually employed in a heterogeneous field, where the field is divided into a number of strata (or parts) and samples are drawn independently from each stratum (Tan, 2005).

Some other sampling methods which can be incorporated into common sampling methods or done as a standalone to sample contaminated soils include the following:

- *Depth profiling*: This is used to measure the vertical distribution of HMs in the soil profile at different depths.

- *Hotspot sampling*: During preliminary investigations, it is suggested to identify and prioritise areas which are suspected to have high concentrations (hot spots) of HMs. Hence, during the sampling period, more samples from these areas can be collected for analysis. The choice of sampling can be varied. For example, both horizontal and vertical sampling can be conducted.
- *Transect sampling*: This type of sampling allows samples to be collected at regular intervals along a transect (linear sampling path).
- *Composite sampling*: This is used when there is mixing of multiple subsamples at each location to form a single sample, thus providing a more integrated representation of the HM content at each location.
- *Incremental sampling*: It is the collection of samples in increments for a detailed analysis of variations in HM concentrations.
- *Rake sampling*: It is the collection of a surface layer of the soil using a rake to assess the HM contamination in areas with loose or disturbed soils.
- *Auger sampling*: It is the collection of soil cores and assessing the HMs contamination down the vertical profile.
- *Root zone sampling*: It is the collection of samples from the root zone of plants in order to understand the potential for phytoremediation and the impact of HMs on vegetation.
- *Pit sampling*: In order to understand soil profiles, pit sampling is carried out. Pits are dug to expose soil profiles, and samples are collected at different depths from the pit walls, thus providing a visual representation of soil horizons and HM distribution.
- *Pore water sampling*: It is the collection of samples of pore water from saturated soils, thus giving an insight into the leaching potential of HMs.

4.3.3 *Sampling containers and tools*

The sampling container's role is to protect the sample from losses due to adsorption or volatilisation or from contamination by foreign substances (Paetz and Wilke, 2005). Hence, cleaning of the sample container is paramount for any sampling programme.

Pretreatment of sampling containers is fundamental, especially when dealing with contaminated soil samples for HMs, so that the risk of contamination is minimised. Preferred sampling containers should be made up of inert materials, such as high-density polyethylene (HDPE) or glass containers. The sampling containers need to be acid washed, typically in a 10% nitric acid (HNO_3) bath, and washed thoroughly with distilled deionised (DDI) water and air-dried in a clean environment. Since the focus of this chapter is on inorganic parameters (HMs), the cleaning procedure of the sampling containers for inorganic substances is given.

The tools needed for contaminated soil sampling for HMs also need to be cleaned before sampling. Preference is given to plastic or stainless-steel sampling tools over metallic tools to further avoid metal contamination.

Various studies have mentioned soil sampling techniques. Studies such as those by Yu *et al.* (2022), Gao *et al.* (2021), Xing *et al.* (2019) and Kamunda *et al.* (2016) have elaborated on the different ways of soil sampling from contaminated sites.

4.4 Laboratory Analysis of Heavy Metals

Soils can be analysed for HMs for a variety of reasons, such as determining the total element content in order to obtain a baseline knowledge of the soil composition, assessing risks to human or environmental health through the analysis of soils collected from industrial and mining sites and assessing the availability of metal content in agricultural soils in order to assess the potential plant deficiency or toxicity of the metals present. In this era of sophisticated technology, a range of analytical techniques and methods are available for the determination of HMs in soils.

4.4.1 *Pretreatment and storage of contaminated soil samples*

Once soil samples contaminated with HMs are collected, pretreatment is needed to preserve the HMs in a representative state and to ensure accurate analysis and interpretation of results.

Sample containers need to be clearly labelled with information, such as sample ID, date, location, depth and any other relevant details. After collection, the soil samples are dried in the air or in an oven at a temperature of 40°C or even freeze-dried (depending on the methods of analysis) (Davidson, 2013; Paetz and Wilke, 2005). Paetz and Wilke (2005) mentioned that drying soil samples in an oven at 40°C is preferred to air drying because the increased speed of drying limits any changes to the soil due to microbial activity. However, high temperatures should be avoided during drying since this may alter the speciation of some HMs, such as mercury (Hg). Dried soil samples are homogenised (either by grinding or milling), and fractions less than 2 mm (subsamples) are sieved for chemical analysis. The subsamples should be stored in clean, labelled containers, away from light, moisture and extreme temperatures.

4.4.2 *Reagents*

High-purity or analytical-grade reagents should be used for the extraction of HMs from contaminated soils. Concentrated acids, including nitric acid (HNO_3), sulphuric acid (H_2SO_4), hydrochloric acid (HCl), hydrofluoric acid (HF) and perchloric acid ($HClO_4$), are used for the digestion of soil samples. Standard reference materials (SRMs) or certified reference materials (CRMs), such as National Institute of Standard and Technology (NIST) soil SRMs 2709, 2710 and 2711, as well as SRMs 2586 and 2587 (Davidson, 2013), are also used for method validation.

4.4.3 *Quality control and assurance*

Quality control is usually performed through the application of standard laboratory and quality control protocols, such as replicate analysis of samples, use of blanks and SRMs (as mentioned in Section 4.4.2) for each metal investigated, percentage recovery and instrument accuracy. The percentage recoveries of the metals need to be in the range of 85–115%, with standard deviations of less than 5% between parallel samples. The glassware used in the sample analysis is rinsed thoroughly with detergent, rinsed with deionised water and soaked in a 10% HNO_3 bath for 24 h and then rinsed again with deionised water before use. Standard solutions of the HM (100 mg/L) to be analysed are prepared from stock solutions (1000 mg/L) of each HM, and then the working standards are prepared. All reagents to be used (such as acids) should be of analytical grade.

4.4.4 *Analytical instrumental methods for analysis of contaminated soils*

Specialised analytical methods are required to accurately quantify the concentration of HMs in contaminated soils. HMs can be determined from contaminated soil samples either directly (in soil samples) or indirectly through sample digestion and analyte extraction. There are several instrumentation techniques that could be used to quantify the levels of metals present and are briefly discussed as follows:

- *Atomic absorption spectroscopy (AAS)*: AAS is an analytical technique used for the analysis of individual elements in a sample. It measures the absorption of light of a specific wavelength by the atoms of the analyte. Once the light is absorbed, the electrons of the atom move from the ground state to an excited state. The amount of light absorbed is measured, and using the Beer–Lambert law, the concentration of the element in the sample can be calculated. The soil sample needs to undergo digestion, and the analyte is extracted into solution.
- *Atomic fluorescence spectrometry (AFS)*: The instrumentation for AFS includes a light source to excite the atoms, an atom cell to convert the sample into gaseous atoms and a detection system to collect the fluorescence radiation. The sample is first converted into gaseous atoms, and the element of interest is excited to a higher electronic energy level by a light source; following excitation, the atoms are deactivated by the emission of a photon, and the measurement is made by the detection of radiation emitted by the atom as it relaxes back from the excited state to the ground state. The soil sample needs to undergo digestion, and the analyte is extracted into solution.
- *Graphite furnace atomic absorption spectroscopy (GFAAS)*: GFAAS atomises the sample in a graphite furnace and measures the absorption of light by the analyte atoms. This technique has a higher sensitivity when compared to AAS and is suitable for trace metal studies. Sample preparation is needed, especially for solid samples.
- *Inductively coupled plasma mass spectrometry (ICPMS)*: ICPMS is a multi-element analytical method used to measure the concentration of several elements simultaneously, where the sample solution is nebulised into plasma. The ions generated in the plasma are directed into a quadrupole mass analyser (Davidson, 2005). The mass-to-charge ratios of the ions are measured to quantify the concentration of HMs. Some common elements which can be analysed by ICPMS include Al, As, Cd, Cr, Cu, Mn, Ni, Pb, Zn, B, Ba, Cs, Fe, Se, Sr, Ti, U and V (Jorgensen *et al.*, 2005). ICPMS has high sensitivity and a wide linear dynamic range. However, this instrument is expensive, requires the expertise of skilled operators and needs solid samples to be digested before analysis.
- *Inductively coupled plasma optical emission spectroscopy (ICPOES)*: ICPOES ionises the sample in an inductively coupled plasma source but detects emitted light instead of mass-to-charge ratios. The

ICPOES has high sensitivity with a wide elemental coverage and low operating costs. Sample digestion is needed for solid samples.

- *Hydride generation atomic absorption spectroscopy (HGAAS)*: In the HGAAS technique, certain HMs (such as arsenic and selenium) are converted into volatile hydrides, which are measured using AAS.
- *Anode stripping voltammetry (ASV)*: ASV measures the current generated when HM ions are reduced at a working electrode after preconcentration. The ASV technique has high sensitivity, with low detection limits.
- *X-ray fluorescence (XRF) spectroscopy*: The XRF technique measures the characteristic fluorescent X-rays emitted by elements in the sample when irradiated with X-rays, allowing elemental quantification. Irradiation with X-rays ejects inner electrons from atoms, thus allowing outer electrons to "fall" into the vacancies and release energies as X-ray fluorescence spectra, with the characteristic wavelengths observed for respective elements (Davidson, 2005). This is a non-destructive rapid analysis technique and is suitable for on-site measurements, especially for surface analysis, requiring minimal sample preparation. XRF is a powerful technique for multi-element analysis and determines the concentrations of the analyte directly.
- *Laser ablation inductively coupled plasma mass spectrometry (LA-ICPMS)*: This technique is regarded as one of the most versatile methods for the compositional analysis of solid materials, such as soils or biological samples, thus allowing the determination of HM distribution. It is highly sensitive and simple, and solid samples can be analysed with only minor sample preparation.
- *Sequential extraction analysis (SEA)*: This technique allows the sequential extraction of HMs into different chemical fractions (referred to as metal speciation); thus, the fractions represent their potential bioavailability and mobility in the soils.

4.4.5 *Heavy metal extraction from contaminated soil samples*

HMs can be extracted from soils using traditional (wet chemistry) digestion methods through the use of strong acids, such as HF, HNO_3, aqua regia (HCl and HNO_3), H_2SO_4 or $HClO_4$, before being determined by specific instrumentation, as discussed above. Usually, 0.5–1 g samples are digested using the respective acids at temperatures around 60–100°C until clear solutions are formed; the digestate is then allowed to cool, filtered and diluted to a volume of 50 mL with DDI water and analysed via the specific instruments as per the requirements of the method (Parvez *et al.*, 2023).

The literature provides a vast number of analytical methods for the analysis of HMs in soils. These methods focus on the acid digestion of soil samples by using a hot plate or overnight digestion in Pyrex beakers covered with watch glass or microwave digestion. Studies such as those by Xing *et al.* (2020), Gao *et al.* (2021), Yu *et al.* (2022) and Parvez *et al.* (2023) have discussed the various ways of HM analysis in contaminated soils.

The USEPA Method 3050 is one of the most common, well-established and thoroughly tested methods for determining the concentrations of metals in contaminated soils. USEPA 3050 analyses the contaminants via total elemental analysis, expressing the concentrations of the metals in mg/kg of soil (USEPA, 1996). USEPA 3051A is also used for HM analysis for soils but involves microwave-assisted acid digestion of soil samples. These methods can be found here: https://www.epa.gov/sites/default/files/2015-06/documents/epa-3050b.pdf; https://www.epa.gov/sites/default/files/2015-12/documents/3051a.pdf.

Depending on the instrumentation sensitivity and detection limits of the different elements, different methods of analysis are available for HM determination. Furthermore, the choice of the analytical method also depends on several factors, including the specific HM of interest, sample matrix, detection limits and the availability of resources.

4.5 Conclusion

Soils represent a crucial medium through which life on Earth is sustained. A healthy, productive soil system ensures food security through sustainable agricultural practices, thus enhancing human and ecosystem health. However, soil contamination and pollution are inevitable due to the escalating rates of urbanisation, industrialisation, agricultural production and waste generation globally. Hence, soil sampling at contaminated sites becomes a critical process for assessing the extent, types and concentrations of pollutants, such as HMs, and evaluating environmental regulatory compliance standards.

In an effective soil sampling process, the site is characterised to understand its historical uses and contamination sources. Sampling plans are then developed to representatively capture contamination variations across different areas. After collecting samples, laboratory analyses determine contaminant concentrations, especially for HMs. Risk assessments, using the soil geochemical indices, are also carried out to determine the productivity of soils as well as to assess the risks to human and ecosystem health, which allow informed decisions to be made for remediation techniques (if any) or the urgency of intervention.

References

Adnan, M., Xiao, B., Ali, M. U., Xiao, P., Zhao, P., Wang, H., and Bibi, S. (2024). Heavy metals pollution from smelting activities: A threat to soil and groundwater. *Ecotoxicology and Environmental Safety*, 274, 116189. Available at: https://doi.org/10.1016/j.ecoenv.2024.116189.

Ahamad, M. I., Song, J., Sun, H., Wang, X., Mehmood, M. S., Sajid, M., Su, P., and Khan, A. J. (2020). Contamination level, ecological risk, and source identification of heavy metals in the hyporheic zone of the Weihe River, China. *International Journal of Environmental Research and Public Health*, 17, 1070. Available at: https://doi.org/10.3390/ijerph17031070.

Al-Dahar, R., Rabee, A. M., and Mohammed, R. J. (2023). Calculation of soil pollution indices with elements in residential areas of Baghdad city. *Revista Bionatura*, 8(1), 43. Available at: https://dx.doi.org/10.21931/RB/2023.08.01.43.

Ali, M. J. A. and Al-Tamimi, O. (2018). Ecological indices of the heavy metals in the soil of Shewasoor Sub – Basin, Kirkuk NE Iraq. *Open Science Journal*, 3(1), 1–19.

Alloway, B. J. (2013). *Heavy Metals in Soils: Trace Metals and Metalloids in Soils and Their Bioavailability*. 3rd ed. Environmental Pollution, Vol. 22. DOI: 10.1007/978-94-007-4470-7_1. Dordrecht: Springer.

Almeida, L. C., da Silva, J. B., dos Santos, I. F., de Carvalho, V. S., Santos, A. D., Hadlich, G. M., and Ferreira, S. L. C. (2020). Assessment of toxicity of metals in river sediments for human supply: Distribution, evaluation of pollution and sources identification. *Marine Pollution Bulletin*, 158, 111423.

Arisekar, U., Shakila, R. J., Shalini, R., Jeyasekaran, G., Keerthana, M., Arumugam, N., Almansour, A. I., and Perimal, K. (2022). Distribution and ecological risk assessment of heavy metals using geochemical normalization factors in the aquatic sediments. *Chemosphere*, 294, 133708.

Baldwin, D. R. and Marshall, W. J. (1999). Heavy metal poisoning and its laboratory investigation. *Annals of Clinical Biochemistry*, 36(3), 267–300.

Bhardwaj, S., Soni, R., Gupta, S. K., and Shukla, D. P. (2020). Mercury, arsenic, lead and cadmium in waters of the Singrauli coal mining and power plants industrial zone, Central East India. *Environmental Monitoring and Assessment*, 192, 1–20.

Bocardi, J. M. B., Pletsch, A. L., Melo, V. F., and Quinaia, S. P. (2020). Quality reference values for heavy metals in soils developed from basic rocks under tropical conditions. *Journal of Geochemical Exploration*, 217, 106591. Available at: https://doi.org/10.1016/j.gexplo.2020.106591.

Broughton, E. (2005). The Bhopal disaster and its aftermath: A review. *Environmental Health*, 4(6). Available at: https://doi.org/10.1186/1476-069x-4-6.

Cachada, A., Pereira, M. E., Da Silva, E. F., and Duarte, A. C. (2012). Sources of potentially toxic elements and organic pollutants in an urban area subjected to an industrial impact. *Environmental Monitoring and Assessment*, 184, 15–32.

Chaudhry, H., Vasava, H. B., Chen, S., Saurette, D., Beri, A., Gillespie, A., and Biswas, A. (2024). Evaluating the soil quality index using three methods to assess soil fertility. *Sensors*, 24(3), 864. Available at: https://doi.org/10.3390/s24030864.

Chen, X., Wong, C. U. I., and Zhang, H. (2023). Analysis and pollution evaluation of heavy metal content in soil of the Yellow River Wetland Reserve in Henan. *PeerJ*, 11, e16454. Available at: https://doi.org/10.7717/peerj.16454.

Cheng, K., Tian, H. Z., Zhao, D., Lu, L., Wang, Y., Chen, J., Liu, X. G., Jia, W. X., and Huang, Z. (2014). Atmospheric emission inventory of cadmium from anthropogenic sources. *International Journal of Environmental Science and Technology*, 11, 605–616. DOI: 10.1007/s13762-013-0206-3.

Cui, Y., Bai, L., Li, C., He, Z., and Liu, X. (2022). Assessment of heavy metal contamination levels and health risks in environmental media in the northeast region. *Sustainable Cities and Society*, 80, 103796. Available at: https://doi.org/10.1016/j.scs.2022.103796.

Davidson, C. M. (2013). Methods for the determination of heavy metals and metalloids in soils. In Alloway, B. J. (Ed.), *Heavy Metals in Soils: Trace Metals and Metalloids in Soils and Their Bioavailability*. 3rd ed. Environmental Pollution, Vol. 22. DOI: 10.1007/978-94-007-4470-7_1. Dordrecht: Springer.

de Vries, W., Schütze, G., Lots, S., Meili, M., Römkens, P., de Temmerman, L., and Jakubowski, M. (2003). Critical limits for cadmium, lead and mercury related to ecotoxicological effects on soil organisms, aquatic organisms, plants, animals and humans. In *Proceedings of the Expert Meeting on Critical Limits for Heavy Metals and Methods for Their Application*, Berlin, 2–4 December 2002. Available at: https://edepot.wur.nl/52354.

Duffus, J. H. (2002). "Heavy metals" a meaningless term? (IUPAC Technical Report). *Pure and Applied Chemistry*, 74, 793–807.

FAO and UNEP. (2021). Global Assessment of Soil Pollution: Report. Rome. Available at: https://doi.org/10.4060/cb4894en.

FAO/WHO. (2011). Joint FAO/WHO Food Standards Programme Codex Committee on Contaminants in Foods, pp. 64–89.

Ferreira, S. L. C., da Silva Junior, J. B., Santos, I. F., *et al.* (2022). Use of pollution indices and ecological risk in the assessment of contamination from chemical elements in soils and sediments – practical aspects. *Trends in Environmental Analytical Chemistry*, 35, e00169.

Fraser, B. (2009). La Oroya's legacy of lead. *Environmental Science & Technology*, 43, 5555–5557.

Gao, Z., Dong, H., Wang, S., Zhang, Y., Zhang, H., Jiang, B., and Liu, Y. (2021). Geochemical characteristics and ecological risk assessment of heavy metals in surface soil of Gaomi City. *International Journal of Environmental Research and Public Health*, 18(16), 8329. Available at: https://doi.org/10.3390/ijerph18168329.

Hakanson, L. (1980). An ecological risk index for aquatic pollution control. A sedimentological approach. *Water Research*, 14, 975–1001.

Harada. (1995). Minamata disease: Methylmercury poisoning in Japan caused by environmental pollution. *Critical Reviews in Toxicology*, 25(1), 1–24. Available at: https://doi.org/10.3109/10408449509089885.

Hu, Y. and Cheng, H. (2013). Application of stochastic models in identification and apportionment of heavy metal pollution sources in the surface soils of a large-scale region. *Environmental Science and Technology*, 47, 3752–3760. Available at: https://doi.org/10.1021/es304310k.

Huang, Y., Li, T., Wu, C., He, Z., Japenga, J., Deng, M., and Yang, X. (2015). An integrated approach to assess heavy metal source apportionment in peri-urban agricultural soils. *Journal of Hazardous Materials*, 299, 540–549. Available at: https://doi.org/10.1016/j.jhazmat.2015.07.041.

Hubbard, A. (2005). Heavy metals in the environment: Origin, interaction and remediation. *Journal of Colloid and Interface Science*, 291, 307.

Inaba, T., Kobayashi, E., Suwazono, Y., Uetani, M., Oishi, M., Nakagawa, H., and Nogawa, K. (2005). Estimation of cumulative cadmium intake causing Itai – itai disease. *Toxicology Letters*, 159, 192–201.

ISO. (2015). ISO 11074:2015. Available at: http://www.iso.org/cms/render/live/en/sites/isoorg/contents/data/standard/05/92/59259.html.

Izydorczyk, G., Mikula, K., Skrzypczak, D., Moustakas, K., Witek-Krowiak, A., and Chojnacka, K. (2021). Potential environmental pollution from copper metallurgy and methods of management. *Environmental Research*, 197, 111050.

Jahan, S. and Strezov, V. (2018). Comparison of pollution indices for the assessment of heavy metals in the sediments of seaports of NSW, Australia. *Marine Pollution Bulletin*, 128, 295–306.

Jiang, G.-B., Shi, J.-B., and Feng, X.-B. (2006). Mercury pollution in China. *Environmental Science and Technology*, 40(12), 3672–3678.

Jorgensen, K. S., Jarvinen, O., Sainio, P., Salminen, J., and Suortti, A. M. (2005). Quantification of soil contamination. In Margesin, R. and Schinner, F. (Eds.), *Manual of Soil Analysis: Monitoring and Assessing Soil Bioremediation*. Germany: Springer.

Kabata-Pendias, A. and Pendias, H. (2001). *Trace Metals in Soils and Plants*. 3rd ed. Boca Raton, Florida, USA: CRC Press, 413 p.

Kampa, M. and Castanas, E. (2008). Human health effects of air pollution. *Environmental Pollution*, 151(2), 362–367. Available at: https://doi.org/10.1016/j.envpol.2007.06.012.

Kamunda, C., Mathuthu, M., and Madhuku, M. (2016). Health risk assessment of heavy metals in soils from Witwatersrand Gold mining basin, South Africa. *International Journal of Environmental Research and Public Health,* 13, 663. Available at: https://doi.org/10.3390/ijerph13070663.

Kerur, S. S., Bandekar, S., Hanagadakar, M. S., Nandi, S. S., Ratnamala, G. M., and Hegde, P. G. (2021). Removal of hexavalent chromium – Industry treated water and wastewater: A review. *Materials Today: Proceedings*, 42, 1112–1121.

Khan, S., Cao, Q., Zheng, Y. M., Huang, Y. Z., and Zhu, Y. G. (2008). Health risks of heavy metals in contaminated soils and food crops irrigated with wastewater in Beijing, China. *Environmental Pollution*, 152(3), 686–692.

Koner, S., Chen, J. S., Rathod, J., Hussain, B., and Hsu, B. M. (2023). Unravelling the ultramafic rock-driven serpentine soil formation leading to the geo-accumulation of heavy metals: An impact on the resident microbiome, biogeochemical cycling and acclimatized eco-physiological profiles. *Science of the Total Environment*, 216, 114664.

Le Roux, G., De Vleeschouwer, F., Weiss, D., Masson, O., Pinelli, E., and Shotyk, W. (2019). Learning from the past: Fires, architecture and environmental lead emissions. *Environmental Science and Technology*, 53(15), 8482–8484. Available at: https://doi.org/10.1021/acs.est.9b03869.

Li, C., Zhou, K., Qin, W., Tian, C., Qi, M., Yan, X., and Han, W. (2019). A review on heavy metals contamination in soil: Effects, sources and remediation techniques. *Soil and Sedimentation Contamination: An International Journal*, 28, 380–394.

Li, X., Jiao, W., Xiao, R., Chen, W., and Liu, W. (2017). Contaminated sites in China: counter measures of provincial governments. *Journal of Cleaner Production*, 147, 485–496. Available at: https://doi.org.10.1016/j.jclepro.2017.01.107.

Li, Z., Ma, Z., van der Kuijp, T. J., Yuan, Z., and Huang, L. (2014). A review of soil heavy metal pollution from mines in China: Pollution and health risk assessment. *Science of the total Environment*, 468–469, 843–853. Available at: https://doi.org.10.1016/j.scitotenv.2013.08.090.

Ling, W., Shen, Q., Gao, Y., Gu, X., and Yang, Z. (2007). Use of bentonite to control the release of copper from contaminated soils. *Australian Journal of Soil Research*, 45(8), 618–623.

Liu, R. P., Xu, Y. N., Zhang, J. H., Wang, W. K., and Elwardany, R. M. (2020). Effects of heavy metal pollution on farmland soils and crops: A case study of the Xiaoqinling Gold Belt, China. *China Geology*, 3, 402–410.

Liu, Y., Yang, Y., Li, C., Ni, X., Ma, W., and Wei, H. (2018). Assessing soil metal levels in an industrial environment of Northwestern China and the phytoremediation potential of its native plants. *Sustainability*, 10, 2686. Available at: https://doi.org/10.3390/su10082686.

Lombi, E. and Gerzabek, M. H. (1998). Determination of mobile heavy metal fraction in soil: results of a pot experiment with sewage sludge. *Communications in Soil Science and Plant Analysis*, 29(17–18), 2545–2556.

Lone, M. I., He, Z. L., Stoffella, P. J., and Yang, X. E. (2008). Phytoremediation of heavy metal polluted soils and water: Progresses and perspectives. *Journal of Zhejiang University Science B,* 9, 210–220. Available at: https://doi.org/10.1631/jzus.B0710633.

Lv, J., Liu, Y., Zhang, Z., Zhou, R., and Zhu, Y. (2015). Distinguishing anthropogenic and natural sources of trace elements in soils undergoing recent 10-year rapid urbanization: A case of Donggang, Eastern China. *Environmental Science and Pollution Research International*, 22, 10539–10550.

Ma, C. Y., Cai, D. J., and Yan, H. (2013). Soil Cd pollution and research progress of treatment techniques. *Henan Chemical Industry*, 30(16), 17–22.

Mao, Y., Wang, M., Wei, H., Ging, N., Wang, F., and Zhu, C. (2023). Heavy metal pollution and risk assessment of vegetables and soil in Jinhua City of China. *Sustainability*, 15(5), 4241. Available at: https://doi.org/10.3390/su15054241.

Mansoor, S., Kour, N., Manhas, S., Zahid, S., Wani, O. A., Sharma, V., Wijaya, L., Alyemeni, M. N., Alsahli, A. A., and El-Serehy, H. A. (2020). Biochar as a tool for effective management of drought and heavy metal toxicity. *Chemosphere*, 271, 129458.

McLaughlin, M. J., Hamon, R. E., McLaren, R. G., Speir, T. W., and Rogers, S. L. (2000a). Review: A bioavailability-based rationale for controlling metal and metalloid contamination of agricultural land in Australia and New Zealand. *Australian Journal of Soil Research*, 38(6), 1037–1086.

McLaughlin, M. J., Zarcinas, B. A., Stevens, D. P., and Cook, N. (2000b). Soil testing for heavy metals. *Communications in Soil Science and Plant Analysis*, 31(11–14), 1661–1700.

Mehr, M. R., Shakeri, A., Amjadian, K., Poshtegal, M. K., and Sharifi, R. (2021). Bioavailability, distribution and health risk assessment of arsenic and heavy metals in agricultural soils of Kermanshah Province, West of Iran. *Journal of Environmental Health Science and Engineering*, 1–14.

Monte, C. D., Rodrigues, A. P. D., de Freitas, A. R., Braz, B. F., Freire, A. S., Cordeiro, R. C., Santelli, R. E., and Machado, W. T. V. (2021). Ecological risks associated to trace metals of contaminated sediments from a densely urbanized tropical eutrophic estuary. *Environmental Monitoring and Assessment*, 193(12), 767.

Moukhchan, F., March, J. G., and Cerda, V. (2013). Distribution of trace metals in marine sediments of the Bay of Palma de Mallorca (Mallorca Island, Spain). *Environmental Monitoring and Assessment*, 185(1), 695–706.

Mueller, L., Schindler, U., Mirschel, W., Graham-Shepherd, T., Ball, B. C., Helming, K., Rogasik, J., Eulenstein, F., and Wiggering, H. (2010). Assessing the productivity function of soils. A review. *Agronomy for Sustainable Development*, 30, 601–614. Available at: https://doi.org/10.1051/agro/2009057.

Muller, G. (1969). Index of geo-accumulation in sediments of the Rhine River. *Geological Journal*, 2(3), 108–118.

Münzel, T., Hahad, O., Daiber, A., and Landrigan, P. J. (2023). Soil and water pollution and human health: What should cardiologists worry about? *Cardiovascular Research*, 119(2), 440–449. DOI: 10.1093/cvr/cvac082.

Napoletano, P., Colombo, C., Di Iorio, E., Memoli, V., Panico, S. C., Ruggiero, A. G., Santorufo, L., Maisto, G., and De Marco, A. (2021). Integrated approach for quality assessment of technosols in experimental mesocosms. *Sustainability*, 13, 9101. Available at: https://doi.org/10.3390/su13169101.

Napoletano, P., Guezgouz, N., Iorio, E. D., Colombo, C., Guerriero, G., and Marco, A. D. (2023). Anthropic impact on soil heavy metal contamination in riparian ecosystems of northern Algeria. *Chemosphere*, 313, 137522.

Nunez, O., Fernandez-Navarro, P., Martin-Mendez, I., Bel-Lan, A., Ruperez, J. F. L., and Lopez-Abente, G. (2017). Association between heavy metal and metalloid levels in topsoil and cancer mortality in Spain. *Environmental Science and Pollution Research International*, 24(8), 7413–7421. Available at: https://doi.org/10.1007/s11356-017-8418-6.

Nuralykyzy, B., Wang, P., Deng, X., An, S., and Huang, Y. (2021). Heavy metal contents and assessment of soil contamination in different land-use types in the Qaidam basin. *Sustainability*, 13, 12020. Available at: https://doi.org/10.3390/su132112020.

Obeng-Gyasi, E. (2019). Sources of lead exposure in various countries. *Reviews on Environmental Health*, 34(1), 25–34.

Okedeyi, O. O., Dube, S., Awofolu, O. R., and Nindi, M. M. (2014). Assessing the enrichment of heavy metals in surface soil and plant (*Digitaria eriantha*) around coal-fired power plants in South Africa. *Environmental Science and Pollution Research*, 21, 4686–4696. Available at: https://doi:10.1007/s11356-013-2432-0.

Okrent, D. (1999). On intergenerational equity and its clash with intragenerational equity and on the need for policies to guide the regulation of disposal of wastes and other activities posing very long time risks. *Risk Analysis*, 19, 877–901.

Oves, M., Khan, M. S., Zaidi, A., and Ahmad, E. (2012). Soil contamination, nutritive value, and human health risk assessment of heavy metals: An Overview. In *Toxicity of Heavy Metals to Legumes and Bioremediation*. Berlin, Germany: Springer, pp. 1–27.

Paetz, A. and Wilke, B. M. (2005). Soil sampling and storage. In *Monitoring and Assessing Soil Bioremediation*. Soil Biology, Vol. 5. Berlin: Springer. Available at: https://doi.org/10.1007/3-540-28904-6_1.

Parvez, M. S., Nawshin, S., Sultana, S., Hossain, M. S., Khan, M. H. R., Habib, M. A., Nijhum, Z. T., and Khan, R. (2023). Evaluation of heavy metal contamination in soil samples around Rampal, Bangladesh. *ACS OMEGA,* 8, 15990–15999. Available at: https://doi.org/10.1021/acsomega.2c7681.

Paz-Kagan, T., Shachak, M., Zaady, E., and Karnieli, A. (2014). A spectral soil quality index (SSQI) for characterizing soil function in areas of changed land use. *Geoderma,* 230, 171–184.

Pejman, A., Nabi Bidhendi, G., Ardestani, M., Saeedi, M., and Baghvand, A. (2015). A new index for assessing heavy metals contamination in sediments: A case study. *Ecological Indicators*, 58, 365–373. Available at: https://doi.org/10.1016/j.ecolind.2015.06.012.

Priya, A. K., Muruganandam, M., Ali, S. S., and Kornaros, M. (2023). Clean-Up of heavy metals from contaminated soil by phytoremediation: A multidisciplinary and eco-friendly approach. *Toxics*, 11(5), 422. DOI: 10.3390/toxics11050422.

Rai, P. K., Lee, S. S., Zhang, M., Tsang, Y. F., and Kim, K. H. (2019). Heavy metals in food crops: health risks, fate, mechanisms, and management. *Environment International*, 125, 365–385. Available at: https://doi.org/10.1016/j.envint.2019.01.067.

Ran, J., Wang, D., Wang, C., Zhang, G., and Zhang, H. (2016). Heavy metal contents, distribution, and prediction in a regional soil-wheat system. *Science of the Total Environment*, 544, 422–431.

Rashid, A., Schutte, B. J., Ulery, A., Deyholos, M., Sanogo, S., Lehnhoff, E. A., and Beck, L. (2023). Heavy metal contamination in Agricultural soil: Environmental pollutants affecting crop health. *Agronomy*, 13(6), 1521. Available at: https://doi.org/10.3390/agronomy13061521.

Rehman, Z. U., Junaid, M. F., Ijaz, N., Khalid, U., and Ijaz, Z. (2023). Remediation methods of heavy metal contaminated soils from environmental and geotechnical standpoints. *Science of the Total Environment*, 867, 161468.

Sellami, S., Zeghouan, O., Dhahri, F., Mechi, L., Moussaoui, Y., and Kebabi, B. (2022). Assessment of heavy metal pollution in urban and peri — urban soil of Setif city (High Plains, eastern Algeria). *Environmental Monitoring and Assessment*, 194(2), 1–17.

Shahid, M., Khalid, S., Abbas, G., Shahid, N., Nadeem, M., Sabir, M., and Dumat, C. (2015). Heavy metal stress and crop productivity. *Crop Production and Global Environmental Issues*, pp. 1–25. Switzerland: Springer International Publishing.

Sharma, N., Sodhi, K. K., Kumar, M., and Singh, D. K. (2021). Heavy metal pollution: Insights into chromium eco-toxicity and recent advancement in its remediation. *Environmental Nanotechnology, Monitoring & Management*, 15, 100388.

State Environmental Protection Administration. (1995). Environmental Quality Standard for Soils (GB15618-1995). State Bureau of Technology Supervision. Available at: https://english.mep.gov.cn/standards_reports/standards/Soil/Quality_Standard3/200710/W020070313485587994018.pdf.

Swarnalatha, K., Letha, J., and Ayoob, S. (2013). An investigation into the heavy metal burden of Akkulam-Veli Lake in South India. *Environmental Earth Sciences*, 68(3), 795–806.

Tan, K. H. (2005). *Soil Sampling, Preparation, and Analysis*. 2nd ed. Boca Raton, Florida: CRC Press.

Tang, X., Shen, C., Shi, D., Cheema, S. A., Khan, M. I., Zhang, C., and Chen, Y. (2010). Heavy metal and persistent organic compound contamination in soil from Wenling: An emerging e-waste recycling city in Taizhou area, China. *Journal of Hazardous Materials*, 173, 653–660. Available at: https://doi.org.10.1016/j.jhazmat.2009.08.134.

Tian, K., Huang, B., Xing, Z., and Hu, W. (2017). Geochemical baseline establishment and ecological risk evaluation of heavy metals in greenhouse soils from Dongtai, China. *Ecological Indicators*, 72, 510–520. Available at: https://doi.org/10.1016/j.ecolind.2016.08.037.

Tomlinson, D. L., Wilson, J. G., Harris, C. R., and Jeffrey, D. W. (1980). Problems in the assessment of heavy-metal levels in estuaries and the formation of a pollution index. *Helgolander Meeresunters*, 33(1), 566–575. Available at: https://doi.org/10.1007/BF02414780.

Uchimiya, M., Bannon, D., Nakanishi, H., McBride, M. B., Williams, M. A., and Yoshihara, T. (2020). Chemical speciation, plant uptake, and toxicity of heavy metals in agricultural soils. *Journal of Agriculture and Food Chemistry*, 68(46), 12856–12869.

Upadhyay, V., Kumari, A., and Kumar, S. (2024). From soil to health hazards: Heavy metals contamination in northern India and health risk assessment. *Chemosphere*, 354, 141697. Available at: https://doi.org/10.1016/j.chemosphere.2024.141697.

USEPA. (1989). Risk Assessment Guidance for Superfund Volume 1: Human Health Evaluation Manual (Part A). Washington, DC, USA: Office of Emergency and Remedial Response.

USEPA. (1996). Method 3050B: Acid Digestion of Sediments, Sludges, and Soils. Revision 2. Washington, DC.

USEPA. (2007). Method 3051A (SW – 846): Microwave Assisted Acid Digestion of Sediments, Sludges and Oils. Revision 1. Washington, DC.

USEPA. (2019). Integrated Risk Information System. Washington, DC.

Usese, A., Chukwu, O. L., Rahman, M. M., Naidu, R., Islam, S., and Oyewo, E. O. (2017). Enrichment, contamination and geo-accumulation factors for assessing arsenic contamination in sediment of a Tropical Open Lagoon, Southwest Nigeria. *Environmental Technology and Innovation*, 8, 126–131.

Wang, L., Cui, X., Cheng, H., Chen, F., Wang, J., Zhao, X., Lin, C., and Pu, X. (2015). A review of soil cadmium contamination in China including a health risk assessment. *Environmental Science and Pollution Research*, 22, 16441–16452. Available at: https://doi.org.10.1007/s11356-015-5273-1.

Wang, P., Chen, H., Kopittke, P. M., and Zhao, F. J. (2019). Cadmium contamination in agricultural soils of China and the impact on food safety. *Environmental Pollution*, 249, 1038–1048. Available at: https://doi.org/10.1016/j.envpol.2019.03.063.

Wang, Z., Lin, K., and Liu, X. (2022). Distribution and pollution risk assessment of heavy metals in the surface sediment of the intertidal zones of the Yellow River Estuary, China. *Marine Pollution Bulletin*, 174, 113286. Available at: https://doi.org/10.1016/j.marpolbul.2021.113286.

WHO. (1996). Permissible Limits of Heavy Metals in Soil and Plants. Geneva, Switzerland: World Health Organization.

WHO/FAO. (2001). Codex Alimentarius Commission. Food Additives and Contaminants. Food Standards Programme, ALINORM 10/12A. Joint FAO/WHO.

Wilson, M. A., Burt, R., Indorrante, S. J., Jenkins, A. B., Chiaretti, J. V., Ulmer, M. G., and Scheyer, J. M. (2008). Geochemistry in the modern soil survey program. *Environmental Monitoring and Assessment*, 139(1–3), 151–171. Available at: https://doi.org/10.1007/s100661-007-9822-z.

Winkler, D., Bidló, A., Bolodár-Varga, B., Erdõ, A., and Horváth, A. (2018). Long — term ecological effects of the red mud disaster in Hungary: Regeneration of red mud flooded areas in a contaminated industrial region. *Science of the Total Environment*, 644, 1292–1303. Available at: https://doi.org/10.1016/j.scitotenv.2018.07.059.

Wong, S., Li, X., Zhang, G., Qi, S., and Min, Y. (2002). Heavy metals in agricultural soils of the pearl river delta, south China. *Environmental Pollution*, 119, 33–44. Available at: https://doi.org/10.1016/S0269-7491(01)00325-6.

Wu, Q., Hu, W., Wang, H., Liu, P., Wang, X., and Huang, B. (2021). Spatial distribution, ecological risk and sources of heavy metals in soils from a typical economic development area, southeastern China. *Science of the Total*

Environment, 780, 146557. Available at: https://doi.org/10.1016/j.scitotenv.2021.146557.

Wuana, R. A. and Okieimen, F. E. (2011). Heavy metals in contaminated soils: A review of sources, chemistry, risks and best available strategies for remediation. *International Scholarly Research Network Ecology*, 402647. Available at: https://doi.org/10.5402/2011/402647.

Xing, W., Liu, H., Banet, T., Wang, H., Ippolito, J. A., and Li, L. (2020). Cadmium, copper, lead and zinc accumulation in wild plant species near a lead smelter. *Ecotoxicology and Environmental Safety*, 198, 110683. Available at: https://doi.org/10.1016/j.ecoenv.2020.110683.

Xing, W., Zheng, Y., Scheckel, K. G., Luo, Y., and Li, L. (2019). Spatial distribution of smelter emission heavy metals on farmland soil. *Environmental Monitoring and Assessment*, 191(2), 115. Available at: https://doi.org/10.1007/s10661-019-7254-1.

Xu, J., Liu, C., Hsu, P. C., Zhao, J., Wu, T., Tang, J., Liu, K., and Cui, Y. (2019). Remediation of heavy metal contaminated soil by asymmetrical alternating current electrochemistry. *Nature Communications*, 10(1), 1–8.

Yan, A., Wang, Y., Tan, S. N., Mohd Yusof, M. L., Ghosh, S., and Chen, Z. (2020). Phytoremediation: A promising approach for revegetation of heavy metal — Polluted land. *Frontiers in Plant Science*, 11, 359. Available at: https://doi.org/10.3389/fpls.2020.00359.

Yang, L. Q., Huang, B., Hu, W. Y., Chen, Y., Mao, M. C., and Yao, L. P. (2013). The impact of greenhouse vegetable farming duration and soil types on phytoavailability of heavy metals and their health risk in eastern China. *Chemosphere*, 103, 121–130.

Yang, Q., Zhang, L., Wang, H., and Martin, J. D. (2022). Bioavailability and health risk of toxic heavy metals (As, Hg, Pb and Cd) in urban soils: A Monte Carlo simulation approach. *Environmental Research*, 214, 113772.

Yu, F., Zhang, J., Li, Z., and Liu, S. (2022). Ecological and health risk assessment of heavy metals in soils from recycled lead smelting sites. *Water*, 14(9), 1445. Available at: https://doi.org/10.3390/w14091445.

Yuan, Z., Luo, T., Liu, X., Hua, H., Zhuang, Y., and Ren, J. (2019). Tracing anthropogenic cadmium emissions from sources to pollution. *Science of the Total Environment*, 676, 87–96.

Zhang, M. K., Liu, Z. Y., and Wang, H. (2010). Use of single extraction methods to predict bioavailability of heavy metals in polluted soils to rice. *Communications in Soil Science and Plant Analysis,* 41(7), 820–831.

Zhao, Q. and Kaluarachchi, J. J. (2002). Risk assessment at hazardous waste-contaminated sites with variability of population characteristics. *Environment International*, 28(1–2), 41–53.

Zhao, H., Wu, Y., Lan, X., Yang, Y., Wu, X., and Du, L. (2022). Comprehensive assessment of harmful heavy metals in contaminated soil in order to score pollution level. *Scientific Reports*, 12, 3552. Available at: https://doi.org/10.1038/s41598-022-07602-9.

Zhou, H., Chen, Y., Yue, X., Ren, D., Liu, Y., and Yang, K. (2023). Identification and hazard analysis of heavy metal sources in agricultural soils in ancient mining areas: A quantitative method based on the receptor model and risk assessment. *Journal of Hazardous Materials*, 445, 130528. Available at: https://doi.org/10.1016/j.jhazmat.2022.130528.

https://doi.org/10.1142/9781800617841_0005

Chapter 5

Cryogenic Core Collection and High-Resolution Sample Analysis to Support Subsurface Remediation

Saeed Kiaalhosseini[*,||], Tom Sale[†,#], Rick Rogers[‡], Maria Irianni-Renno[§], and Rick Johnson[¶]

[*]*ONxpress Transportation Partners, Toronto, Ontario, Canada*

[†]*Department of Civil and Environmental Engineering, Colorado State University, Fort Collins, CO, USA*

[‡]*Drilling Engineers Inc., Fort Collins, CO, USA*

[§]*Department of Civil and Environmental Engineering, Colorado State University, Fort Collins, CO, USA*

[¶]*OHSU/PSU School of Public Health, Oregon Health & Science University, Portland, OR, USA*

[||]*skiaalhosseini@onxpress.com*

[#]*thomas.Sale@colostate.edu*

Abstract

For more than half a century, the management of risks posed by contaminants in soils and groundwater has been a major concern. The novel cryogenic core collection technique presented here, coupled with high-resolution analysis of frozen cores, substantially increases our ability to characterise and manage impacted sites. High-resolution characterisation of representative core samples is critical because subsurface environments are almost always heterogeneous with respect to soil permeabilities, which govern the transport and distribution of contaminants in all phases, redox conditions, physical properties and microbial ecology.

With respect to characterising contaminant releases, soil-groundwater settings can be subdivided into 14 compartments, where up to four contaminant phases (i.e., nonaqueous, aqueous, sorbed and gas) occur in zones with contrasting permeabilities (referred to as high- and low-permeability (*k*) zones) in source zones and downgradient plumes (Sale and Newell, 2011). To address all compartments, "whole" soil samples must be collected from the subsurface. Guiding principles for sample collection and analysis include the following: acquiring samples from known targeted intervals, preserving relevant *in situ* attributes of the samples (e.g., lithology, pore fluids, pore gases, pore structure, volatile contaminants, mineralogy, redox conditions and microbial ecology) and employing sample analytics needed to accurately resolve the relevant attributes of interest.

Collection and analysis of subsurface soil samples have improved greatly over the past half century. Cryogenic core collection represents, for the first time, an opportunity to characterise all 14 of the compartments associated with releases of contaminants to subsurface soils and groundwater. Cryogenic core collection uses conventional hollow-stem auger drilling methods with a modified sample collection system. To freeze samples *in situ*, liquid nitrogen is circulated from the surface through a dual-wall cooling chamber in the core barrel. Freezing is typically accomplished in 4–6 min. The Time required for freezing increases below the water table and with greater depths. Subsequently, the core barrel is pulled to ground surface, where the frozen 2 ¼-inch outer diameter (OD) soil core (in a polyvinyl chloride (PVC) liner) is pulled from the barrel and placed in a cooler on dry ice for shipment to a laboratory. Complete freezing is verified via visual inspection.

In this chapter, insights gained from the collection of ~800 feet of cryogenic core from 52 borehole locations at 12 sites in the US are discussed. This includes a description of the cryogenic core collection technique, a presentation of multi-parameter high-resolution datasets, and a discussion of the advantages and limitations of cryogenic core collection.

5.1 Introduction

Management of risks associated with chemical contaminants in the subsurface is a global challenge, carrying a multi-billion-dollar price tag. This is in large part due to the complexity of the subsurface, which makes it exceptionally difficult to characterise chemical occurrences, distribution, transport and fate. Over the past half century, a wide variety of tools have been developed to improve site characterisation and related site conceptual site model (CSM) that support risk management decisions. In the work described herein, cryogenic core collection, coupled with high-resolution subsampling and analysis of frozen cores, is described as an advanced and highly accurate method to improve subsurface characterisation and support the development of improved and reliable risk management decisions.

From the perspectives of hydrogeology and chemistry, the remediation practice generally recognises that contaminants can exist in (or in a combination of) 14 compartments (or categories) in the subsurface (Sale and Newell, 2010). These compartments are organised into four phases (i.e., nonaqueous, aqueous, sorbed and gas) in four zones (i.e., low-permeability and transmissive source zones and low-permeability and transmissive plume zones, Figure 5.1). The zones can exist in perhaps an infinite arrangement of geometries, depending on the geologic depositional environment in which they were created. In addition to these complexities, spatial variations in biogeochemical conditions can occur to an even greater degree and often at a smaller scale than physical heterogeneities.

Commonly, following initial work to characterise a site through evaluation of data from subsurface borings, contaminant characterisation is founded on the prevalence of water quality data obtained from monitoring wells. Groundwater samples collected via conventional methods from

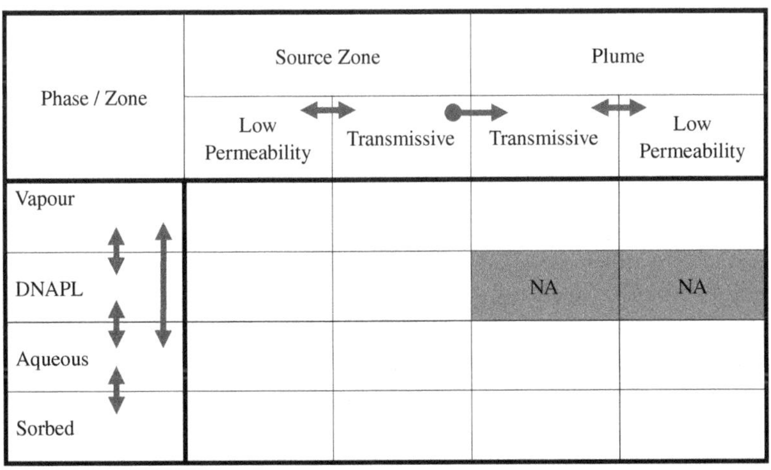

Figure 5.1. Fourteen-compartment model for all contaminant phases in transmissive and low-*k* zones, in source zones and plumes, following Sale and Newel (2010).

monitoring wells (i.e., purging a well and collecting a small sample of the water stream abstracted from the well) provide integrated samples, often biased and weighted towards transmissive zones. Monitoring wells provide vertically averaged data from only two compartments in the 14-compartment model (i.e., an aqueous phase in transmissive zones of source zones and plumes). Risk assessment and site management decisions based on only a few compartments are incomplete and have the potential to result in outcomes that fall short of expectations.

Data collection that can inform all (or at least, a more complete list of) compartments in the 14-compartment model requires collecting subsurface soil samples that meet the following criteria:

- Soil samples must be collected from known targeted vertical intervals.
- The physical, chemical and microbial properties of the soil samples need to be preserved during the sampling and storage process.

Soil samples for environmental characterisation are collected from the subsurface using a variety of widely used techniques that access shallow-to-deep zones below ground, including drilling using hollow stem augers, rotary drilling, direct push, wireline coring and sonic methods. Each of these methods has advantages in different environments; however, all are to some degree subject to the following challenges:

(1) gravels or cobbles larger than the inner diameter (ID) of the sampling system blocking sample collection and limiting recovery (i.e., sample length or sampled interval);
(2) samples falling out of the sample barrel while pulling the sampling system to grade;
(3) fluids draining out of samples during recovery and pore spaces being backfilled with atmospheric gases, including oxygen;
(4) soils liquifying and flowing up into drill systems (augers, drive rods or drive casings) leading to sample biases.

For achieving representative characterisation of biogeochemical properties, high-quality soil cores should avoid, or at least limit the occurrence of, the following conditions during sample recovery as well as during storage and handling:

(1) redistribution of pore fluids,
(2) exposure to oxygen and atmospheric conditions,

(3) volatilisation of contaminants,
(4) degradation of contaminants and biomolecules.

Development of new tools for improved collection of representative soil samples has been an applied research topic for decades. Many researchers have used freezing to overcome the challenges listed above. Durnford *et al.* (1991) used downhole expansion of compressed carbon dioxide (CO_2) gas to partially freeze soils and improve retention of liquids and soils. Similarly, Murphy *et al.* (1996) employed CO_2 freezing in a drive shoe using a wireline coring technique. Johnson *et al.* (2013) used liquid nitrogen to freeze soils *in situ* using a direct-push sampling tool. Unfortunately, none of these methods have been robust enough for broad use.

Building on previous work, we developed novel tooling and techniques to better address the constraints listed above. In the following sections, specific examples of how the new techniques addressed sampling challenges are discussed. In addition, examples of how high-resolution sampling of the cryogenic cores were used to improve site characterisation are discussed. Finally, the strengths and limitations of cryogenic core collection and analysis are evaluated to help practitioners better understand when cryogenic core collection might be appropriate for their sites.

5.2 Methods

This section describes the following: (1) study sites where cryogenic core collection has been applied; (2) cryogenic core collection equipment and procedures (Kiaalhosseini *et al.*, 2016; Sale *et al.*, 2016, 2017); (3) methods to process and analyse frozen cores. The initial development of cryogenic core collection tools and procedures was conducted at Drilling Engineers Inc. and Colorado State University (CSU) facilities in Fort Collins, Colorado, US. The tools and operational procedures were then tested and refined at 12 contaminated field sites across the US.

5.2.1 *Sites*

Approximately 800 linear feet of frozen and non-frozen cores have been collected at 12 sites in the US. Given the evolving practices during method development, minor inconsistencies exist in the symbols used in cryogenic core data plots. The key attributes of the sites are presented in Table 5.1.

Table 5.1. Summary of cryogenic core collection sites.

Site Name	Borehole ID	Depositional Environment	Total Length of Coring, Frozen and Non-Frozen (feet Co)[1]	Approximate Depth to Water Table (ft bgs)[1,2]	Approximate Active Industry Use	Contaminant of Concern (CoC)[3]
Wood Treatment (WT)	WT-1–WT-7	Glacial-fluvial, sand and gravel with interbedded silt layers	62.5	River-dependent; 18–30	1900–1991	PCP in carrier oil, DRO, GRO, methane
Active Refinery (AR)	AR-1 and AR-2	Braided stream deposit, silt and sand overlying coarse sand and gravel	26.2	8–12	1920s–present	Benzene in gasoline, DRO, GRO, methane
Former Oil Field (FOF)	FOF-1–FOF-8	Coastal aeolian sand dune deposits composed of well-sorted medium sand	142	23–32	1955–1990	Intermediate distillates
Former Refinery #1 (FR#1)	FR#1-1–FR#1-7	Glacial-fluvial, sand and gravel with interbedded silt layers	84.5	River-dependent; 11–35	1931–1985	Benzene in gasoline, DRO, GRO, methane
Former Refinery #2 (FR#2)	FR#2-1 and FR#2-2	Fluvial over bank deposit, silt and sand beds	51	River-dependent; 12–17	1904–1985	Benzene in gasoline, DRO, GRO, methane
Former Refinery #3 (FR#3)	FR#3-1–FR#3-4	Braided river alluvium, sediments grade	146	10	1923–1982	Gasoline and diesel fuel

Active Gas Station #1 (AGS#1)	AGS#1-1 to AGS#1-3	Low-energy fluvial	39	10–12	1990–present	Benzene in gasoline, DRO, GRO, methane
Active Gas Station #2 (AGS#2)	AGS#2-1–AGS#2-3	Low-energy fluvial	24	10–23	1960–present	Benzene in gasoline, DRO, GRO, methane
Active Gas Station #3 (AGS#3)	AGS#3-1	Braided stream deposit, silt and sand overlying coarse sand and gravel	8.7	8–12	1970s–present	Benzene in gasoline, DRO, GRO, methane
Petroleum Terminal (PT)	PT-1–PT-3	Low-energy fluvial	35	7–12	1970s–present	Benzene in gasoline, DRO, GRO, methane
Air Force Base (AFB)	AFB-1–AFB-6	Aeolian overlying sandstone, heterogenous layers of conglomerate, sand and silt	85	11.5–25	1900–present	TCE, PCE, DCE, methane
Naval Facility (NF)	NF-1–NF-6	Tidal estuary, fine grained sand and silt	93	6–8	1960–1980	PCE, TCE, DCE, VC

Notes: [1]Imperial units are used in the table and throughout the manuscript to match field practices.

[2]ft bgs: feet below ground surface.

[3]DRO: diesel range organics; DCE: dichloroethane; GRO: gasoline range organics; PCE: tetrachloroethylene; PCP: pentachlorophenol; TCE: trichloroethylene; and VC: vinyl chloride.

5.2.2 *Cryogenic core collection tools and equipment*

A Central Mining Equipment (CME) 75 truck-mounted hollow-stem auger (HSA) drill rig was used at all sites. Standard HSAs, with 4¼-inch ID and 7⅝-inch OD, were employed to advance boreholes (see Figure 5.2).

The cryogenic core barrel consists of a modified 4-inch OD, 5-foot CME Continuous Sample Tube System™ with a "dual-wall cooling barrel" fitted inside (Kiaalhosseini *et al.*, 2016; Sale *et al.*, 2017). To accommodate the liquid nitrogen delivery and exhaust lines for the dual-wall cooling barrel, 2¾-inch holes were drilled in the top of the drive head. The CME drive shoe was also modified to accommodate the dual-wall cooling barrel and an outer layer of insulation.

Figure 5.2. Cryogenic core collection using a Central Mining Equipment (CME) 75 truck-mounted drill rig.

5.2.3 *Dual-wall cooling cylinder*

The dual-wall cooling barrel is 3 feet in length and consists of two overlapping stainless-steel tubes with sealed ends (Kiaalhosseini *et al*., 2016) (Figure 5.3). It is equipped with ¼-inch OD stainless-steel inlet and exhaust tubes. The inlet line carries the liquid nitrogen to the base of the cooling barrel. Delivery of liquid nitrogen to the base of the cooling barrel maximises freezing at the drive shoe and the bottom of the cores. The exhaust line at the top of the cooling cylinder discharges gas- and liquid-phase nitrogen to the ground surface. The entire dual-wall cooling cylinder is covered in insulation, except for the section where it contacts the drive shoe. The insulation accelerates sample freezing, limits the amount of liquid nitrogen required, reduces the time needed to freeze samples and limits the freezing of the cryogenic core barrel to the augers. Clear PVC liners with 2½-inch OD and 2¼-inch ID are placed inside the cooling cylinder for core collection.

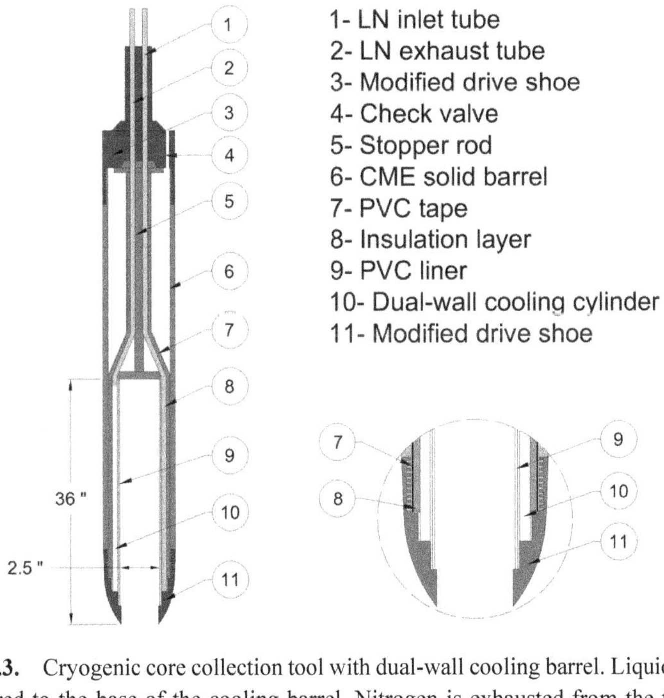

Figure 5.3. Cryogenic core collection tool with dual-wall cooling barrel. Liquid nitrogen is delivered to the base of the cooling barrel. Nitrogen is exhausted from the top of the cooling barrel. Insulation is excluded where the cooling barrel is in contact with the modified drive shoe.

5.2.4 *Coolant delivery and exhaust*

Liquid nitrogen is used as the coolant in the cryogenic core collection system. At atmospheric pressure, liquid nitrogen provides temperatures as low as −196°C. Liquid nitrogen dewars (160 L), generating up to 230 psi, are employed. An insulated, flexible hose with a stainless-steel vacuum jacket is used to connect the liquid nitrogen dewar to insulated stainless-steel downhole delivery lines. Handling of liquid nitrogen follows guidelines as outlined in the vendor's material safety data sheets.

A cryogenic throttle valve and pressure gauge are placed on the exhaust line so that the backpressure could be adjusted to help maximise the boiling of the liquid nitrogen within the cooling barrel (Kiaalhosseini *et al.*, 2016).

5.2.5 *Operational procedures*

Operational procedures for cryogenic core collection are shown in Figure 5.4. The primary steps are as follows:

(1) Advance the HSA and the cryogenic core barrel in 2.5-foot intervals to the targeted depth. The cryogenic core barrel is fixed while the augers rotate.
(2) Deliver liquid nitrogen to the cryogenic core collection tool to freeze the core and stabilise soils at and below the drive shoe. Freezing below the drive shoe occurs through the uninsulated drive shoe, which is in direct contact with the cooling barrel.
(3) Remove the cryogenic core barrel from the HSA by pulling it up to the ground surface.
(4) Remove the soil core (in PVC liner) from the cryogenic core barrel.
(5) Measure core recovery, verify freezing, cap the ends of the core, label the core and place the core on dry ice in a cooler.
(6) Clean and decontaminate the cryogenic core barrel, drain water out of the cryogenic core barrel and insert a new PVC liner into the core barrel.

A notable complication is that, for cores collected below the water table, the core liners can freeze to the wall of the cooling cylinder. This

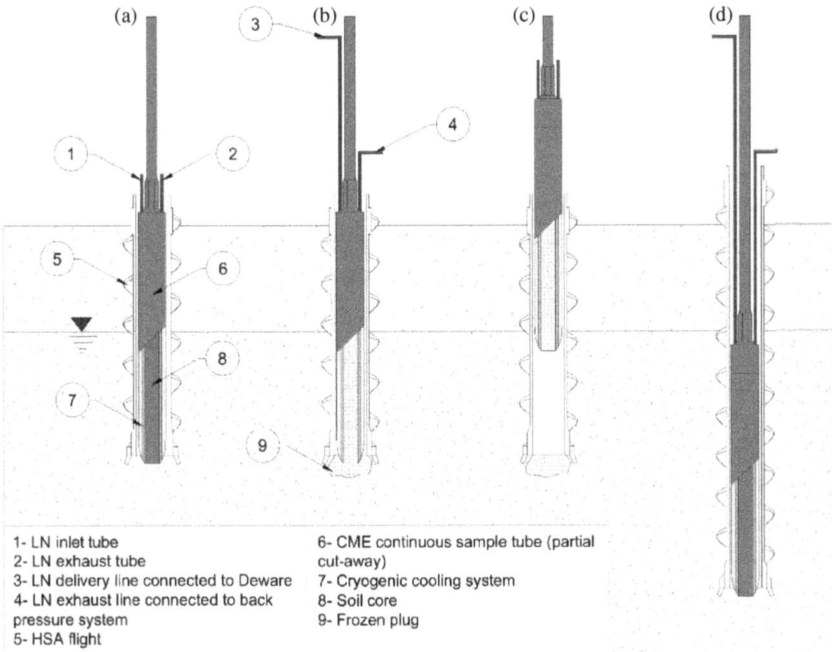

Figure 5.4. Schematic diagram of cryogenic core collection operational procedures: (a) advancement of the HSA and the cryogenic core barrel to the targeted depth, (b) delivery of liquid nitrogen to the cryogenic core collection tool to freeze the core and stabilise soils at and below the drive shoe, (c) retrieval of the cryogenic core barrel from the HSA by pulling it up to the ground surface, (d) returning the decontaminated cryogenic core barrel to the next depth interval.

issue is managed by briefly (~10–30 seconds) running hot water from a pressure washer through the dual-wall cooling cylinder to thaw ice holding the liner in the cooling cylinder (Kiaalhosseini *et al.*, 2016; Sale *et al.*, 2016). After thawing, it is critical to remove all water from the dual-wall cooling system before its next use. Water left in the cooling systems will freeze upon subsequent use, either plugging or damaging the dual-wall cooling cylinder. Water is displaced using pressurised air.

Frozen cores are sent via overnight couriers to laboratories in coolers with excess dry ice to ensure that they remain frozen during shipping. Upon arrival at laboratories, cores are immediately placed in −80°C freezers.

5.2.6 *Core cutting, processing and analyses*

Plans for sub-sampling each core are developed based on core recovery and intervals of interest. Guiding principles for sub-sampling include the following:

(1) avoiding the immediate ends of cores where samples may be biased,
(2) obtaining sub-samples over uniform intervals (i.e., 6 inches or less),
(3) addressing conditions at geological, hydrological and/or contamination interfaces.

To prepare samples for analysis, in most cases, two adjacent 1-inch sections, referred to as "pucks," are cut using a chop saw (Figure 5.5). Puck 1 is quartered into sub-samples A–D for chemical and/or microbial analyses, as summarised in Table 5.2. Puck 2 is used for visual logging and measurement of fluid saturations. Unused sections of core are archived in −80°C freezers and selectively used to determine physical properties, including hydraulic conductivity.

5.2.7 *Analytical methods*

A primary motive for cryogenic coring is the *in situ* preservation of samples. Freezing stabilises pore fluids, stabilises the arrangement of soil

Figure 5.5. Cutting frozen core using a chop saw equipped with a concrete blade.

Table 5.2. General introductions to core processing and analytical methods.

Sample Size	Sub-Sample	Extraction and Analytical Method	Analytical Parameter	Reference
1-inch cut (Puck 1)	Sub-Sample A	Methanol extraction followed by GC- ECD and GC-MS	Organic compounds analysis (e.g., PCE, TCE, DCE, VC)	Sale *et al.* (2016) and Kiaalhosseini (2016)
	Sub-Sample B	Extraction in DI water followed by GC-FID, IC, colourimetry, reference electrodes, mass and volume measurements	Dissolved gases, major anions and cations, metals pH, ORP, bulk density, porosity, water content, f_{oc}	
	Sub-Sample C	qPCR and pyrosequencing	Microbial ecology	Kiaalhosseini (2016), Sale *et al.* (2016), Irianni-Renno *et al.* (2021), and Irianni-Renno (2024)
	Sub-Sample D	Archive	Future analysis	N/A
1-inch cut (Puck 2)	The entire puck	Placed in a capped jar filled with water	Visual logging and photographs under visible and UV	Kiaalhosseini (2016), Sale *et al.* (2016), and Olson *et al.* (2017)
Sections between samples	Entire length or subsections	None	Hydraulic conductivity and grain size distribution	Sale *et al.* (2016)

Notes: GC: gas chromatography; ECD: electron capture detector; MS: mass spectroscopy; FID: flame ionisation detector; IC: ion chromatography; qPCR: quantitative polymerase chain reaction; and f_{oc}: fraction of organic carbon.

grains, limits losses due to volatilisation, limits enzymes that can degrade RNA, constrains the degradation of contaminants and constrains the introduction of oxygen, which might bias microbial ecology and mineralogy.

Given effective preservation, a wide range of chemical, physical and biological measurements can be completed. Most of the methods discussed in the following have been described elsewhere (Sale *et al.*, 2016; Olson *et al.*, 2017). Analytical methods are briefly introduced here.

5.2.7.1 *Organic compounds*

Organic compounds are extracted from sub-sample A using methanol. Methanol extracts are analysed for extractable organic compounds (e.g., gasoline range organics (GROs), diesel range organics (DROs), benzene, tetrachloroethylene (PCE), trichloroethylene (TCE), dichloroethane (DCE) and vinyl chloride (VC)) using gas chromatography with an electron capture detector (GC-ECD), flame ionisation detector (GC-FID) or mass spectrometer (GC-MS).

5.2.7.2 *Dissolved gases*

Dissolved gases are obtained using sub-sample B in deionised (DI) water. Extracted water is purged with air, and the sparged methane, ethane and ethene are detected using a GC-FID.

5.2.7.3 *Aqueous inorganics*

Major anions and cations are obtained using the aqueous extract from sub-sample B and analysed using ion chromatography (IC). Ferrous iron concentrations are measured using a standard colourimetric method. pH and ORP are measured using electrodes shortly after placing samples in DI water.

5.2.7.4 *Bulk density, water content and porosity*

Bulk density, water content and porosity are determined using sub-sample B. Frozen samples are weighed prior to placing them in water (wet weight). Bulk volume is estimated by measuring the volume of water displaced by immersing the samples in water. Finally, the samples are dried in an oven at 80°C until a constant weight is achieved.

5.2.7.5 *Fraction of organic carbon*

Fraction organic carbon (f_{oc}) content is determined using dried sediments from sub-sample B. Samples are analysed for total- and inorganic-carbon fractions. f_{oc} is calculated by differences. Total carbon is measured by combustion, while inorganic carbon is measured by acid digestion.

5.2.7.6 *Hydraulic conductivity testing*

A "core-in-liner" (CIL) method is used to measure hydraulic conductivity (K) (Sale *et al.*, 2016). The CIL method was designed to test K directly without removing the core from the PVC liner and thus without disturbing sediments (i.e., preserving particle arrangement, sorting and layering). Figure 5.6 shows the CIL assembly, which allows for testing the specimen using both constant-flow (ASTM D2434) and/or falling head methods (Kalinski, 2006).

5.2.7.7 *Grain size distribution*

Grain size distributions are determined by performing standard sieve analyses on dried sub-sample A or sub-sample B in accordance with ASTM D422 and ASTM D1140.

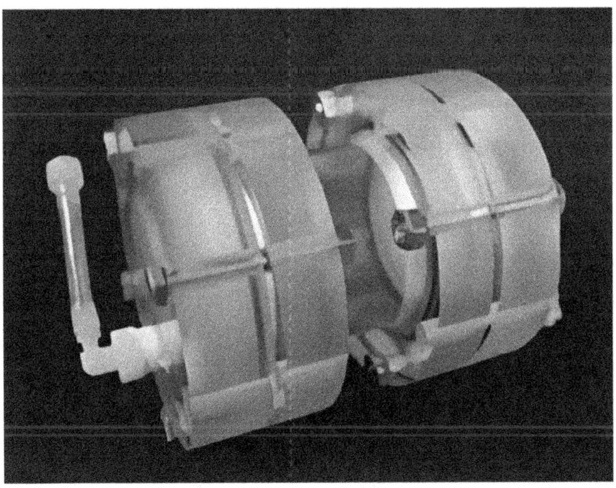

Figure 5.6. Core-in-liner (CIL) assembly.

5.2.7.8 *Microbial analysis*

Stabilisation of RNA via freezing means preserving the integrity of RNA molecules by storing them at very low temperatures, typically in a freezer at −20°C or −80°C, which significantly slows down the activity of enzymes that could degrade the RNA, effectively halting its breakdown and allowing for long-term storage.

Microbial analyses are performed in triplicate on the core samples. DNA is extracted from pretreated samples using commercially available kits, with some modification to the pretreatment process, as described by Sale *et al.* (2016), Kiaalhosseini (2016) and Irianni-Renno (2024). Quantitative polymerase chain reaction (qPCR) assays are used to quantify the bacterial and archaeal 16S ribosomal RNA (rRNA) genes. Pyrosequencing analysis is performed to identify 16S community profiles. Microbial analyses reported in this study were supported by Dr. Susan De Long at CSU, including the use of Dr. De Long's laboratory facilities.

5.2.8 *Calculations*

Table 5.3 summarises the mathematical equations used to estimate key parameters.

5.2.9 *Transmissive and low-k zones*

Transmissive and low-*k* zones in subsurface media are conceptually differentiated as zones with two or more orders of magnitude differences in permeability relative to one another. The terms "transmissive" and "low *k*" are relative to one another, not absolute terms. Aquifer materials from frozen cores were visually logged. Relative differences in permeability are based on soil descriptions and estimates of permeability for specified soil types, as given by Freeze and Cherry (1979). To date, measurements of soil permeability (with conversions to hydraulic conductivity) have been made on frozen cores for one site (AFB) using CIL hydraulic conductivity testing (see Section 5.2.7.6).

To provide insights into contaminant distributions in transmissive and low-*k* zones, subsurface lengths of samples and contaminant masses in these zones were normalised to the total length and total contaminant mass in boreholes, as described in Table 5.4.

Table 5.3. Analytical parameters.

Parameter	Equation		Description
Contaminant mass per area for a single sample for a specific contaminant of concern (CoC) (M/L^2)	$\widehat{\omega}_{\mathrm{CoC}} = \omega\, z\, \rho_b$	(5.1)	ω: mass of a CoC in a sample per dry weight soil (M/M); z: length of sample (L); ρ_b: bulk density (M/L^3)
Total contaminant mass at a borehole location for a specific CoC (M/L^2)	$\widehat{\omega}_{\mathrm{CoC,Total}} = \sum_{i=1}^{n} \widehat{\omega}_{\mathrm{CoC},i}$	(5.2)	$\widehat{\omega}_{\mathrm{CoC},i}$: mass per area for a single sample for a specific CoC (M/L^2); n: number of samples at a given borehole
Total mass of all resolved contaminants in place at a borehole location (M/L^2)	$\widehat{\omega}_{\mathrm{SUM\ CoCs}} = \sum_{i=1}^{n} \widehat{\omega}_{\mathrm{CoC,Total},i}$	(5.3)	n: number of contaminated samples in a given borehole
Bulk density (M/L^3)	$\rho_b = M_{\mathrm{dry}} / V_t$	(5.4)	M_{dry}: dry mass of soil (M); V_t: total volume of sample (L^3)
Porosity (L^3/L^3)	$\phi = 1 - (\rho_b/\rho_{\mathrm{minerals}})$	(5.5)	ρ_{minerals}: average density of soil particles – we use quartz (2.65 g/cm^3)
Fluid saturation	$S_{\mathrm{NAPL}} = \left(\mathrm{TPH}\dfrac{M_{\mathrm{dry}}}{\rho_{\mathrm{NAPL}}}\right) \Big/ \left(V_t - \left(\dfrac{M_d}{\rho_{\mathrm{quartz}}}\right)\right)$	(5.6)	TPH: concentration of total petroleum hydrocarbons; ρ_{NAPL}: density of NAPL;
	$S_w = V_1 - S_{\mathrm{NAPL}}$	(5.7)	V_1: liquid (water + LNAPL) volume in sample
	$S_g = 1 - (S_{\mathrm{NAPL}} + S_w)$	(5.8)	
Specific LNAPL volume	$sp_{\mathrm{LNAPL}} = \sum_{i=1}^{n} z_i \phi_i S_{\mathrm{NAPL}_i}$	(5.9)	
Fluid content	$\theta = S\phi$	(5.10)	S: fluid saturation, which can be S_{NAPL}, S_w or S_g

Table 5.4. Fractions.

Fraction	Equation	Description
Fraction of sample that is transmissive and low-*k* zone in borehole (*L*/*L*)	$f_{Tr} = L_{Tr} / L_{Samples}$ (5.13) $f_{LK} = L_{LK} / L_{Samples}$ (5.14) $f_{LK} + f_{Tr} = 1$ (5.15)	L_{Tr}: length of core that is defined as transmissive; L_{LK}: length of core that is defined as transmissive; $L_{Samples}$: total length of core
Fraction of specific CoC in transmissive and low-*k* zone in borehole	$f_{CoC,Tr} = \hat{\omega}_{CoC,Tr} / \hat{\omega}_{CoC,Total}$ (5.16) $f_{CoC,LK} = \hat{\omega}_{CoC,LK} / \hat{\omega}_{CoC,Total}$ (5.17) $f_{CoC,Tr} + f_{CoC,LK} = 1$ (5.18)	See Equations (5.1) and (5.2) in Table 5.3

5.3 Results

Results discussed in this section include the following: (1) core recovery and occurrences of contaminants with and without *in situ* freezing, (2) NAPL distribution, (3) gas distribution (4) distributions of contaminants in transmissive and low-*k* zones, (4) use of frozen cores to resolve physical properties of soils and (5) use of frozen cores to resolve microbial ecology.

5.3.1 *Core recovery and occurrence of contaminants with and without freezing*

Results of core recovery and occurrence of contaminants with and without freezing during core collection at former refinery #1 (Site FR#1) and a former oil field (Site FOF) are presented in the following section.

5.3.1.1 *Former refinery (FR#1)*

Three locations at a former refinery (Site FR#1) were investigated. At the first location, three boreholes were advanced using *in situ* freezing (FR#1-1–FR#1-3), and one borehole was advanced without freezing (FR#1-4). The remaining two locations were investigated with the freezing technique only (FR#1-5 and FR#1-6/FR#1-7, with FR#1-7 covering a section missed in FR#1-6). Contaminants of concern (CoCs) at the first location (FR#1-1–FR#1-4) included methane, benzene and total petroleum hydrocarbons (TPH). The site is underlain by a glacial valley train aquifer beneath a major river. Frozen and non-frozen cores were collected using vertically adjacent 2.5-foot sample intervals over 10 feet, spanning the water table (at about 24–34 ft bgs).

The frozen cores at the first location yielded mean core recoveries (recovered core lengths versus drives of 2.5 feet) of 90 ± 2% (Table 5.5) and specific LNAPL volumes of 0.03 ± 0.01 feet (Figure 5.7). Variability in the triplicate values reflects differences between boreholes and the repeatability of cryogenic core collection and analytical techniques. Recovery and specific LNAPL volumes at the other two locations were 93% and 97% and 0.10 and 0.08 feet, respectively. Intervals missed with freezing are attributed to cobbles blocking recovery (Figure 5.8(c)). In contrast to the cores frozen *in situ* (Figure 5.8(b)), without freezing, the cohesionless soils fell out of the sample sleeves, and the average recovery was only about 27% (Table 5.5, Figure 5.8(a)).

Importantly, the cobble frozen to the drive shoe in Figure 5.8(c) demonstrates freezing below the drive shoe. Temporary freezing below the drive shoe provides soil strength that limits inflow of non-representative "flowing sand" to the open space created upon removal of the core barrel from within the hollow stem augers. As discussed in the following, flowing sands can compromise the ability to collect accurately depth-resolved core samples.

Table 5.5. Key attributes and statistics for frozen and non-frozen cores collected at Site FR#1.

Core ID	Recovery (%)	Specific LNAPL Volume (feet)	Comment
FR#1-1	89	0.02	Frozen Cores-Location 1
FR#1-2	93	0.02	Frozen Cores-Location 1
FR#1-3	88	0.05	Frozen Cores-Location 1
Mean	90	0.03	–
Standard deviation	2	0.01	–
FR#1-4	27	–	Non-Frozen-Location 1
FR#1-5	93	0.10	Frozen-Location 2
FR#1-6 and FR#1-7	97	0.08	FR#1-7 covers an interval missed by FR#1-6-Location 3

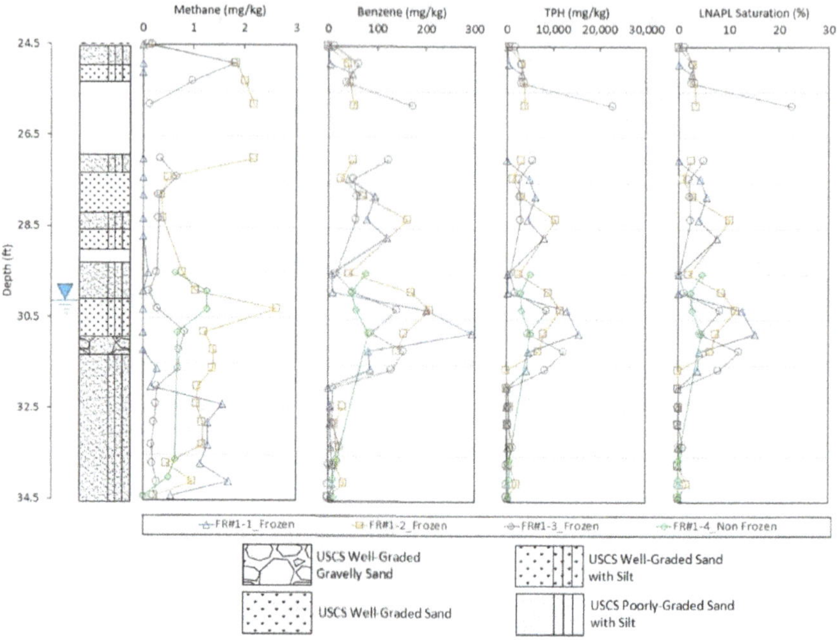

Figure 5.7. Recovery, contaminants of concern and LNAPL saturation with and without freezing at Site FR#1.

Figure 5.8. Site FR#1 coring in glacial outwash: (a) empty sample barrel without freezing, (b) frozen core held in PVC sample liner by ice, and (c) cobble frozen to drive shoe blocking recovery.

Following Table 5.6, benzene (a primary risk driver) is present in transmissive (83.7 ± 5.8%) and low-*k* zones (23 ± 18.8%). Low levels of TPH (up to 25,000 mg/kg) and low NAPL saturations (up to 10%) are encountered in transmissive zones above and below the water table. In general, NAPLs are absent in low-*k* zones.

Table 5.6. Occurrences of contaminants of concern in transmissive and low-*k* zones at Site FR#1.

Parameter	Zone	FR#1-4 Non-Frozen Cores	FR#1-1 Frozen Cores	FR#1-2 Frozen Cores	FR#1-3 Frozen Cores	Mean Values of Frozen Cores	Standard Deviation of Frozen Cores
Benzene (mg/kg dry soil)	Transmissive	66.2	81.8	90.2	79.0	83.7	5.8
	Low-*k*	12.6	15.0	18.0	23.3	18.8	4.2
TPH (mg/kg dry soil)	Transmissive	4,038.1	4,681.3	5,211.0	5,614.0	5,168.8	467.8
	Low-*k*	479.0	614.6	603.2	1393.3	870.4	452.9
LNAPL Saturation (%)	Transmissive	3.6	4.4	4.8	5.2	4.8	0.4
	Low-*k*	0.0	0.5	0.3	1.1	0.6	0.4

5.3.1.2 *Former oil field (FOF)*

Cryogenic core collection, with and without freezing, was applied at a former oil field site (Site FOF) impacted by releases of an intermediate petroleum distillate. The site is underlain by aeolian sand dune deposits composed of well-sorted medium sand overlying fine-grained fluvial overbank deposits. Eight boreholes were advanced within a 100 by 30 feet area. Five of the boreholes were advanced using cryogenic core collection techniques (FOF-1–FOF-5), and three were advanced without *in situ* freezing (FOF-6–FOF-8). At FOF, the collection of non-frozen cores involved pushing a standard 4-foot direct-push sampling tool beyond a lead hollow stem auger flight. Non-frozen cores were placed in a cooler on dry ice immediately after collection. Cores were collected from 31 to 44 ft bgs. Depths to the water table were approximately 34 ft bgs at all drill locations.

At the FOF site, without the stabilising effect of freezing below the drive shoe, flowing sands were prevalent at or below the water table, as indicated by sampling tools encountering sand inside the auger flights above the targeted sampling interval. A consequence of flowing sand was that 100% or greater recovery was achieved when driving sample tools through non-representative flowing sands in auger flights and then advancing through the targeted sample intervals.

For cryogenic boreholes FOF-1 and FOF-5, core collection operations went smoothly with the recovery of 2.5-foot cores every 30 minutes (min).

At a 30-min frequency freezing, flowing sands were limited between sample runs. Recovery at locations where flowing sands were controlled via freezing was close to 100%. Due to collapsed liners and other issues at borehole locations FOF-2, FOF-3 and FOF-4, more than an hour passed between the collection of some frozen cores below the water table. In those cases, thawing of the frozen formation, which was the product of the application of the cryogenic core collection technique, led to sand flowing into the auger flights.

Figure 5.9 and Table 5.7 document the observed smear zone heights (vertical LNAPL intervals), specific LNAPL volumes and maximum LNAPL saturations with and without freezing. Apparent smear zone heights are similar in collected cores with and without freezing. Apparent specific LNAPL volumes and maximum LNAPL saturations are four times larger in cores collected without freezing.

Only considering cores from boreholes FOF-1 and FOF-5, where cryogenic core collection "went well," apparent specific LNAPL volumes and maximum LNAPL saturations are 6 and 13 times greater without

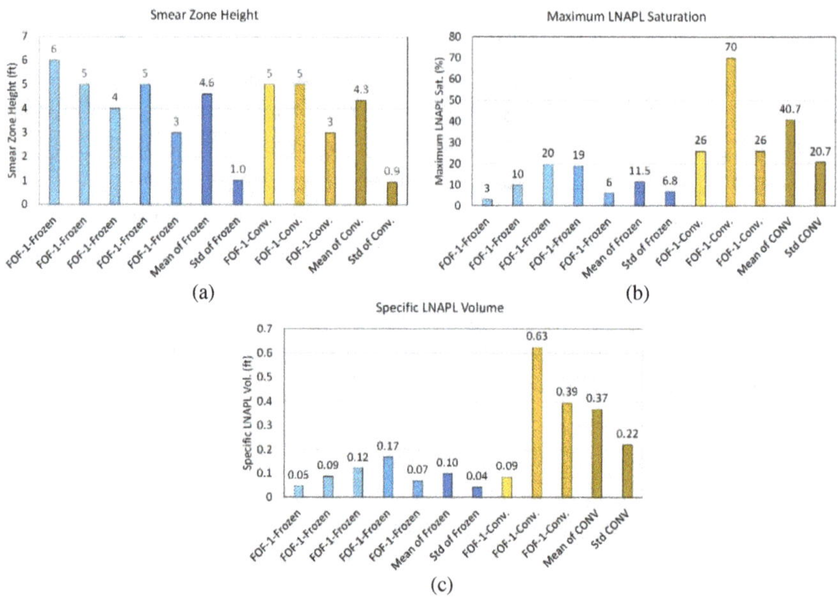

Figure 5.9. Results of laboratory analysis of frozen and non-frozen cores collected at Site FOF: (a) observed heights of LNAPL in soils (smear zone), (b) maximum LNAPL saturations and (c) specific LNAPL volumes.

Table 5.7. Results summary of analysing frozen and non-frozen cores collected at Site FOF.

Parameter	Type of Core	Mean	Standard Deviation	Ratio of Mean Non-Frozen/ Frozen Cores
Smear zone height	Frozen cores	4.6	1.0	0.94
	Non-frozen cores	4.3	0.9	
Specific LNAPL volume (feet)	Frozen cores	0.01	0.04	3.53
	Non-frozen cores	0.37	0.22	
Maximum LNAPL saturation (%)	Frozen cores	11.5	6.8	3.72
	Non-frozen cores	40.7	20.7	

freezing. Non-frozen sample biases are attributed to encountering and collecting expanded "flowing sand," where expansion of the samples is accommodated by inflow of LNAPL and water.

5.3.2 *High-resolution site characterisation*

Because cryogenic core collection effectively collects soils from targeted intervals and preserves attributes of subsurface media, it can be an important part of high-quality, high-resolution (6-inch intervals or less) characterisation of sites. Examples of high-resolution data from a chlorinated solvent site and three hydrocarbon sites are presented in the following section.

5.3.2.1 *Air force base site (AFB)*

Six cryogenic core holes were advanced using cryogenic core collection techniques at an active Air Force Base (Site AFB) located in a semiarid area in the US. Site AFB is impacted by chlorinated solvents, including PCE, TCE, cis-DCE and VC. The geology of the site consists of highly heterogeneous alluvial and aeolian (sand and loess) deposits with intermittent caliche beds. Three boreholes (AFB-1, AFB-2 and AFB-3) were located across the facility in a one-mile square area. Data from these core holes are described in this section. Three additional boreholes (AFB-4, AFB-5 and AFB-6) are located at a single spill site. Data from these boreholes are described in the following section, addressing distributions of contaminants in transmissive and low-*k* zones. Depths to the water table range from 11 to 25 ft bgs. Frozen cores collected from 15 to 40 ft bgs were analysed for organic and inorganic compounds, including PCE, TCE, methane, ferrous iron, chloride, nitrite, sulphate, ORP, pH and total organic carbon (TOC).

Figures 5.10–5.12 present results from core holes AFB-1–AFB-3. Except for intervals where concrete-like caliche beds were encountered, core recovery at this site is almost 100%. Per the second data columns in Figures 5.10–5.12, TCE is the primary CoC. Lower concentrations of PCE follow occurrences of TCE.

Elevated methane concentrations occur in finer, less permeable sediments. Consistent with standard chemistry principles:

- TCE concentrations are inversely related to methane concentrations.
- Reduced ferrous iron concentrations increase with larger methane concentrations.
- Oxidised sulphur, in the form of sulphate, is less prevalent in intervals with higher methane concentrations.

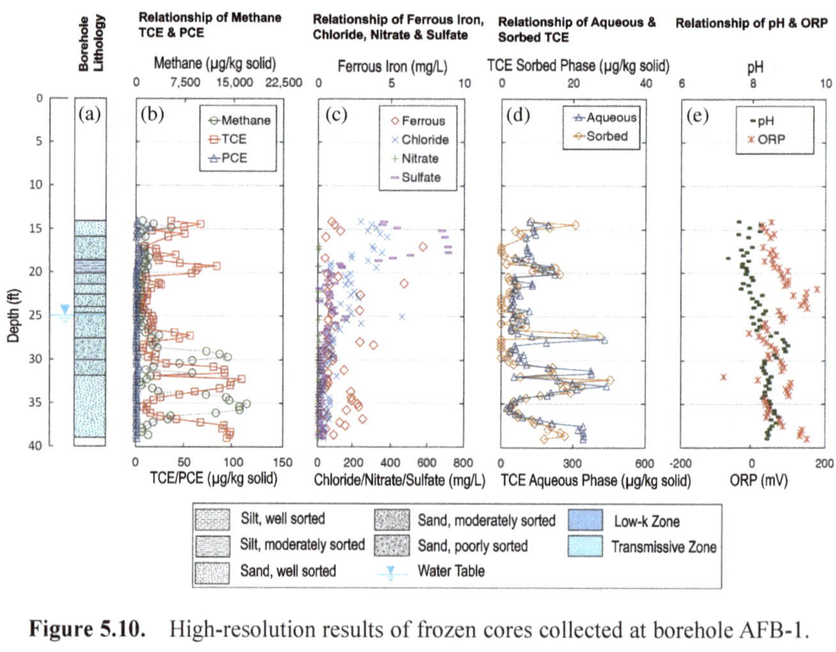

Figure 5.10. High-resolution results of frozen cores collected at borehole AFB-1.

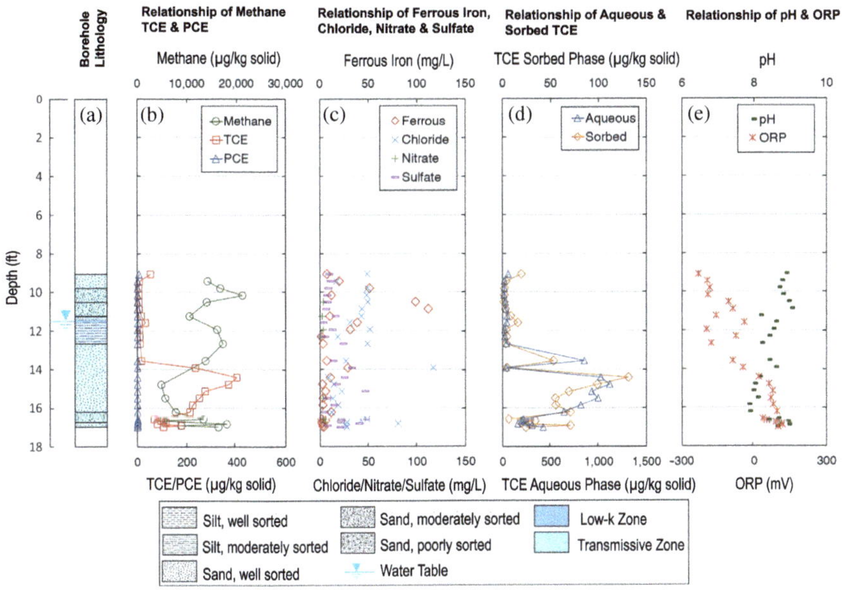

Figure 5.11. High-resolution results of frozen cores collected at borehole AFB-2.

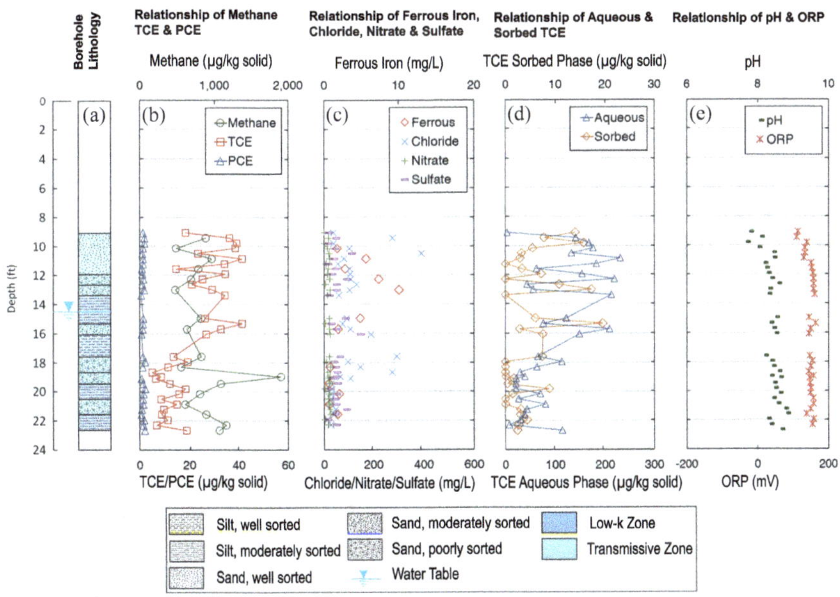

Figure 5.12. High-resolution results of frozen cores collected at borehole AFB-3.

- Chloride concentrations, a potential product of natural reductive dechlorination, are elevated in intervals where TCE and reducing conditions (higher methane, low ORP, elevated ferrous iron) occur.
- Overall, nitrate concentrations are largely uniform in frozen cores.

Aqueous and sorbed-phase TCE concentrations are based on measured fractions of organic carbon (f_{oc}), measured total concentrations of TCE and partitioning coefficients. pH values are largely constant and slightly alkaline (pH near 8). ORP values are lower where methane concentrations are higher.

5.3.2.2 *Petroleum terminal (PT) and active gas stations (AGS#1 and AGS#2)*

Cryogenic cores collected from a petroleum terminal (Site PT) and two active gas stations (Sites AGS#1 and AGS#2) are discussed in this section. The sites have historical releases of gasoline and diesel fuel and are underlain by glacial till. At each site, frozen cores were collected upgradient of sources, in source zones (release areas) and downgradient of sources. Depths of coring varied between 0 and 25 ft bgs. Depths to the water table range from 5 to 12 ft bgs. Parameters of interest include benzene, DRO, GRO, TPH, methane, LNAPL saturations and specific LNAPL volumes.

Figures 5.13–5.15 present results from the three sites. Average core recovery from the three sites was 85%. The primary constraint to recovery was large cobbles blocking movement of soils into the core collection tool. Darker sediment colours were observed in source zones and downgradient plumes. Colours are attributed to biogeochemical transformations of naturally occurring (i.e., upgradient) oxidised soil minerals (i.e., red iron oxides) to reduced minerals with darker colours (e.g., black metal sulphides). The data suggest that soil minerals provide solid-phase electron acceptors supporting natural source zone depletion (NSZD) in source zones and natural attenuation (NA) in plumes.

DRO and GRO were encountered in source zones adjacent to the water table. The sum of DRO and GRO is reported as TPH. Specific LNAPL volumes in source zones were 0.19 feet at Site PT, 0.02 feet at Site AGS#1 and 0.03 feet at Site AGS#2. Benzene, a primary human health risk driver, is only encountered in source zones following DRO and GRO. Methane is encountered in source zones and downgradient plumes.

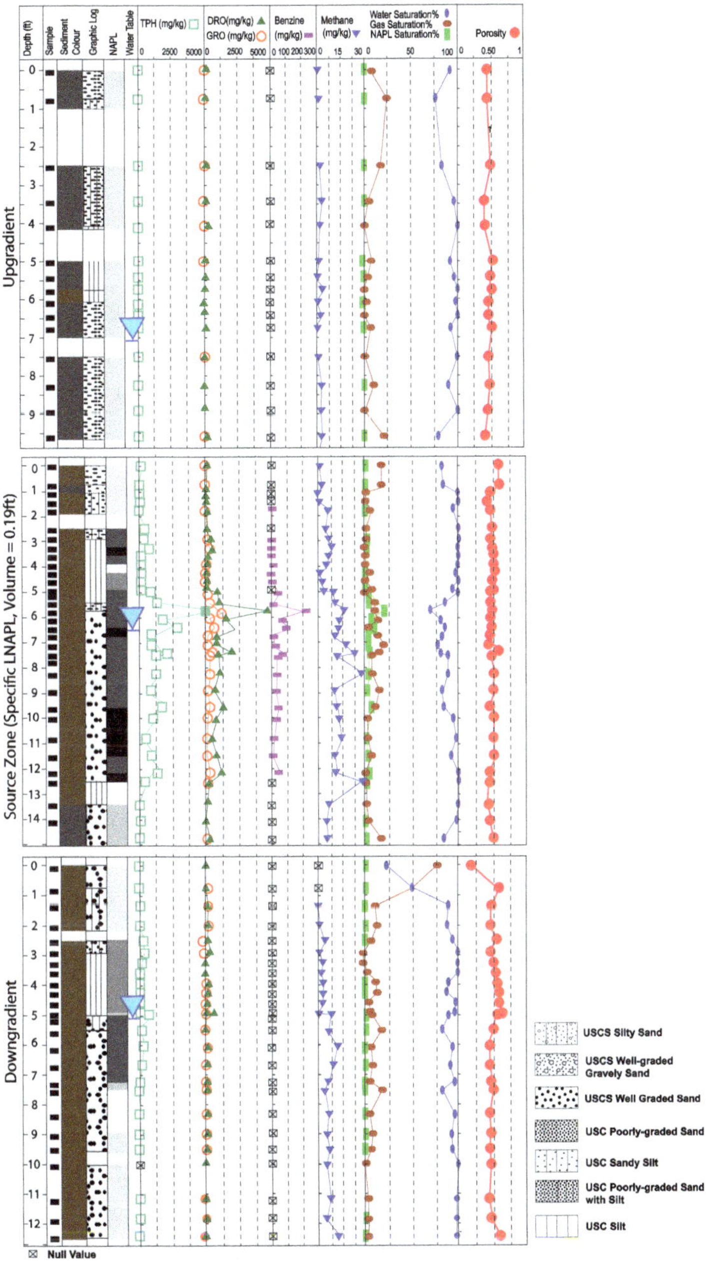

Figure 5.13. High-resolution data collected by analyses of frozen cores collected from Site PT.

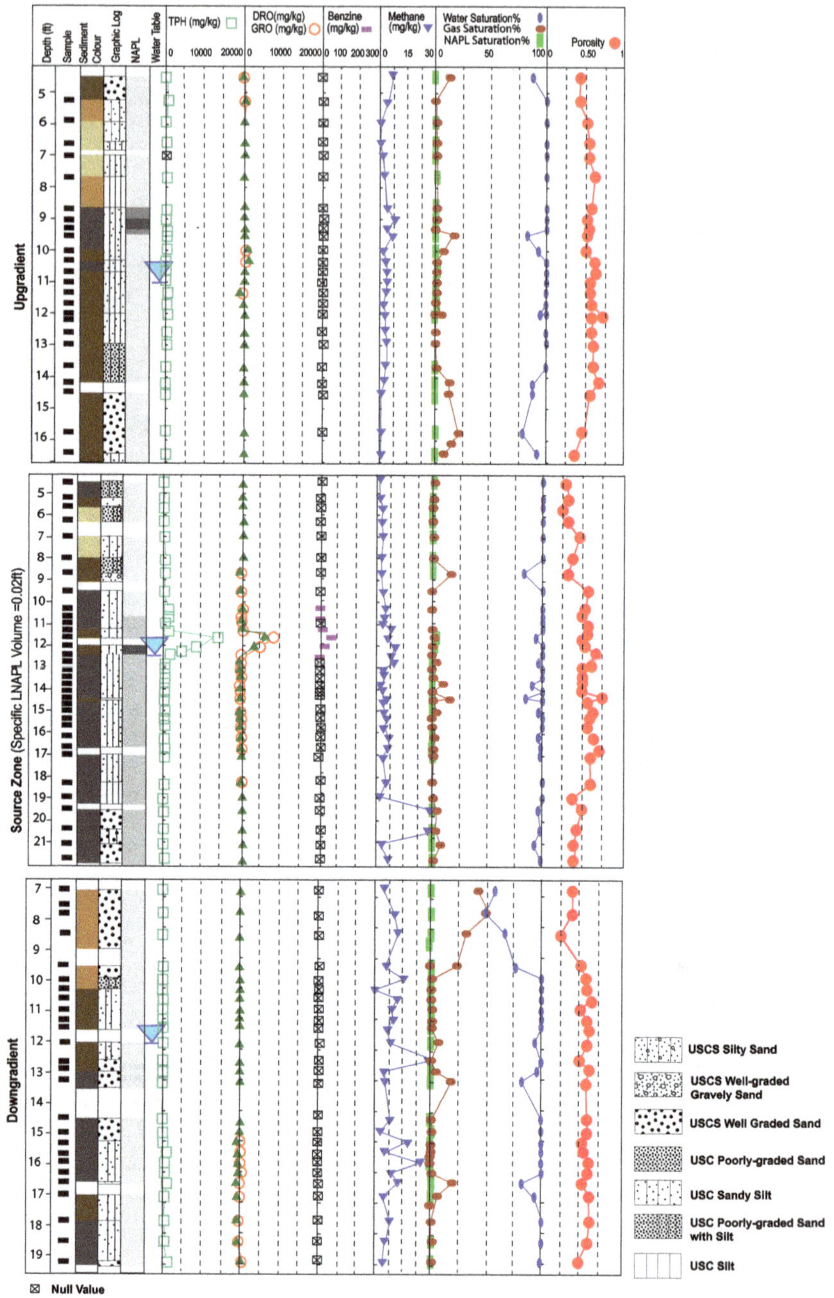

Figure 5.14. High-resolution data collected by analyses of frozen cores collected from Site AGS#1.

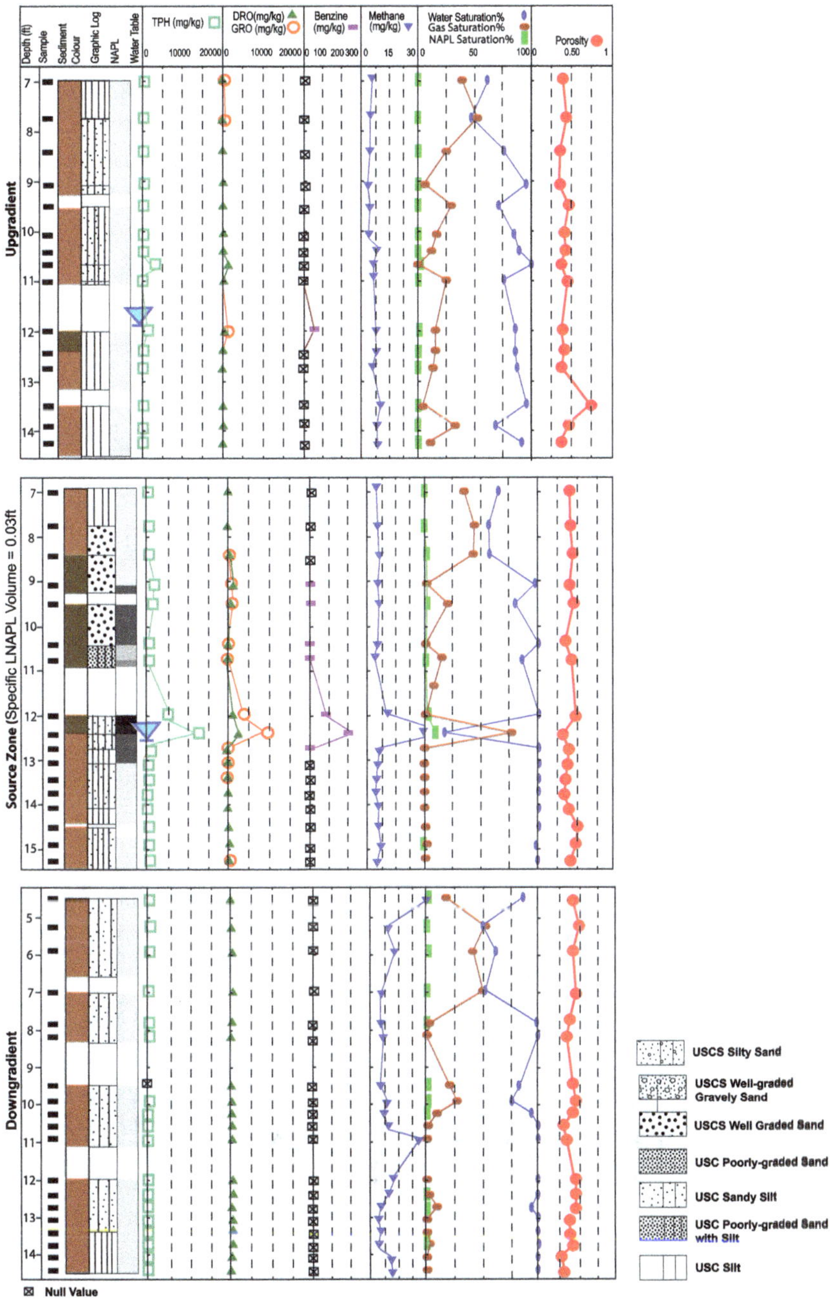

Figure 5.15. High-resolution data collected by analyses of frozen cores collected from Site AGS#2.

Methane suggests ongoing biogeochemical transformations of CoCs to carbon dioxide and methane via methanogenesis. Similarly, elevated gas saturations below the water table in source zones and plumes provide a line of evidence for ongoing natural assimilation of CoCs. Overall data in Figures 5.13–5.15 depict scenarios where the primary risks posed by historic releases of petroleum hydrocarbons are, in large part, mitigated by ongoing natural assimilation processes.

5.3.3 *Distributions of contaminants in transmissive and low-k zones*

Soil–groundwater systems impacted by releases can be conceptualised as being composed of transmissive and low-*k* zones. Contaminants move through transmissive zones via advection. Low-*k* zones store and release contaminants from and to transmissive zones via diffusion and slow advection (Chapman and Parker, 2005; Parker *et al.*, 2008; Sale *et al.*, 2013). Releases from low-*k* zones are commonly referred to as back diffusion. Based on the plots presented in Figures 5.10–5.15, transmissive and low-*k* zones range from thin interbeds at the scale of inches to larger beds ranging up to 10 feet.

Figure 5.16 presents fractions of low-*k* and transmissive zones as well as fractions of contaminants in low-*k* and transmissive zones (Roads,

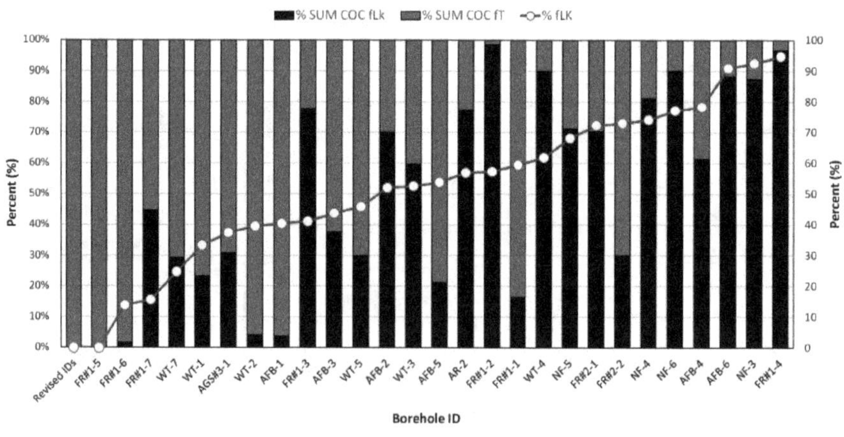

Figure 5.16. Circles indicate the fraction of low-*k* zones in all frozen cores collected from 29 boreholes. Bars indicate the fractions of total contaminant mass in transmissive and low-*k* zones.

2020). The primary purpose of the data is to quantitatively support the theme that impacted soils and groundwater are almost always composed of transmissive and low-*k* zones.

Data from all frozen cores collected at the sites described in Table 5.1 are used in Figure 5.16. Applying the generalisation of transmissive and low-*k* zones, the white circles in Figure 5.16 present percentages of low-*k* zones in each core. Estimates of transmissive and low-*k* zones at each borehole (Roads, 2020) are based on visual logging of sub-samples from frozen cores as defined above. Percentages of low-*k* zones range from 0 to 92%, with a mean value of 50%. Correlations between visual soil descriptions and measured permeabilities for the AFB site are addressed in Table 5.9 (in subsequent text).

Data bars in Figure 5.16 present ranked estimates of fractions of contaminants in low-*k* zones. All borehole locations with low-*k* zones have contaminants in low k zones. Fractions of contaminants in low-*k* zones range from 0 to 99%. For eight of the 29 borehole locations, more than half of the contaminant mass is present in low-*k* zones. As a first-order approximation, sites with larger fractions of low-*k* zones have greater fractions of total contaminant mass in low-*k* zones. Exceptions include sites with small fractions of low-*k* zones with high fractions of total contaminants stored in low-*k* zones. These sites are likely mature releases wherein contaminants in low-*k* zones are likely to be a persistent source of low levels of contamination in transmissive zones.

5.3.4 *Physical properties*

The following section documents the results of physical properties measured using frozen cores collected from the chlorinated solvent site (AFB) and a petroleum site (FR#3).

5.3.4.1 *Bulk density, water content, porosity and fraction of organic carbon*

Results for bulk density (ρ_b), water content (w), porosity and fraction of organic carbon (f_{oc}) estimated through the analysis of frozen cores collected from Site AFB are shown in Figure 5.17. Values for ρ_b range from 1.1 to 2.4 g/cm^3, with an average of 1.8 g/cm^3. Figure 5.18 indicates that water content varies between 8% and 52%, with an average of 21%. As shown in Figure 5.19, porosity of saturated samples ranges between 8%

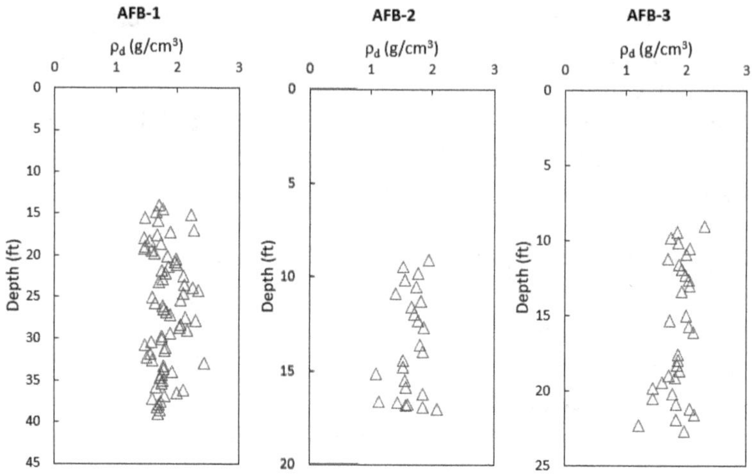

Figure 5.17. Dry bulk density of frozen cores collected from Site AFB.

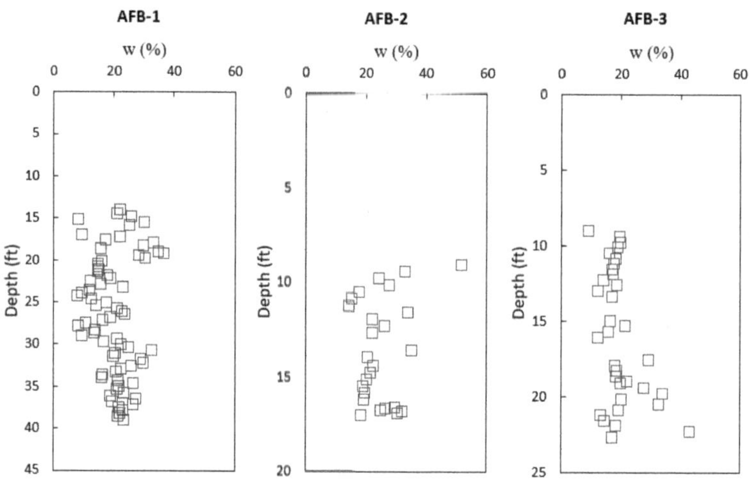

Figure 5.18. Water content of frozen cores collected from Site AFB.

and 52% with an average of 21%. f_{oc} is estimated between 0 and 0.004, with an average of 0.0008, as shown in Figure 5.20.

5.3.4.2 *Fluid saturation and fluid content*

Fluid saturation and fluid content were determined for four boreholes from a former refinery (Site FR#3). The site is underlain by a braided

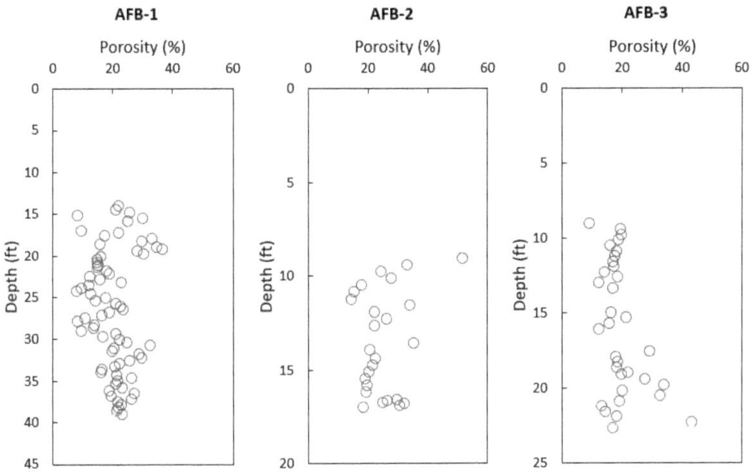

Figure 5.19. Porosity of frozen cores collected from Site AFB.

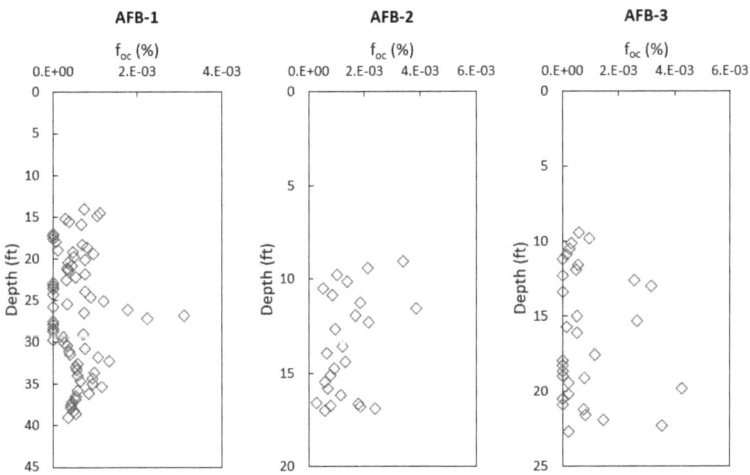

Figure 5.20. f_{oc} of frozen cores collected from Site AFB.

stream deposit. With depth, sediments grade from fine-grained overbank sands and silts to gravels. Four core holes (FR#3-1–FR#3-4) were advanced in a 10-by-10-foot area using the cryogenic core collection technique. Cores were collected from the ground surface to 14 ft bgs. The water table at Site FR#3 was approximately 4 ft bgs. Core sub-samples were analysed for fluid porosities, fluid saturations and specific LNAPL volumes (using TPH data from Kiaalhosseini *et al.*, 2016).

Figure 5.21 presents fluid saturations, fluid contents and porosities from frozen cores collected at Site FR#3. Fluid porosities, saturations and specific LNAPL volumes are presented in Table 5.8. Lower porosity values reflect samples where large pieces of solid aggregate comprised a large fraction of the sample. As with Sites PT, AGS#1 and AGS#2, large gas content/saturations below the water table are attributed to natural depletion of LNAPL (i.e., NSZD) via ongoing biologically mediated assimilation processes (Irianni-Renno *et al.*, 2016).

5.3.5 *Hydraulic conductivity*

The results of hydraulic conductivity testing on the frozen core sections collected at Site AFB are summarised in Table 5.9. The results indicate three orders of magnitude variation in the measured hydraulic conductivities (i.e., from 7.8×10^{-3} to 3.5×10^{-6} cm/s).

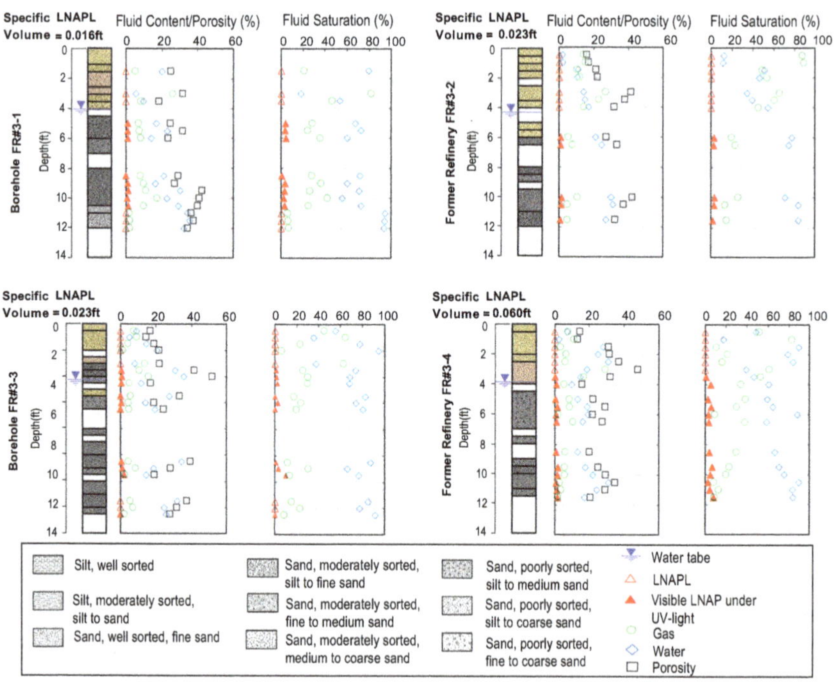

Figure 5.21. Results of fluid saturation, fluid content and porosity estimated through analysis of frozen cores collected at Site FR#3.

Table 5.8. Median porosity, fluid content and fluid saturation of frozen cores collected at Site FR#3.

	Median Results (%)	
Parameter	**Above Water Table**	**Below Water Table**
Porosity	23.7	29.2
Water content/saturation	14.5/49.5	22.1/74.2
Gas content/saturation	10.8/50.5	5.8/22.9
LNAPL content/saturation	0.2/0.6	0.9/2.8

Table 5.9. Results of laboratory hydraulic conductivity measurements of frozen cores collected from Site AFB using the CIL technique.

Borehole Location	Depth (feet)	Description	Water Content (g/g)	*K*, CF[1] (cm/s)	*K*, FH[2] (cm/s)
AFB3	15.34	Silt, well-sorted	0.205	–	5.2×10^{-5}
AFB-3	19.08	Sand, moderately sorted, fine-to-medium grained	0.208	7.8×10^{-3}	1.2×10^{-3}
AFB-3	20.53	Silt, moderately sorted, with fine sand	–	–	3.5×10^{-6}
AFB-3	21.94	Silt, moderately sorted, with fine sand	–	–	2.2×10^{-5}
AFB-2	9.80	Sand, well-sorted, fine-grained	0.240	2.3×10^{-3}	–
AFB-2	10.53	Sand, poorly-sorted, fine-to-coarse grained	0.180	1.6×10^{-3}	–
AFB-2	10.90	Sand, moderately sorted, silt-to-fine grained	0.149	7.3×10^{-4}	–

Notes: [1] Constant flow testing methods ($K > 10^{-4}$ cm/s). [2] Falling head testing methods ($K < 10^{-4}$ cm/s).

5.3.6 *Microbial analysis*

Figure 5.22 presents microbial analysis results for frozen cores collected at Site AFB. Microbial analyses were facilitated via support from Dr. Susan De Long at CSU, including use of Dr. De Long's laboratory resources at CSU. Cryogenic preservation is widely considered a "gold standard" for preservation of biological samples (Jungare *et al.*, 2022).

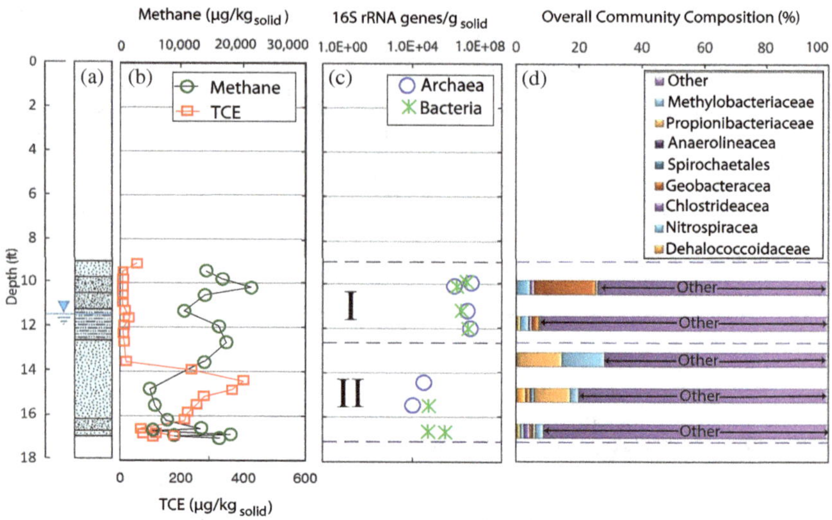

Figure 5.22. Microbial analysis of frozen cores collected from AFB-2: (a) visual soil log, (b) methane and TCE concentrations, (c) the number of 16s rRNA genes per gram of core for archaea and bacteria as a function of methane and TCE and (d) the overall microbial community composition where each horizontal bar represents a specific depth and is divided into colored segments, each representing a different microbial family. The x-axis represents the percentage of the total community composed by each family.The dominant family across all depths appears to be "Other", highlighting that a significant portion of the microbial community may not belong to the specifically identified families.

Envisioned advantages of *in situ* versus *ex situ* freezing of samples for microbial analyses include the following:

- stabilising pore fluids *in situ*, as opposed to drainage of pore fluids as commonly occurs absent *in situ* freezing;
- limited entry of atmospheric oxygen;
- flash (~ 5 min) *in situ* freezing, as opposed to slower *ex situ* freezing in coolers with dry ice.

Figure 5.22 presents visual soil logging data, methane and TCE concentrations, the number of 16s rRNA genes per gram of solids for archaea and bacteria and the overall community composition of microbial ecology at location AFB-3. Two distinct intervals were identified: (1) a moderately sorted silt (low-*k* zone) with low TCE, high methane and elevated levels of 16s rRNA genes and (2) a fine-grained, well-sorted sand (transmissive zone) with high TCE, low methane concentrations and lower levels of the

16s rRNA genes. Figure 5.22(d) describes the overall microbial community at this location based on five samples analysed. The data suggest distinct microbial communities (I) in the low-*k* zone and (II) in the transmissive zone.

5.4 Discussion

Cryogenic core collection provides, perhaps for the first time, a mechanism for resolving the distribution of contaminants in all 14 compartments introduced in Figure 5.1, as well as associated redox conditions, physical properties and microbial ecology of the subsurface. The holistic characterisation of contaminants, redox conditions, physical properties and microbial ecology provides a foundation for sound decisions regarding management of sites (Sale and Newell, 2011). In contrast, solely relying on data from monitoring wells, two of the 14 relevant compartments, can lead to flawed decisions. Furthermore, relying on non-representative soil samples can lead to incomplete characterisation of sites and poorly informed remedial decisions. The following section outlines the strengths and limitations of cryogenic core collection.

5.4.1 *Advantages of cryogenic coring*

In situ freezing of soil cores provides opportunities to (1) improve core recovery, (2) stabilise pore fluids, (3) limit losses of volatile CoCs, (4) preserve redox conditions, (5) preserve microbial communities and (6) preserve physical properties of soils.

5.4.1.1 *Core recovery*

In situ freezing of cores prevents soils from falling out of sample liners during retrieval to the surface. Although only one dataset directly shows that this was presented here (Site FR#1, 83% recovery of frozen core, 27% recovery of unfrozen core), based on other drilling, we have found this to be a common occurrence in non-cohesive formations (e.g., sands and gravels).

5.4.1.2 *Control of "flowing sands"*

During the core recovery process, the weight of overburden soils and water inside augers is significantly reduced, and non-cohesive soils can flow from the formation into the empty augers. As documented at the

FOF site, "flowing sands" can lead to the collection of non-representative samples. In the case of Site FOF, expanded flowing sands appear to lead to overestimation of specific LNAPL volumes by a factor of ~3.5 (Table 5.7). Processes driving the flow of low cohesion soils into auger flights are explored in Figure 5.23. Notably, little information has been published regarding sample biases introduced by flowing sands or practical solutions for flowing sands.

Cryogenic core collection tools described here are designed to temporally freeze and thereby stabilise soils below the lead auger. Design details include (1) excluding insulation between the drive shoe and the cooling barrel to improve thermal contact with the soils at the tip of the drive shoe

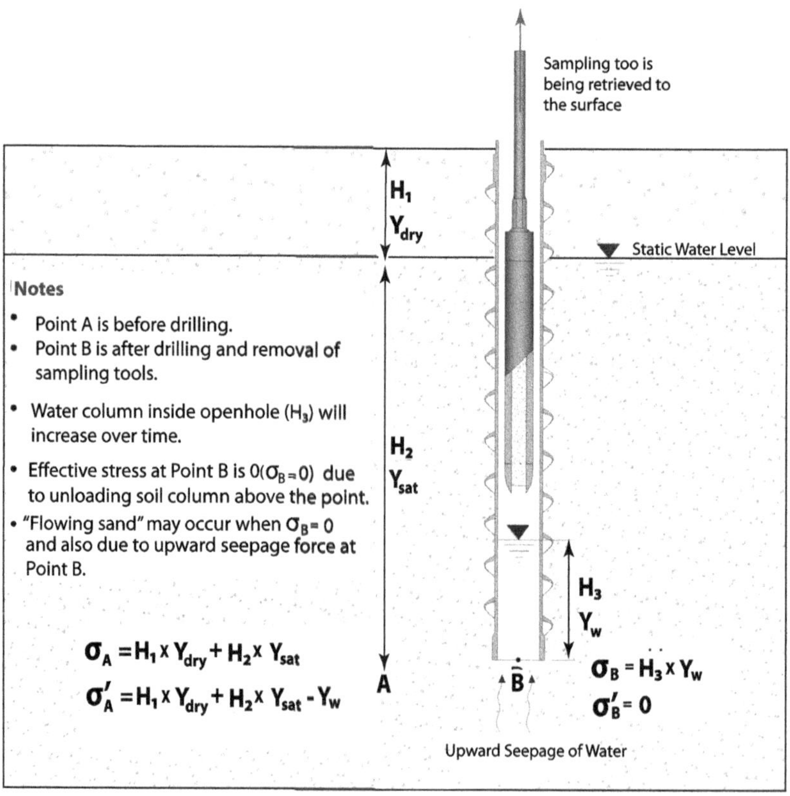

Figure 5.23. Forces driving flowing low-cohesion soils into auger flights. σ and σ' are total and effective stresses (Terzaghi, 1925), respectively. γ are densities for dry soils (γ_{dry}), water-saturated soil (γ_{sat}) and water (γ_w). Effective stresses at point A (before drilling) greater than effective stress at point B (after drilling, $\sigma'_A > \sigma'_B$) can lead to low cohesion media flowing into auger flights.

(Figure 5.24); (2) delivery of liquid nitrogen directly to the drive shoe, with upward flow to the rest of the barrel, again to focus cooling at the drive shoe; (3) extending the duration of liquid nitrogen delivery in intervals with a potential for flowing sands.

A potential downside to freezing the formation near the drive shoe is that the core barrel may become frozen in place (to the formation and augers), and the augers may need to be held in place while pulling the rods and the cryogenic core barrel out of the augers (Figure 5.25). Holding rods

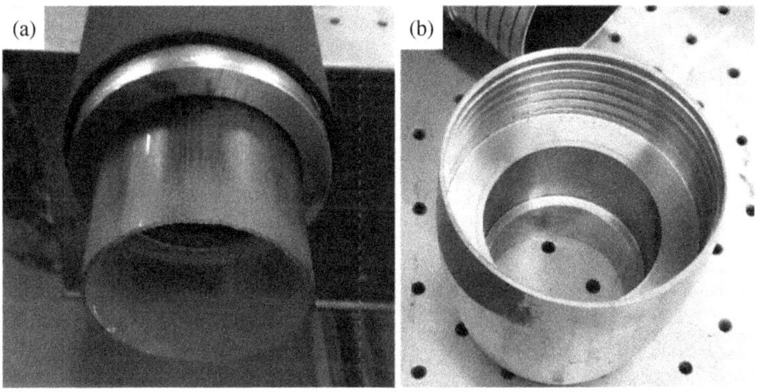

Figure 5.24. (a) PVC core liner in insulated dual-wall cooling barrel. Insulation is absent where the cooling barrel seats inside the drive shoe. (b) Drive show cut to accommodate the cooling barrel and PVC liner.

Figure 5.25. (a) Braces used to hold augers down while pulling the cryogenic coring tool out after freezing. (b) Clamps used to draw frozen cores in liners out of the dual-wall cooling barrel. (c) Pulling the cores in liners out of the dual wall cooling barre.

in tension for up to 5 min may be required to allow the frozen tooling to release from the formation and auger flights.

As described above for Site FOF, delays in returning the cryogenic core barrel for subsequent sampling can lead to invasion of flowing sands and elevated estimates of specific LNAPL volumes. To overcome this, it is recommended that a second cryogenic core barrel be available so it can be used to collect the next sample while the frozen core is recovered from the first core barrel.

5.4.1.3 *Stabilisation of pore fluids*

Meaningful estimates of fluid saturations (i.e., water, NAPLs and gases), dissolved gases, redox conditions and microbial ecology are achieved via *in situ* freezing of cores. Rapid freezing and subsequent storage of cores at -80°C stabilises pore fluids by (1) freezing water in soils, (2) increasing the viscosity of NAPLs and (3) entrapping dissolved gases and NAPLs in frozen pore water.

5.4.1.4 *Preservation of biogeochemical conditions*

Freezing limits the inflow of atmospheric gases (e.g., oxygen) that occur when pore liquids drain out or redistribute during core recovery. Drainage of pore fluids can also lead to losses of volatile compounds and changes in the composition of the soil gas, including microbially derived gases. The introduction of atmospheric oxygen into cores is also likely to impact oxidation-reduction conditions in samples.

Note that the advantages of cryogenic core collection are greatest for less consolidated soil with a potential for drainage and/or flowing sands. Unfortunately, it can be difficult to know in advance the nature of the soils that will be encountered. Furthermore, in comparing coring with and without *in situ* freezing, Kearney *et al.* (2024) observe that sands are missed without freezing in media dominated by fine-grained cohesive sediments.

5.4.1.5 *Preservation of physical properties of soils*

Preserving *in situ* soil structure can be essential to determining physical properties of soil, including the distributions of permeability. Acquiring estimates of in situ physical properties is a central tenet for advancing sound SCMs and sound remedies.

5.4.2 *Insights gained through cryogenic coring*

The following section summarises insights gained through subsurface investigations using high-resolution analysis of cores obtained via cryogenic core collection, including improved recovery, acquisition of representative soil samples, and preservation of soil properties.

5.4.2.1 *Fractions of transmissive and low-k zones*

As the data in Figures 5.7, 5.10–5.15 and 5.21 show, impacted sites are composed of interbedded transmissive and low-*k* zones. Low-*k* zones include interbedded layers where the thicknesses of beds range from inches to feet. The greatest fractions of low-*k* zones are seen in cores from Sites FR#2, AR#1, AFB and NF, corresponding to fluvial overbank, braided stream and tidal estuary deposits, as listed in Table 5.1.

According to Figure 5.16, all sites with low-*k* zones have contaminants within those zones, and eight of the 29 frozen core locations have more than 50% of the total contaminant mass in low-*k* zones. This is important because in many instances contaminants in low-*k* zones will act as persistent long-term sources of contaminants to transmissive zones. In addition, flushing contaminants out of low-*k* zones or delivery of treatment reagent to low-*k* zones is difficult, and only a few remedies (i.e., *in situ* heating, Davis, 1997, and *in situ* soil mixing, Olson *et al.*, 2012) have the potential to address contaminants in low-*k* zones).

5.4.2.2 *LNAPL distribution*

LNAPL quantities and distributions are key elements in understanding risks posed by petroleum-contaminated sites. Cryogenic core collection represents a substantial improvement in our ability to determine both the quantity and distributions of NAPLs.

As an example of how specific LNAPL volume data can be used, the mean specific volume value from the data reported for Site FR#1 of 0.05 feet can be used to estimate the remaining longevity of LNAPL at the site. Assuming a constant NSZD rate through time of 0.003–0.006 feet/year (1,000–2,000 gallons/acre/year, Kulkarni *et al.*, 2022), the estimated longevity of the LNAPL is in the range of 8–16 years. Without high-quality specific LNAPL volume estimates, there is little basis for anticipating the longevity of LNAPL bodies.

5.4.2.3 *Natural assimilation processes*

Most contaminants are assimilated by natural physical and biogeochemical processes through time. Consequently, natural assimilation is commonly a primary factor controlling risks to human health and the environment. However, demonstration of natural assimilation requires multiple lines of evidence. Cryogenic core collection provides a promising approach to providing the multiple lines of evidence needed for assessment of assimilation remedies.

5.4.2.4 *Petroleum hydrocarbons*

Building on data from Sites PT, AGS#1 and AGS#2 in Figures 5.13–5.15, methane is encountered at three sites in source zones along with LNAPL constituents including DRO, GRO and benzene. Methane is also encountered in downgradient plumes, while it is absent in the upgradient of source zones. The presence of methane in NAPL source zones supports ongoing NSZD via conversion of LNAPL constituents to methane and carbon dioxide via methanogenesis. The presence of methane in downgradient plumes, absent DRO, GRO and benzene, supports ongoing methanogenic NA in the downgradient plumes.

Further support for degradation-related assimilation is seen in soil colours. Soils are darker in source zones and plumes as compared to those in upgradient locations, with the shifts to darker colours likely due to NSZD- and NA-related reduced mineral precipitation in source zones and plumes.

5.4.3 *Limitations*

While cryogenic core collection, as described, has numerous benefits, the method is not without limitations. The following describes notable limitations.

5.4.3.1 *Additional costs and time*

The steps involved in cryogenic core collection (connecting lines for LN delivery and exhaust, handling LN dewars, *in situ* freezing and recovering frozen core) take more time than conventional hollow stem auger and direct-push soil sampling methods. On average, cryogenic core collection takes two to three times more time, and consequently it costs more per foot of core than non-cryogenic methods. On the other hand, given the high cost of flawed site management decisions based on flawed SCMs, cryogenic core collection can be a basis for great cost savings.

5.4.3.2 *Liquid nitrogen dewars*

Cryogenic core collection work described here was carried out using 160 L liquid nitrogen dewars. Typically, one 160 L dewar is needed for every 20–30 feet of frozen core. The performance of individual dewars under field conditions can vary, and at times inadequate delivery of liquid nitrogen has slowed core collection and constrained complete *in situ* freezing of cores.

5.4.3.3 *Reduced benefits in fine-grained consolidated cohesive soils*

The advantages of cryogenic core collection can be reduced in fine-grained consolidated cohesive soils. Conventional core collection with placement of samples on dry ice at grade may provide similar results (Kearny *et al.*, 2024).

5.4.3.4 *Alteration of mineralogy*

Work presented by Axe *et al.* (2020) suggests that cryogenic sampling processes can alter iron mineral coatings in contaminated sediments. Care is needed in applying cryogenic core collection methods at sites where soil mineralogy is an issue.

5.4.3.5 *Application of cryogenic core collection at PFAS sites*

Per- and polyfluoroalkyl substances (PFAS) have emerged as primary CoCs in soils and groundwater. Challenges include long-chained PFAS attached to the air/water interfaces presented by the thin films of water around sand grains in the vadose zone and potentially at the capillary fringe. It is anticipated that the ability to stabilise pore fluids in the vadose and groundwater zones could lead to consequential new insights regarding the distributions of PFAS in soils and groundwater, better CSMs and better solutions. To the authors knowledge, as of 2024, cryogenic core collection methods have not been applied at a PFAS site. It seems likely that cryogenic core collection methods will be applied at a PFAS site in 2025.

5.5 Conclusions

A major constraint to managing contaminated soils and groundwater has been the incomplete resolution of the distribution of contaminants in the subsurface according to the 14-compartment model. Furthermore, a clear understanding of biogeochemical conditions in all compartments is critical to resolving the fate of contaminants in soils and groundwater. Well-informed characterisations of the distribution of contaminants and biogeochemical conditions in all compartments provide a basis for well-informed holistic CSMs and a sound basis for site management decisions.

As discussed, *in situ* freezing has multiple benefits. First, *in situ* freezing facilitates better collection of soils from targeted intervals by improving recovery and controlling flowing sands. Second, freezing preserves pore fluids, contaminants in all phases (i.e., nonaqueous, aqueous, sorbed and gas), redox conditions, physical properties and microbial ecology. Lastly, freezing enables a high-resolution multiple-lines-of-evidence basis for resolving natural assimilation processes, which are increasingly being recognised as playing a primary role in mitigating risks associated with releases to soils and groundwater.

Acknowledgements

Primary funding for the work presented in this manuscript came from SERDP Project ER-1740, sponsors of the University Consortium for Field Focused Groundwater Research and the State of Colorado's OEDIT programme.

Supporting research faculty at CSU include Drs. Mitch Olson, Susan DeLong, Maria Irianni-Renno and Jens Blotevogel. Supporting students at CSU include Eric Roads, Ray Buzeta and Cathelyne Powers. Consulting firms supporting fieldwork include TriHydro Inc., TRC, AECOM, GES and ROUX and Associates.

Fabrication of cryogenic core collection systems was performed at CSU's Engineering Research Center (ERC) Machine Shop by Bart Rust, Horatio (Junior) Graza and Matthew Szydlowski.

Drilling services were provided by Rick Rogers, Andy Lozano and Dustin Coy, with Drilling Engineers Inc.

References

ASTM Standard D2434. Standard Test Methods for Measurement of Hydraulic Conductivity of Coarse-Grained Soils. West Conshohocken, PA: ASTM International. DOI: 10.1520/D2434-68R06. Available at: www.astm.org.

ASTM Standard D1140. Standard Test Methods for Determining the Amount of Material Finer than 75-μm (No. 200) Sieve in Soils by Washing. West Conshohocken, PA: ASTM International. DOI: 10.1520/D1140-17. Available at: www.astm.org.

ASTM Standard D422. Standard Test Method for Particle-Size Analysis of Soils. West Conshohocken, PA: ASTM International. DOI: 10.1520/D0422-63R07. Available at: www.astm.org.

Axe, L., Hua, H., Dyer, J., Irianni-Renno, M., Landis, R., Sale, T. C., and Yin, X. (2020). Impacts of cryogenic sampling processes on iron mineral coatings in contaminated sediment. *Science of the Total Environment*. Available at: https://doi.org/10.1016/j.scitotenv.2020.142796.

Chapman, S. W. and Parker, B. L. (2005). Plume persistence due to aquitard back diffusion following dense nonaqueous phase liquid removal or isolation. *Water Resource Research*, 41(12), W12411.

Davis, E. L. (1997). How Heat Can Enhance In-situ Soil and Aquifer Remediation: Important Chemical Properties and Guidance on Choosing the Appropriate Technique. US EPA/540/S-97/502.

Durnford, D., Brookman, J., Billica, J., and Milligan, J. (1991). LNAPL distribution in a cohesionless soil: A field investigation and cryogenic sampler. *Ground Water Monitoring and Remediation*, 11(3), 115–122.

Freeze, R. A. and Cherry, J. A. (1979). *Groundwater*. New Jersey, USA: Prentice Hall Inc.

Irianni-Renno, M. (2024). Tools for characterization and monitoring natural source zone. PhD Dissertation, Colorado State University, Fort Collins, Colorado State, USA.

Irianni-Renno, M., Akhbari, D., Olson, M. R., Byrne, A. P., Lefèvre, E., Zimbron, J., Lyverse, M., Sale, T. C., and De Long, S. K. (2016). Comparison of bacterial and archaeal communities in depth-resolved zones in an LNAPL body. *Applied Microbiology and Biotechnology*, 100, 3347–3360.

Irianni-Renno, M., Sale, T. C., and De Long, S. K. (2021). Advanced methods for RNA recovery from petroleum impacted soils. *Methods X*, 8, 101503.

Johnson, R. L., Brow, C. N., Johnson, R. O., and Simon, H. M. (2013). Cryogenic core collection and preservation of subsurface samples for biomolecular analysis. *Groundwater Monitoring and Remediation*, 33(2), 38–43. DOI: 10.1111/j.1745-6592.2012.01424.x.

Jungare, K. A., Radha, R., and Sreekanth, D. (2022). Cryopreservation of biological samples — A short review. *Materials Today Proceedings*, 51(Part 3), 1637–1641.

Kalinski, M. E. (2011). *Soil Mechanics Lab Manual*. 2nd ed. John Wiley & Sons, 111 River Street, Hoboken, NJ 07030.

Kearney, K. J., Blount, T., Palmer, E., Arvizu, R., Sanford, R., Valocchi, A., Schaefer, C., and Werth, C. (2024). Assessment of cryogenic coring to preserve vertical distributions of trichloroethylene and reaction products in an aquitard. *Groundwater Monitoring and Remediation*, 44(2), 118–128.

Kiaalhosseini, S. (2016). Third generation site characterization, cryogenic core collection, nuclear magnetic resonance and electrical resistivity. Ph.D. Dissertation, Department of Civil and Environmental Engineering, Colorado State University, Fort Collins, Colorado, USA.

Kiaalhosseini, S., Johnson, R., Rogers, R., Irriani-Renno, M., Olson, M., Lyverse, M., and Sale, T. (2016). Cryogenic core collection from unconsolidated subsurface media. *Groundwater Monitoring & Remediation*, 36(4), 41–49.

Kulkarni, P. R., Walker, K. L., Newell, C. J., Askarani, K. K., Li, Y., and McHugh, T. E. (2022). Natural source zone depletion (NSZD) insights from over 15 years of research and measurements: A multi-site study. *Water Research*, 119170.

Murphy, F. and Herkelrath, W. N. (1996). A sample-freezing drive shoe for a wire line piston core sampler. *Ground Water Monitoring and Remediation*, 16(3), 86–90. DOI: 10.1111/j.1745-6592.1996.tb00143.

Olson M., Clayton, W., Sale, T. C., De Long, S., Irianni-Renno, M., and Johnson, R. (2017). Evaluating Long-Term Impacts of Soil-Mixing Source-Zone Treatment using Cryogenic Core Collection. ESTCP Project ER-201587, US DoD.

Olson, M. R., Sale, T. C., Shackelford, C. D., Bozzini, C., and Skeean, J. (2012). Chlorinated solvent source-zone remediation via ZVI-clay soil mixing: 1-year results. *Groundwater Monitoring & Remediation*, 32(3), 63–74.

Parker, B. L., Chapman, S. W., and Guilbeault, M. A. (2008). Plume persistence caused by back diffusion from thin clay layers in a sand aquifer following TCE source-zone hydraulic isolation. *Journal of Contaminant Hydrology*, 102, 86–104.

Roads, E. (2020). Analysis of contaminant mass in place in transmissive and low-K zones. M.Sc. Thesis, Department of Civil and Environmental Engineering, Colorado State University, Fort Collins, Colorado, USA.

Sale, T., Kiaalhosseini, S., Olson, M., Johnson, R., and Rogers, R. (2016). Third-Generation (3G) Site Characterization: Cryogenic Core Collection and High-Throughput Core Analysis. An Addendum to Management of Contaminants Stored in Low Permeability Zones: A State-of-the-Science Review, SERDP Project ER-1740.

Sale, T. and Newell, C. J. (2010). Impacts of source management on chlorinated solvent plumes. In *In Situ Remediation of Chlorinated Solvent Plumes*, pp. 185–216. SERDP and ESTP Remediation Technology Monograph Series, New York, USA: Springer.

Sale, T. and Newell, C. J. (2011). A Guide to Selecting Remedies for Subsurface Releases of Chlorinated Solvents. ESTCP Project ER-200530, U.S. Department of Defense Environmental Security Technology Certification Program.

Sale, T. C., Johnson, R. L., Rogers, R., and Kiaalhosseini, S. (2017). Cryogenic core collection, US Patent 20170362908 A1, December 21, 2017.

Sale, T. C., Parker, B., Newell, C., and Devlin, J. (2013). Management of Contaminants Stored in Low Permeability Zones-A State of the Science Review. DTIC Document, Not Peer Reviewed/Refereed.

Terzaghi, K. (1925). *Erdbaumechanik auf Bodenphysikalischer Grundlage.* Vienna: Dueticke.

https://doi.org/10.1142/9781800617841_0006

Chapter 6

Leveraging Machine Learning Models for Enhanced Contaminated Site Assessment

Alvin Lal[*,†,¶], **Junaid Sadeque**[‡], **Ashneel Sharan**[§], **Asadi Srinivasulu**[†], **Yanju Liu**[*,†], **and Ravi Naidu**[*,†]

[*]*Global Centre for Environmental Remediation (GCER), College of Engineering, Science and Environment, the University of Newcastle, Callaghan, Australia*

[†]*The Cooperative Research Centre for Contamination Assessment and Remediation of the Environment (crcCARE), the University of Newcastle, Callaghan, Australia*

[‡]*AECOM, Los Angeles, CA, USA*

[§]*College of Science and Engineering, James Cook University, Queensland, 4811, Australia*

[¶]*alvin.lal@newcastle.edu.au*

Abstract

This chapter delves into the transformative potential of machine learning (ML) models in environmental management, emphasising their role in evaluating and identifying harmful substances in specific areas to mitigate environmental and health risks. Traditional methods often fall short in real-time monitoring and handling complex data, limiting their effectiveness. This review highlights how ML enhances site assessment and monitoring by presenting innovative solutions for real-time data analysis and complex data management, thus revolutionising environmental management. ML techniques effectively address significant drawbacks of traditional assessment methods, such as inefficient remediation due to underutilised data, difficulties in predicting contaminant movement and inadequacies in risk assessments. The research underscores the critical role of ML in adopting new approaches to assessment and remediation, significantly contributing to addressing environmental and health challenges. In addition, we present a local case study in which we demonstrate the role of ML models in assessing the influence of various soil and groundwater properties on PFAS concentration levels in soil. In summary, this chapter highlights that integrating ML technologies improves data analysis, prediction accuracy and overall remediation efficiency. Future recommendations suggest expanding the applicability of ML models, incorporating real-time sensor data and fostering collaboration to develop user-friendly tools. These strategies aim to effectively tackle environmental and health challenges associated with contaminated sites, ensuring a safer and healthier environment.

6.1 Introduction

Environmental contamination presents significant risks to both ecosystems and human health, making comprehensive site assessments essential to identify and mitigate these dangers (Ritter, 2002). Assessing contaminated sites involves scrutinising specific areas to detect and measure harmful substances, ensuring that appropriate measures are implemented to manage potential risks. While traditional assessment methods are valuable, they often face challenges with real-time monitoring and the complexity of environmental data, resulting in inefficiencies in remediation efforts and risk evaluation. For example, traditional site assessments often involve extensive field sampling and laboratory analyses, which can be costly. Multiple rounds of sampling may be required to obtain a comprehensive understanding of contamination, increasing overall costs (Back, 2001). In addition, traditional assessments can take a long time to complete due to the need for extensive data collection, laboratory analysis and report generation. This extended timeline can delay decision-making and remediation efforts (Reible and Demnerova, 2012). As environmental issues become increasingly complex, there is a pressing need for advanced tools capable of meeting the demands of contemporary environmental management.

Recently, machine learning (ML) has emerged as a powerful tool for transforming the assessment and remediation of contaminated sites. ML models, which are based on artificial intelligence (AI) algorithms, provide innovative solutions for real-time monitoring and the analysis of complex data, effectively addressing the limitations of traditional methods (Tyagi and Chahal, 2020). While applications of ML models in other fields are extensively documented, their application for contaminated site assessment is quite rare. This research chapter investigates the potential of ML to assist in contaminated site assessment and eventually revolutionise environmental management. We demonstrate that by integrating advanced ML techniques, contaminated site assessors can enhance data analysis, increase prediction accuracy and optimise remediation processes, thereby contributing to a safer and healthier environment.

This study highlights the transformative impact of ML in contaminated site assessment, showcasing its potential to provide innovative solutions to significant environmental and health challenges. We specifically explore what ML models are, review their applications in various site assessment contexts, examine how they are developed and present a local case study demonstrating ML models' effectiveness in assisting with contaminated site assessment.

6.2 Understanding Contaminated Site Assessment

Contaminated site assessment is an essential process that involves examining and identifying harmful substances in specific areas to mitigate environmental and health risks. This process aims to detect, measure and understand the distribution and concentration of pollutants, ensuring that appropriate actions are taken to manage the associated risks. Traditionally, this involves extensive sampling and testing during site characterisation, followed by the selection and design of remediation technologies based on the collected data. However, as described in Section 6.1, traditional methods often struggle with the complexities of real-time monitoring and handling intricate environmental data, leading to inefficiencies in remediation efforts and risk assessment.

ML models have proven essential in overcoming challenges associated with traditional methods. These models have transformed the field of contaminated site assessment and have recently been successfully applied in various case studies. ML models with their predictive capabilities provide innovative solutions for real-time monitoring and complex data analysis and illustrations. By utilising ML models, site assessors can improve the accuracy of contamination predictions, optimise remediation strategies and enhance overall environmental management (Zhong *et al.*, 2021). This transformative approach not only increases the efficiency of detecting and managing contaminants but also significantly contributes to mitigating the environmental and health challenges posed by contaminated sites. The adoption of these advanced techniques highlights the critical role of modern technology in delivering forward-thinking solutions for effective contaminated site assessment and remediation.

6.2.1 *Stages in contaminated site assessment*

Contaminated site assessment has four crucial stages for identifying, evaluating and managing sites that pose environmental and health risks due to the presence of hazardous substances (NSW EPA, 2011). These four stages are preliminary site investigation (PSI), detailed site investigation (DSI), site remedial action plan (RAP) and site validation and ongoing monitoring (SVOM). A more detailed description of these four stages is given in Figure 6.1.

These stages collectively ensure a systematic and comprehensive approach to managing contaminated sites, protecting public health and preserving the environment.

6.3 What Are ML Models and Their Applications in Contaminated Site Assessment?

6.3.1 *What are ML models?*

ML is pivotal in advancing environmental science, offering robust tools for analysing large datasets and complex relationships. As a branch of AI, ML involves developing algorithms and models that enable computers to learn from data and make predictions or decisions. In other words, an ML model is a computational system designed to recognise patterns, make predictions or perform decisions based on data. Unlike traditional algorithms, which follow explicitly programmed instructions, ML models learn from data, adapting their behaviour to improve performance over

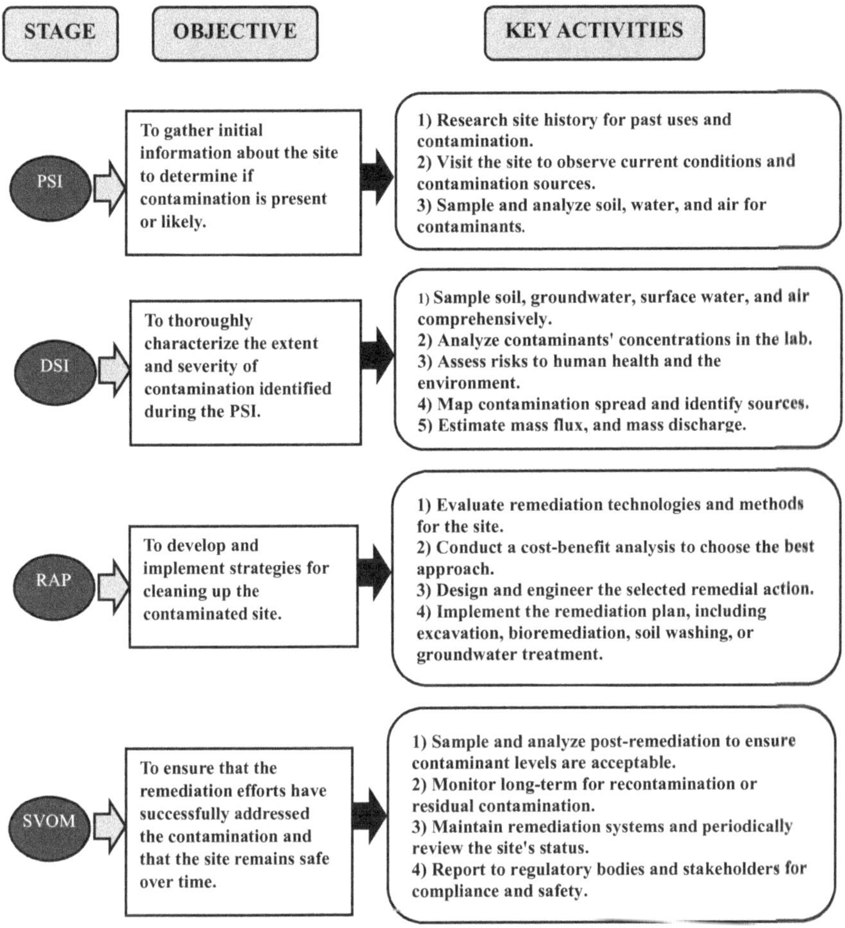

Figure 6.1. Key features of the four stages in a contaminated site assessment process.

time. The essence of an ML model lies in its ability to generalise from historical data to make accurate predictions or decisions on new, unseen data. The core components of an ML model are described in Figure 6.2.

6.3.2 *ML benefits for contaminated site assessments*

ML models offer several benefits for contaminated site assessments, transforming how environmental data are analysed, interpreted and used to make decisions. Some key benefits are described in Figure 6.3.

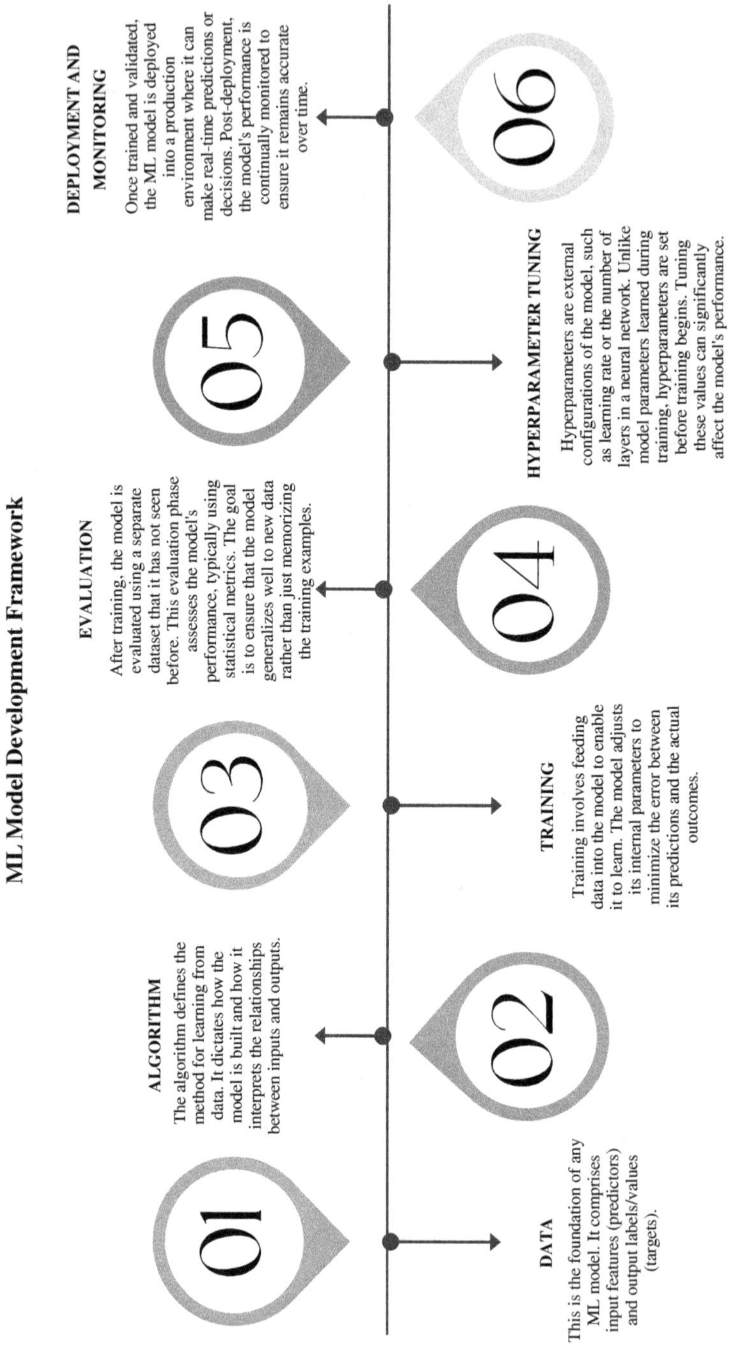

Figure 6.2. ML model development framework.

Figure 6.3. Key benefits of ML models in contaminated site assessments.

In addition to these general benefits, we have reviewed various case studies in which ML models have proven beneficial for contaminated site assessment. A detailed summary of the application of various ML techniques in improving the assessment of environmental contamination across diverse case studies is presented (see Table 6.1) to establish the role and benefits of ML models.

6.4 Case Study: Preliminary Investigation of PFAS in Irrigation Areas at a Contaminated Site

6.4.1 *Approach and purpose*

The case study area was a fenced-off irrigation area with native vegetation, where treated effluents from a sewage treatment plant were discharged. We conducted a preliminary study to assess and monitor the presence of PFAS in soils subjected to effluent irrigation (spray). The research team conducted a field visit and analysed field samples for

Table 6.1. Applications of ML in contaminated site assessment.

Case Study No.	Location	Problem Definition	ML Model Used	Findings	Reference
1	LUCAS topsoil dataset (25 European Union Member States)	Topsoil arsenic (As) contamination threatens the ecological environment and human health. However, traditional methods for As identification rely on onsite sampling and chemical **analysis, which are cumbersome, time-**consuming and costly. Here, we developed a method combining visible near-infrared spectra and deep learning to predict topsoil As content.	Neural network model	The results showed that the optimum fully connected neural network model had high robustness and generalisation (R^2 values of 0.688 and 0.692 on the validation and testing sets). **Using the model, the relative As contents at regional and** global scales were estimated and the human populations that might potentially be affected were determined.	Wu *et al.* (2024)
2	Thailand	The present study used machine learning models to locate risk areas of As contamination in Rayong coastal aquifers, Thailand	Random Forest (RF), support vector machine (SVM) and artificial neural network (ANN)	The result of the risk map obtained from the ML models revealed that the deep aquifer in the northern part of the Rayong basin has a higher risk for people to be exposed to As.	Sumdang *et al.* (2023)

3	Industrial Park in North China	Developed an ML model to finely characterise the distribution of soil contaminants using sparse-biased drilling data.	RF and ANN	The study revealed that contaminants migrated downwards with soil water under gravity-driven conditions and penetrated through the subsoil to reach the saturated aquifer, forming a contamination plume with groundwater flow. The feature importance analysis revealed that soil particle size was dominant for the migration of As, while the migration of 1,2-dichloroethane was highly depended on vertical permeability coefficients of the soil.	Liu *et al.* (2024)
4	Brazil	This study employed ML techniques to predict soil quality indicators at the Fundão dam in Mariana, Brazil	Decision tree	The model accurately predicted soil quality indicators in the area and suggested that the model has the potential to revolutionise the monitoring efforts conducted by regulatory agencies in these regions.	Lima *et al.* (2024)

(*Continued*)

Table 6.1. (*Continued*)

Case Study No.	Location	Problem definition	ML model used	Findings	Reference
5	China	To develop a novel framework for the preliminary risk assessment of industrial enterprise sites in the Yangtze River Delta, which is one of the fastest economic development and most prominent contaminated regions in China.	RF	The preliminary risk ranking of industrial enterprise sites by the random forest showed that high risks (probabilities) were mainly distributed in Shanghai, southern Jiangsu and northeastern Zhejiang from 2000 to 2015.	Jiang *et al.* (2022)
6	Iran	To develop a machine learning-based framework to map groundwater quality in an unconfined aquifer in the north of Iran	Extreme gradient boosting (XGB), RF, SVM, ANN, *k*-nearest neighbour (KNN) and Gaussian classifier model (GCM)	Accuracy assessment showed that ML algorithms provided high accuracy in predicting groundwater quality. However, RF was selected as the optimum model given its higher accuracy (overall accuracy, precision and recall=0.92; ROC=0.95).	Sahour *et al.* (2023)

the presence of PFAS. It was indicated that certain PFAS constituents were present in both surface and subsurface soils. While the highest concentrations and most of the PFAS were identified in the surface soils (0.15 m), significant amounts were observed in the vadose zone, demonstrating its vertical leaching. These findings suggest that the vertical movement of PFAS in soils may go deeper. We were further interested in investigating the retention and release of PFAS constituents in surface and subsurface soils; in particular, there are current research gaps for PFAS transport in the vadose zone. Hence, in Phase II of the project, we wanted to use ML models to build a generic tool to help understand the relationship between the volume of influent (simulated rainwater mimicking irrigation water) and the transport and leaching of PFAS in soils at different water saturations, using dynamic column studies.

6.4.2 *Conceptual model*

Upon surface deposition and subsequent infiltration into the vadose zone, numerous transport mechanisms may exist that govern the fate and transportation of PFAS (Figure 6.4). PFAS dissolution in infiltrating water from precipitation, irrigation, runoff or stormwater leads to their leaching and downward migration, where they may be taken up by plant or soil organisms, until they eventually reach the aquifer and form groundwater plumes. In parallel, PFAS may sorb to soil or accumulate at air–water interfaces as well (Sharifan *et al*., 2021). The amount of PFAS retained in the vadose zone also relies on the composition of the soil and its moisture level.

6.4.3 *ML-based prediction using previous analysis data*

The preliminary analysis results for the PFAS concentration and soil properties were used to determine any correlation between surface soil properties and PFAS concentration (Figure 6.5). The analysis aims to identify the most influential soil characteristics affecting PFAS levels using regression and random forest algorithms.

Data for soil samples (0.15 m) from various locations within the spray field were included for the analysis, including moisture content, pH, electrical conductivity (EC), total organic carbon (TOC), total nitrogen (TN) and soil texture (sand, silt or clay). PFAS compounds identified

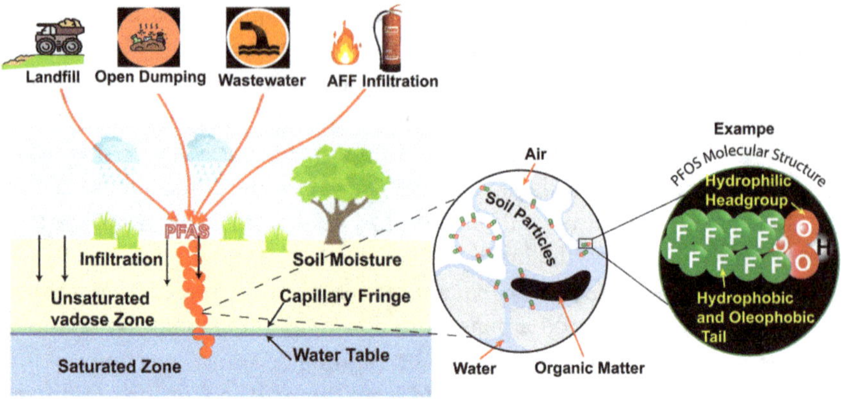

Figure 6.4. Schematic diagram of the sources of PFAS (AFFF-impacted fire training and fire response sites, industrial sites, landfills and wastewater treatment plants) and the transport and migration in the subsurface. The first circular insert shows a zoomed-in cross-section of a configuration of soil grains, air and water showing the distribution of PFAS in the vadose zone and the relevant transport processes. The second insert shows a PFAS molecular structure (PFOS is used as an example), showing a hydrophobic and oleophobic tail and a hydrophilic head group.

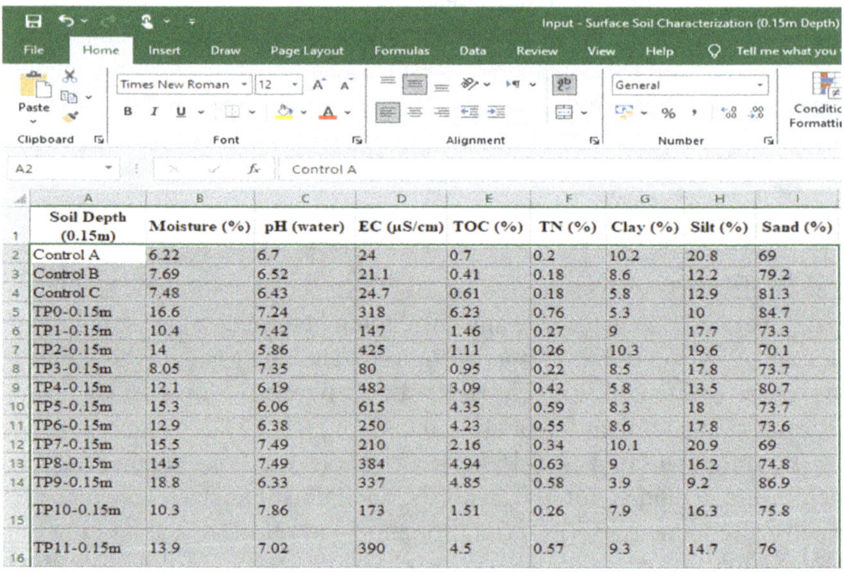

Soil Depth (0.15m)	Moisture (%)	pH (water)	EC (μS/cm)	TOC (%)	TN (%)	Clay (%)	Silt (%)	Sand (%)
Control A	6.22	6.7	24	0.7	0.2	10.2	20.8	69
Control B	7.69	6.52	21.1	0.41	0.18	8.6	12.2	79.2
Control C	7.48	6.43	24.7	0.61	0.18	5.8	12.9	81.3
TP0-0.15m	16.6	7.24	318	6.23	0.76	5.3	10	84.7
TP1-0.15m	10.4	7.42	147	1.46	0.27	9	17.7	73.3
TP2-0.15m	14	5.86	425	1.11	0.26	10.3	19.6	70.1
TP3-0.15m	8.05	7.35	80	0.95	0.22	8.5	17.8	73.7
TP4-0.15m	12.1	6.19	482	3.09	0.42	5.8	13.5	80.7
TP5-0.15m	15.3	6.06	615	4.35	0.59	8.3	18	73.7
TP6-0.15m	12.9	6.38	250	4.23	0.55	8.6	17.8	73.6
TP7-0.15m	15.5	7.49	210	2.16	0.34	10.1	20.9	69
TP8-0.15m	14.5	7.49	384	4.94	0.63	9	16.2	74.8
TP9-0.15m	18.8	6.33	337	4.85	0.58	3.9	9.2	86.9
TP10-0.15m	10.3	7.86	173	1.51	0.26	7.9	16.3	75.8
TP11-0.15m	13.9	7.02	390	4.5	0.57	9.3	14.7	76

Figure 6.5. Soil depth (0.15 m) characteristics analysis.

include PFHxS, 6:2FTS, PFOA, PFOS and PFDA. A correlation matrix (Figure 6.2) was generated to identify relationships between soil properties and PFAS concentrations. Feature importance was evaluated using a random forest model. Linear regression and random forest models were employed to predict PFAS concentrations based on soil characteristics. Model performance was assessed using mean squared error (MSE) and R-squared (R^2) values.

The correlation matrix revealed significant positive correlations between certain soil properties, such as TOC and moisture content, and overall PFAS concentrations. This suggests that these soil characteristics may influence PFAS retention in surface soils. The random forest model identified TOC and EC as the most influential predictors of PFAS concentrations. These features were found to be critical in understanding the distribution and retention of PFAS in the study area. The study found that the random forest model outperformed the linear regression model, providing better predictive accuracy and a closer fit to the actual data.

The variability in soil properties, such as moisture content and EC, reflects diverse soil conditions across different sampling points. These differences potentially influence the distribution and retention of PFAS compounds. The random forest model's superior performance highlights its effectiveness in identifying complex interactions between soil properties and PFAS concentrations. The model's ability to rank feature importance provides valuable insights into the factors driving PFAS contamination, aiding in the development of targeted soil management and remediation strategies. A detailed investigation of PFAS retention and release kinetics in surface and subsurface soils would further enhance the prediction using *Kd* values and soil properties. In addition, the correlation matrix (Figure 6.6) highlights significant positive correlations between specific soil properties, such as TOC and moisture content, with overall PFAS concentrations, indicating that these soil characteristics may affect PFAS retention in surface soils.

In general, the findings from this case study emphasise the importance of incorporating advanced ML models into environmental assessments to enhance prediction accuracy and remediation strategies. The research provides a foundational understanding of PFAS distribution in the study area and suggests further detailed investigations, such as evaluating PFAS retention and release kinetics and developing cost-effective techniques for PFAS management.

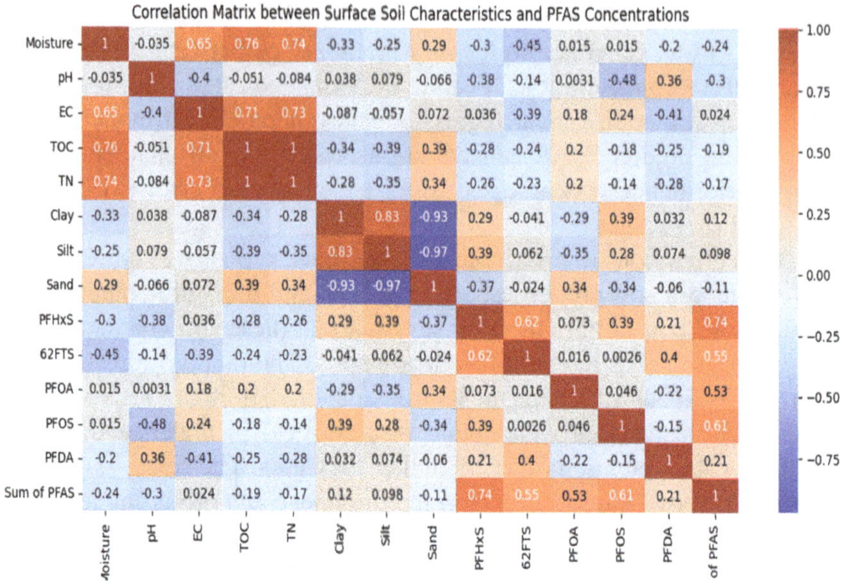

Figure 6.6. Correlation matrix between surface soil characteristics and PFAS concentrations.

6.5 Integration with Remote Sensing and IoT

The integration of remote sensing and Internet of Things (IoT, referring to a network of interconnected physical devices that communicate and exchange data with each other through the internet or other communication networks) technologies with ML models greatly enhances the assessment and remediation of contaminated sites. Remote sensing, through satellite imagery and aerial surveys, provides extensive spatial data, allowing effective monitoring of large areas. When combined with IoT devices, such as ground sensors and drones, real-time data collection becomes more detailed and accurate. This combination enables continuous monitoring and rapid detection of environmental changes. For instance, IoT sensors in the soil can measure contaminant levels, moisture content and other critical parameters, transmitting this data in real time to ML models for analysis. According to Zhang *et al.* (2022), integrating IoT with remote sensing improved the detection accuracy of soil contaminants by 25%, showcasing the potential of this approach to enhance environmental management practices. Moreover, applying ML algorithms to data gathered from remote sensing and IoT devices allows for advanced predictive analytics and efficient data processing. Sophisticated ML models, such as neural networks and random forests, can analyse extensive datasets, identifying patterns and predicting future contamination trends with high precision. This capability is vital for proactive risk management and effective remediation planning. For example, in 2021, the Environmental Protection Agency (EPA) used a combination of satellite data and ground-based sensors in Flint, Michigan, to predict and mitigate lead contamination in water supplies. This integrated approach not only provided accurate contamination forecasts but also enabled the timely implementation of remediation measures, protecting public health.

Looking ahead, the fusion of ML with remote sensing and IoT is expected to drive further advancements in environmental monitoring and site assessment. Emerging technologies such as 5G networks and edge computing will enable faster data transmission and real-time processing, thereby improving the responsiveness and accuracy of monitoring systems. Additionally, the increasing availability of high-resolution satellite imagery and advanced sensors will offer richer datasets for ML models to analyse. Recent studies project that the global market for IoT in environmental monitoring will reach US$22.6 billion by 2027 (Kumar *et al.*, 2022). By leveraging these advancements, future environmental management practices can become more dynamic, data-driven and effective in tackling the complex challenges associated with contaminated sites.

Figure 6.7 depicts the integrated process of utilising remote sensing and IoT technologies with ML models to improve contaminated site assessment. It includes elements such as satellite imagery, aerial surveys, ground sensors and drones for data collection, highlighting the extensive spatial and real-time data acquisition capabilities. The diagram shows how IoT devices transmit data in real time to ML models, which then conduct predictive analytics and efficient data processing using advanced algorithms such as neural networks and random forests. Arrows in the diagram illustrate the data flow from collection to analysis and remediation planning, emphasising the enhancement in detection accuracy through these integrated technologies. It also highlights the role of 5G networks and edge computing in boosting data transmission speed and real-time processing capabilities, thereby contributing to more effective environmental management practices.

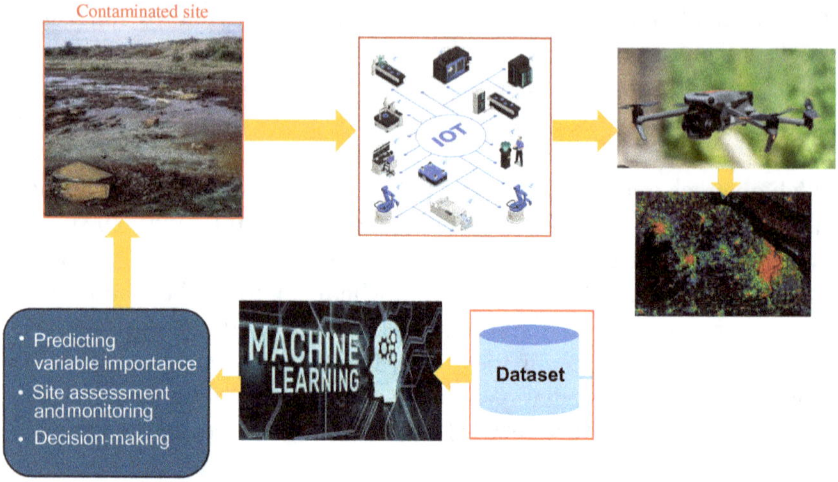

Figure 6.7. Integration of remote sensing and IoT with ML for contaminated site assessment.

6.6 Future Trends and Scope

The future of utilising ML models for improved contaminated site assessment and continuous monitoring is characterised by significant advancements and transformative potential. As technology progresses, there is an increasing focus on expanding the applicability of ML models within environmental science. The integration of real-time sensor data is anticipated to play a pivotal role, providing a more dynamic and responsive approach to monitoring. Collaborative efforts aimed at developing user-friendly tools will enhance accessibility, fostering the widespread use of ML techniques in the field. The recognition of ML's effectiveness in addressing challenges related to data complexity, variability and real-time monitoring suggests a path towards extensive adoption. Ongoing research will drive the refinement and adaptation of ML models, leading to more detailed and nuanced assessments of contaminated sites. Future efforts will likely prioritise comprehensive collaborations, interdisciplinary research and the integration of emerging technologies, positioning ML as a crucial tool for proactive and efficient environmental management. The evolving landscape of assessing and monitoring contaminated sites is marked by emerging trends and advancements that promise significant potential for the future. The transformative impact of ML models on these practices represents a major breakthrough, addressing the limitations of traditional assessment methods. Research underscores ML's ability to provide data-driven insights, conduct real-time analysis and enable adaptive management, thereby greatly enhancing the effectiveness and forward-thinking nature of environmental management. Despite these positive developments, challenges such as ensuring the adaptability of ML models to diverse environmental contexts, handling various contaminant types and managing the complexities of real-time monitoring remain. Overcoming these challenges is essential for the ongoing success of ML applications in contaminated site assessment and monitoring. As the field advances, continuous research and implementation efforts should concentrate on refining ML algorithms, exploring the integration of real-time sensor data and fostering collaboration among interdisciplinary teams. This proactive approach will support the continuous evolution and effectiveness of ML in protecting the environment and public health.

6.7 Conclusion and Recommendations

ML models have demonstrated significant potential in enhancing the efficiency and effectiveness of contaminated site assessments. As evidenced by the case studies discussed, ML technologies not only facilitate more accurate evaluations but also streamline the overall assessment process. These advancements are expected to play an increasingly crucial role in site assessments in the near future. The integration of ML into environmental assessments provides a robust framework for analysing complex datasets and identifying patterns that may not be immediately apparent through traditional methods. For instance, the local case study we performed illustrates how ML can effectively analyse various soil and groundwater properties, shedding light on their influence on PFAS concentrations. This detailed insight is invaluable for understanding the distribution and impact of contaminants, which is essential for developing targeted remediation strategies. Overall, ML is proving to be a beneficial tool that not only simplifies the assessment process but also enhances its accuracy and reliability. As technology continues to evolve, ML's role in site assessments will become increasingly integral, making it easier to address and manage contaminated sites effectively .

References

Back, P.-E. (2001). Sampling strategies and data worth analysis for contaminated land. A literature review.

Guo, B., Zeng, J., and Brusseau, M. L. (2020). A mathematical model for the release, transport, and retention of per- and polyfluoroalkyl substances (PFAS) in the vadose zone. *Water Resources Research*, 56(2), e2019WR026667. Available at: https://doi.org/https://doi.org/10.1029/2019WR026667.

Jiang, Y., You, Q., Chen, X., Jia, X., Xu, K., Chen, Q., Chen, S., Hu, B., and Shi, Z. (2022). Preliminary risk assessment of regional industrial enterprise sites based on big data. *Science of The Total Environment*, 838, 156609.

Kumar, S. V., Mary, G. A. A., Uthirasamy, R., Selvi, S. A., and Glorindal, G. (2022). Design of field to fork framework: Integrating IoT into traditional farming. In *2022 International Conference on Smart Technologies and Systems for Next Generation Computing (ICSTSN)*.

Lima, H. S., de Oliveira, G. F. V., dos Santos Ferreira, R., de Castro, A. G., Silva, L. C. F., de Souza Ferreira, L., dos Santos Oliveira, D. A., da Silva, L. F., Kasuya, M. C. M., and de Paula, S. O. (2024). Machine learning-based soil quality assessment for enhancing environmental monitoring in iron ore mining-impacted ecosystems. *Journal of Environmental Management*, 356, 120559.

Liu, S., Yang, X., Shi, B., Liu, Z., Yan, X., Zhou, Y., and Liang, T. (2024). Utilizing machine learning algorithm for finely three-dimensional delineation of soil-groundwater contamination in a typical industrial park, North China: Importance of multisource auxiliary data. *Science of the Total Environment*, 911, 168598.

NSW EPA. (2011). Guidelines for Consultants Reporting on Contaminated Sites. NSW EPA.

Reible, D. and Demnerova, K. (2012). *Innovative Approaches to the On-Site Assessment and Remediation of Contaminated Sites*, Vol. 15. Springer Science & Business Media. Dordrecht: Springer.

Ritter, L., Solomon, K., Sibley, P., Hall, K., Keen, P., Mattu, G., and Linton, B. (2002). Sources, pathways, and relative risks of contaminants in surface water and groundwater: a perspective prepared for the Walkerton inquiry. *Journal of Toxicology and Environmental Health Part A*, 65(1), 1–142.

Sahour, S., Khanbeyki, M., Gholami, V., Sahour, H., Kahvazade, I., and Karimi, H. (2023). Evaluation of machine learning algorithms for groundwater quality modeling. *Environmental Science and Pollution Research*, 30(16), 46004–46021.

Sharifan, H., Bagheri, M., Wang, D., Burken, J. G., Higgins, C. P., Liang, Y., Liu, J., Schaefer, C. E., and Blotevogel, J. (2021). Fate and transport of per- and polyfluoroalkyl substances (PFASs) in the vadose zone. *Science of the Total Environment*, 771, 145427. Available at: https://doi.org/https://doi.org/10.1016/j.scitotenv.2021.145427.

Sumdang, N., Chotpantarat, S., Cho, K. H., and Thanh, N. N. (2023). The risk assessment of arsenic contamination in the urbanized coastal aquifer of Rayong groundwater basin, Thailand using the machine learning approach. *Ecotoxicology and Environmental Safety*, 253, 114665.

Tyagi, A. K. and Chahal, P. (2020). Artificial intelligence and machine learning algorithms. In *Challenges and Applications for Implementing Machine Learning in Computer Vision*. IGI Global, Hershey, PA, pp. 188–219.

Wu, M., Qi, C., Derrible, S., Choi, Y., Fourie, A., and Ok, Y. S. (2024). Regional and global hotspots of arsenic contamination of topsoil identified by deep learning. *Communications Earth & Environment*, 5(1), 10.

Zhong, S., Zhang, K., Bagheri, M., Burken, J. G., Gu, A., Li, B., Ma, X., Marrone, B. L., Ren, Z. J., and Schrier, J. (2021). Machine learning: new ideas and tools in environmental science and engineering. *Environmental Science & Technology*, 55(19), 12741–12754.

https://doi.org/10.1142/9781800617841_0007

Chapter 7

Quantifying Exchange: An Overview of Groundwater–Surface Water Measurement Methodologies

M. M. Cremeans

Geosyntec Consultants, Overland Park, KS

mackenzie.cremeans@geosyntec.com

Abstract

Groundwater–surface water interactions (GWSWIs) are a crucial part of the hydrogeologic cycle. While often overlooked in assessments of contaminant fate and transport, assessing and remediating the water involved in GWSWIs can be a vital aspect of assessing and remediating contaminated sites. To capture as much contaminant mass as possible, thereby ensuring that the total burden of contaminants in the environment does not increase with time, we must consider the assessment of this part of contaminant fate and transport to design the most effective solutions.

Currently, various promising methods exist to quantify these flows, such as mini-piezometers, streambed point velocity probes, temperature profilers, seepage meters and passive flux meters. Although the tools and methods for quantifying GWSWIs are not perfect, many studies have and will continue to benefit from fate and transport assessments that include this component of flow. Each method should be approached with caution, given the various limitations and assumptions involved with measurements taken at the groundwater–surface water interface. This chapter seeks to review the literature, alongside case studies, exploring the application of various measurement tools to measure GWSWIs, their pitfalls, possible corrections and successful assessments of contamination.

7.1 Introduction

Documentation of human reliance on groundwater dates to Ancient Rome, if not earlier. Roman aqueducts relied on groundwater in the form of springs. The flow was often augmented by dug tunnels known as *adits*. Further, writings from Plato and Aristotle suggest an understanding of groundwater and infiltration as a water source (Deming, 2019). Even so, the modern study of groundwater is most often attributed to Henry Darcy's work, first published in 1856 (Darcy, 1856; Loheide, 2018). While Darcy's empirical formula, which is still used today, was groundbreaking for groundwater science, it does not include a discussion of how groundwater interacts with surface water. However, discussion of groundwater–surface water interactions first appears in the literature in 1877 (21 years later) and, like the study of groundwater, has been ongoing (Boussinesq, 1877; Kaminsky *et al.*, 2023). Even so, an understanding of groundwater–surface water interactions or, of equal importance, the mechanisms that occur in the exchange zone has not been fully recognised. The goals of this chapter are to: (1) define groundwater-surface water interactions, (2) describe their broad importance in the global water cycle, and (3) provide an overview of some common methods of quantifying groundwater-surface water interactions, including advantages and pitfalls of each method, and (4) present some general site assessment scenarios that have benefited from the measurement of groundwater-surface water interactions.

7.1.1 *What are groundwater–surface water interactions?*

Groundwater and surface water systems are connected in most landscapes, and this connection can be studied on a variety of scales. As with any scientific topic, the choice of a relevant scale depends on the question to be answered. However, groundwater–surface water interactions are generally studied at either the local-to-reach scale or the regional-to-catchment

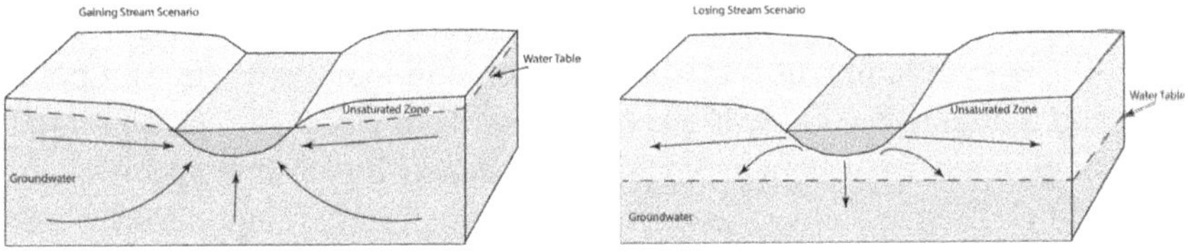

Figure 7.1. Left: a generalised diagram of a gaining stream. Right: a generalised diagram of a losing stream.

Source: Modified from Winter *et al.* (1998).

scale. In a general sense, the term "groundwater–surface water interactions" describes the passage of water from groundwater to surface water, or vice versa, across a boundary (such as the bed of a river, lake, stream, wetland or ocean) (Figure 7.1).

While this task might seem simple at first, the diagram above does not consider many of the complexities affecting groundwater–surface water exchange, including topography, climate, lithologic heterogeneities, different depositional environments and discernment of bed hydraulics (and accompanying geochemistry). In field settings, groundwater–surface water interactions can be particularly difficult to characterise because the boundary, or interface, often has high spatial and temporal variability in flow, heat exchange and biogeochemical activity (Larned *et al.*, 2015). Consequently, data collection must account for the variability at the interface, which can be significant in dynamic environments such as oceans, rivers and streams. Aside from continuous erosion and deposition, other factors (such as weather, biota activity and human influences) can affect the spatial and temporal variability of groundwater–surface water interactions. As such, non-continuous (or single-measurement) data collection represents a "snapshot" of a system that can only be relied upon to describe the specific conditions that were present during data collection. Continuous (or multi-measurement) data collection can provide a more detailed and dynamic picture of the system as it changes.

7.1.2 *Why are groundwater–surface water interactions important?*

Groundwater–surface water interactions have been investigated in a wide variety of contexts including water supply (Winter, 2007), ecosystem health (Lewandowski *et al.*, 2019), water quality (Kaminsky *et al.*, 2023;

Lewandowski *et al.*, 2019), contaminant fate and transport (Caschetto *et al.*, 2013), contaminant remediation (Ren and Packman, 2004), operations and maintenance evaluations of remedial structures (Cremeans *et al.*, 2019b), evolutionary change (Van Damme *et al.*, 2009), rare-species conservation (DiStefano *et al.*, 2009), military history (Younger, 2012) and the potential existence and nature of life on Mars (Gooseff *et al.*, 2010). Though groundwater–surface water interactions can be discussed broadly, across many disciplines, the focus of this chapter is on quantifying groundwater and contaminant fluxes to address issues of water supply, water quality and contaminant fate and transport.

From a water supply standpoint, groundwater has long been documented as a contributor to lakes (e.g., Dimova *et al.*, 2013), estuaries (e.g., Null *et al.*, 2012), oceans (Wilson *et al.*, 2024) and streamflow, both at headwaters and to maintain baseflow (Winter, 2007). This, of course, is important to the calculation of sustainable water supply for all uses (e.g., agriculture, industry and domestic supply). Water quality can be affected by mass flux generated by groundwater exchange, and the area surrounding the interface (known as the exchange, hyporheic or reaction zone in streams and the subterranean estuary in coastal areas) also serves as a "bioreactor," which can change water quality, regardless of flow direction (Kaminsky *et al.*, 2023; Lewandowski *et al.*, 2019; Moore, 1999). Groundwater–surface water interactions are important to contaminant fate and transport for similar reasons. The mass flux of contaminants across the interface zone can lead to complexities in remediation design due to biogeochemical changes in the reaction zone and cross-contamination between groundwater and surface water sources. However, depending on the contaminant, microbial presence and physical conditions, reactivity in the exchange zone can help attenuate contaminants.

For this chapter, the scale of interest is the local-to-reach scale, and as a result, the presented methods will largely focus on the measurement of small-scale interactions.

7.2 Methods

Fortunately, many methods exist to quantify groundwater–surface water interactions (so many, in fact, that one chapter is not enough to cover them all). As such, this chapter presents a subset of groundwater–surface water interaction measurement methods, focusing primarily on applications in rivers and streams, including mini-piezometers, tracer tests, streambed point velocity probes, temperature profilers and seepage meters. Throughout this section, application, best practices and pitfalls will be discussed for each of these methods.

7.2.1 *Mini-piezometers*

Mini-piezometers rely on a Darcian approach that is widely used and accepted, easy to implement and cost-effective. Depending on the study design, mini-piezometers are also versatile, allowing for continuous or non-continuous measurements, high-resolution characterisation and use in a variety of geologic settings. The benefit of continuous measurements, in this context, is that groundwater–surface water interactions can be measured *as they change*, which occurs frequently (as discussed above). Contrastingly, non-continuous methods offer a snapshot of groundwater–surface water interactions under one set of conditions. While this approach is more cost-effective, the data interpretation may not yield results that apply to other site conditions.

However, like all Darcian methods, large uncertainties are associated with estimates of hydraulic conductivity and, secondarily, porosity. Additional errors can stem from the measurement of gradients over small distances (such as vertically through a streambed), leading to unreliable data (Devlin and McElwee, 2007; Harvey *et al.*, 2013; Post and von Asmuth, 2013). These errors can be mitigated in a couple of ways. To obtain a relatively accurate estimate of hydraulic conductivity, the application of mini-piezometers could also include conducting slug tests within them or collocated drive point piezometers and analysing the subsequent response (Baxter and Hauer, 2003; Rosenberry and LaBaugh, 2008). Of course, slug testing is subject to its own limitations and should be carefully designed according to the literature's recommendations. While further discussion of slug testing is beyond this chapter's scope, Butler (2019) contains a thorough compendium of practical advice for the design, implementation and analysis of slug tests. Overcoming errors in gradient measurements over small distances is not truly possible, but errors can be reduced by using the most accurate measurement methods available (transducers, instead of a manual tape, for example). However, this type of improvement to the method also increases the cost of implementation for mini-piezometers.

As noted previously, many scientists have relied on mini-piezometers to characterise groundwater–surface water interactions (Lee and Cherry, 1979; Cey *et al.*, 1998; Malard *et al.*, 2001; Hughes, 2006; Acuña and Tockner, 2009; Brodie *et al.*, 2009; Cremeans *et al.*, 2019a; Halder *et al.*, 2023). Mini-piezometers are built with small-diameter pipes (such as open-ended, clear polyvinyl chloride pipes) and are installed directly at the interface at regular intervals (Figure 7.2). Smaller-diameter pipes are preferred, as they are easier to drive into the subsurface. However, if col-located slug tests are part of the study plan, the diameter of the pipe will need to be large enough to accommodate both a transducer and a slug. In this case, pneumatic slug testing is preferred due to the elimination of a physical slug (allowing the pipe diameter to be much smaller). Stilling wells are recommended in applications with flowing water to allow for a more accurate measurement of surface water levels (with less disturbance from natural turbulence). Before measuring hydraulic heads, piezometers should be allowed to equilibrate for 24 h. Measurements should be taken

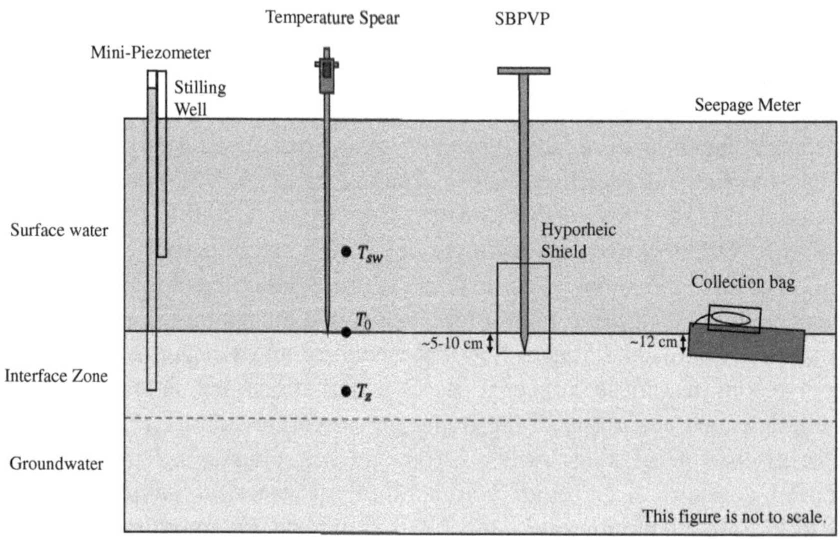

Figure 7.2. From left to right, schematic views of a mini-piezometer, temperature spear, streambed point velocity probe and seepage meter.

Note: The depths of installation for mini-piezometers and temperature spears are relatively flexible, while SBPVPs and seepage meters are generally installed at the depths indicated in the figure.
Source: Modified from Cremeans *et al.* (2019).

using the highest-precision tool available. Replicate measurements should also be performed to verify data quality. After data collection, seepage velocities at the interface can be calculated using the following adaptation of Darcy's law:

$$v = -\frac{Ki}{n},$$

where v is the seepage velocity (L/T), K is the hydraulic conductivity (L/T), i is the hydraulic gradient (unitless) and n is the effective porosity (unitless). As noted above, typically, K is estimated from the literature based on sediment type or determined by slug testing, while n can be determined empirically or through a literature-based estimation. i is measured by the mini-piezometers and calculated as $\Delta H/\Delta L$. Baxter and Hauer (2000) and Baxter *et al.* (2003) provide an overview of using piezometers to quantify groundwater–surface water interactions.

7.2.2 *Large-scale tracer tests*

Tracer tests are another common method of evaluating groundwater–surface water interactions and have been used to quantify hyporheic exchange in rivers (Bourke *et al.*, 2014; Frei and Gilfedder, 2015), as well as the flow of groundwater to rivers, lakes and oceans (Cook *et al.*, 2006; Garcia-Solsona *et al.*, 2010; Krabbenhoft *et al.*, 1990). Though injected tracers are numerous and useful, non-injected tracers are also commonly used and are often preferred in environments with fragile ecosystems (or, on occasion, at sites where more than one study is being conducted). While injected tracers must be placed into a system and then traced, non-injected tracers are already present in the system and can be leveraged to quantify groundwater–surface water interactions (Figure 7.3).

Injected tracers include conservative compounds, such as NaCl, and reactive compounds, such as resazurin (González-Pinzón *et al.*, 2015, for example). Non-injected tracers include any traceable differences between groundwater and surface water (such as alkalinity, silica, geogenic radioactive tracers, pre-existing contaminants, etc.) (Rodgers *et al.*, 2004; Cook *et al.*, 2018). While tracer techniques provide integrated data about the hydrologic system, including its transient surface and subsurface storage zones, often the tracer is not recaptured at 100% of its actual, or theoretical, original volume which means some flow remains

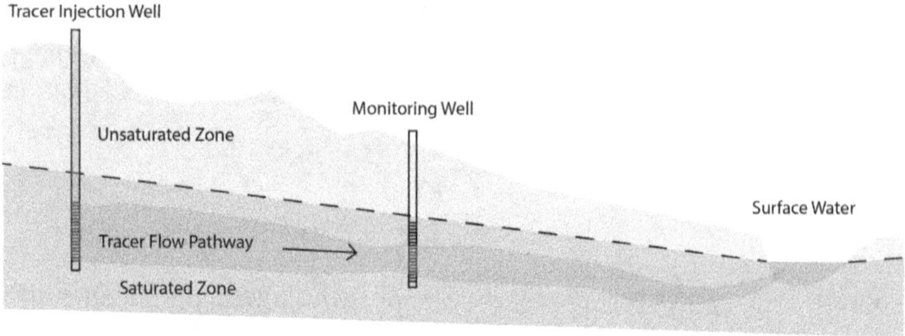

Figure 7.3. A simplified example of an injected tracer test, including an intermediate monitoring well and a discharge zone into surface water.

undetected (e.g., Harvey and Wagner, 2000; Ward *et al.*, 2013; Schmadel *et al.*, 2013). Additionally, complexities in modelling geochemical mixing to quantify flux can be daunting and difficult to overcome. However, properly conducted tracer tests remain a reliable method to determine zones of recharge and discharge.

To properly conduct an injected tracer test, several important considerations should be made prior to the test. First, the selection of tracer type and tracer quantity should be thoroughly examined. Manufactured tracers can have vastly different properties, and a careful review of the manufacturer's instructions should be part of the pre-planning process. Second, injected tracer tests rely heavily on the selection of the appropriate sampler type. The most consistent results have been found with activated charcoal samplers (compared to all other types) (Aley, 2019). Third, effective injected tracer tests require a study design that adequately accounts for the collection of all the tracer (or as much as reasonably possible), ensures timely analysis of samples and encompasses enough collection area or detailed sampling to provide a credible result.

Related to the use of large-scale tracer tests, groundwater–surface water interactions can be considered as bulk withdrawals from or additions to surface water bodies or as small-scale movements of water, characterised in high-resolution detail. Large-scale tracer tests are, perhaps obviously, better suited to the former. However, tracer test methods can be adapted to higher-resolution testing regimens as well. A comprehensive overview of large-scale dye tracer testing methods can be found in the work of Aley (2019).

7.2.3 *Temperature profiling*

Temperature profiling offers a point measurement alternative to large-scale tracer testing. While the latter relies either on visual detection of tracer arrival or geochemical analysis to determine tracer arrival, the former relies on the temperature contrast between groundwater and surface water to interpret where discharge and recharge zones are present. Heat, or temperature contrast, is a viable tracer because groundwater temperature is generally more stable than surface water temperature. Groundwater remains at a similar temperature on daily and annual timescales, while surface water temperature can vary significantly depending on weather conditions, sunlight exposure and human activities. As such, the difference in groundwater and surface water temperatures can be used to identify and quantify zones of recharge and discharge (Bredehoeft and Papadopulos, 1965; Kalbus *et al.*, 2006; Krause *et al.*, 2012; Irvine *et al.*, 2017b; Coluccio, 2018).

Advantageously, temperature-based methods offer a fast, relatively inexpensive and efficient way to delineate and estimate water flux using tools such as temperature spears and fibre-optic distributed temperature sensing cables (Becker *et al.*, 2004; Schmidt *et al.*, 2006, 2007; Lautz, 2010; Bhaskar *et al.*, 2012; Irvine *et al.*, 2016, 2017a; Selker *et al.*, 2006a, 2006b; Tyler *et al.*, 2009; Hausner *et al.*, 2011; Van de Giesen *et al.*, 2012). These tools can collect point measurements or time-series data collection. The former provides a "snapshot" of groundwater–surface water interactions at the time of data collection, while the latter provides a valuable picture of temporal variation, which is important for evaluating dynamic environments as previously discussed (Hatch *et al.*, 2006; Keery *et al.*, 2007; Shanafield *et al.*, 2011). Aside from qualitative delineation of groundwater–surface water interactions, temperature contrasts can be used to quantitatively model the magnitude of flow using the heat transport equation (and its associated analytical and numerical solutions). These mathematical models describe a combination of advective heat transport and conductive heat transport (Domenico and Schwartz, 1998; Carslaw and Jaeger, 1959; Suzuki, 1960; Bredehoeft and Papadopolus, 1965; Stallman, 1965; Turcotte and Schubert, 1982; Silliman *et al.*, 1995).

Use of these solutions requires measurement of a temperature profile beneath the stream.

Schmidt *et al.* (2007) propose a simple, one-dimensional analytical solution that can be easily employed in Excel. Cremeans *et al.* (2019) adapted this solution to site-specific parameters in the following way:

$$q_z = -\frac{k_{fs}}{p_f c_f \cdot z} \cdot \ln \frac{T_z - T_L}{T_0 - T_L},$$

where q_z is the Darcy flux in the vertical direction (m/s), κ_{fs} is the thermal conductivity of the solid–fluid system (J $s^{-1}m^{-1}K^{-1}$), $p_f\, c_f$ is the volumetric heat capacity of the fluid (J m^{-3} K^{-1}), z is the depth of measurement (m), T_z is the temperature at the depth of z in the interface zone (°C), T_L is the temperature of the groundwater which is fixed for all calculations (°C) and is generally taken from a nearby well (not shown in Figure 7.2, as a result) and T_0 is the temperature at $z = 0$ (°C) or the surface of the interface zone. Surface water temperature measurements should also be collected at each point (T_{sw}) to observe the potential for temperature measurement drift over the sampling period. However, the calculation only relies on three measurements, plus estimates of the thermal conductivity of the solid–fluid system and the volumetric heat capacity of the fluid (which can be estimated from values in the literature) to determine zones of recharge and discharge (Figure 7.2).

While this method is simple to execute, the results are only qualitatively reliable in some cases. The qualitative interpretation of temperature data follows a general pattern: Gaining streams (which are dominated by groundwater flow) have relatively stable sediment temperatures, while losing streams have more variable sediment temperatures (Winter *et al.*, 1998; Constantz, 1998; Constantz and Stonestrom, 2003). Quantitatively, fluxes determined by temperature profiling have been shown to be unreliable in some cases, with a consistent order-of-magnitude (or more) underestimation of fluxes (Schmidt *et al.*, 2007; Jensen and Engesgaard, 2011; Schulz, 2017; Cremeans *et al.*, 2019a). Reasons for this underestimation include contrasts that were naturally too small to adequately characterise flow, changes in ambient temperature over the course of the sampling and depth limitations for establishing a profile. A comprehensive review of heat as a tracer can be found in Anderson (2005).

7.2.4 *Streambed point velocity probes*

The streambed point velocity probe (SBPVP) is a relatively recent addition to the methods used for the evaluation of groundwater–surface water interactions. Though the name suggests only use in streambeds, the tool is suitable for use in environments with collapsible sediments, which could include coastal sediments, sea beds and lake beds, for example. Similarly to the temperature profiler, the SBPVP relies on tracer technology to measure the direction and magnitude of flow. Specifically, the SBPVP is a reusable tool that can be installed directly into the interface to measure the point velocity of groundwater–surface water interactions at the location of installation at the time of measurement. Once installed, the SBPVP can be used to quantify the average linear velocity of groundwater in a streambed based on a small injection of water (with a different resistivity than the ambient water) followed by collection and analysis of specific conductance breakthrough curves at the point scale. This method of testing can be conceived as a "mini-tracer test" on the surface of the probe, which, advantageously, eliminates the sources of error related to Darcy's law calculations (i.e., hydraulic conductivity and porosity estimates are not necessary). Laboratory and field testing has suggested that, in unconsolidated sandy sediments, the SBPVP may be a viable tool for measuring groundwater–surface water interactions moving at least 18 cm/day up to 2,700 cm/day (Cremeans and Devlin, 2017; Cremeans *et al.*, 2018). Schulz (2017) demonstrated that the SBPVP was also viable in some poorly sorted, coarsely grained and organically rich settings.

To use an SBPVP in the field, the probe must be fully assembled and firmly pressed 5–10 cm into the sediment at the interface (Figure 7.2). The viability of the test relies heavily on the collapse of sediment against the probe surface, which is not possible in all geological settings. For installation purposes, the water cannot be deeper than the length of the pipe holding the probe. If the water is shallower than the hyporheic shield height, a siphon must be applied to avoid stagnation within the shield. After the SBPVP equilibrates (which can take up to an hour), the tracer can be injected onto the probe surface, where it will travel over detectors on the probe surface, producing a breakthrough curve. In most cases, tracer volumes range from 0.1 to 1 mL, with concentrations between 1 and 2 g/L NaCl, as reported by Schillig *et al.* (2014). Data are most easily processed using a pre-built Excel macro, VelProbePE (Schillig, 2012).

Reduction of noise in the data is vital to the proper use of the SBPVP. The first consideration is the selection of the correct tracer fluid. Specifically, the chosen tracer must show enough electrical conductivity contrast with the ambient water to produce a strong signal while avoiding a significant contrast in density (which would otherwise affect the movement of the tracer on the probe's surface). A second consideration is that these tests are highly susceptible to noise from unsteady tracer injection and movement of the probe caused by currents or passing watercraft. Such movement should be reduced as much as possible to ensure clean data. Best practices for using the SBPVP have been discussed by Osorno (2021).

7.2.5 *Seepage meters*

Seepage meters were originally invented to monitor water losses from irrigation canals (Israelsen and Reeve, 1944). However, by the late 1970s, the design and methods were adapted to quantify groundwater–surface water interactions in shallow water environments (Lee, 1977). Seepage meters are a reusable tool that can be installed directly into a stream, river, lake or seabed to measure groundwater–surface water interactions over a particular cross-sectional area. Generic seepage meter construction includes the top of a plastic or metal barrel and a plastic water collection bag. These two containers are connected via a piece of tubing (and are sealed to stop the infiltration of surface water). The data from a seepage meter are volume measurements of the contents of the bag, which either collects or loses water depending on the direction of flow at the interface. The change in water volume in the bag (over time per area) is used to estimate seepage flux. The design proposed by Lee (1977) relies on manual operation of the meter; other designs have automated seepage meter measurement (e.g., Taniguchi and Fukuo, 1993; Paulsen *et al.*, 2001; Rosenberry and Morin, 2004).

To collect a seepage meter measurement, the meter must be installed at the interface (a depth of around 12 cm is recommended) (Figure 7.2). After 24 h equilibration, the collection bag can be installed on the meter to begin a test. In flowing water, Murdoch and Kelly (2003) and Rosenberry (2008) noted head losses from the movement of water over the collection bag, resulting in groundwater–surface water interaction measurements that were lower than expected. Therefore, placement of the collection bag in a container is suggested to keep the water surrounding the bag stationary. Additionally, to avoid missing a recharge area (and to mitigate any resistance from the bag itself), filling the bag with a small amount of water is recommended prior to beginning a test. Once the bag is connected, great care should be taken to avoid disturbing the interface near the seepage meter. Depending on the rate of flow associated with groundwater–surface water interactions, the test lengths can vary. When the test concludes, a measurement of the final volume of the collection bag is taken, and this value is used to calculate the seepage:

$$v = \frac{V}{A \cdot t \cdot n},$$

where V is the change in the volume of water collected in the bag, A is the cross-sectional area of the meter, t is the time over which the volume was collected and n is the porosity of sediment. Because this calculation includes time, seepage fluxes from seepage meter data should be considered time-averaged flow measurements.

Advantageously, seepage meters are inexpensive and simple to construct (like mini-piezometers) and have been effectively used in many field studies (Rosenberry, 2005, 2008; Rosenberry and LaBaugh, 2008; Kennedy *et al.*, 2010; Solder *et al.*, 2016). However, as with the other methods, proper adherence to the established methods is vital for reducing the bias in seepage meter results. These biases are particularly prevalent in early construction designs and when seepage meters are deployed in deep or fast-moving water. Specifically, as detailed by Rosenberry *et al.* (2005), seepage meter data have been associated with underestimations of flow and can depend on the application of calibration tests and correction factors (usually ranging from 1.05 to 1.7). Additionally, seepage meters are difficult to deploy within hard, or cobbly, interfaces (Erickson, 1981; Brock *et al.*, 1982; Woessner and Sullivan, 1984; Blanchfield and Ridgeway, 1996; Zamora, 2006). A comprehensive discussion of best practices for seepage meter use can be found in the work of Murdoch and Kelly (2003).

7.3 Application of Groundwater–Surface Water Methods to Site Assessment

The application of a remedial strategy is often only successful if the preceding site assessment is sufficiently detailed. At sites with groundwater–surface water exchange, neglecting the connection between groundwater and surface water could mean missing a significant volume of constituents of interest (as well as a significant conduit of transport). For contaminants that are geochemically reactive, additional complications can arise related to changes in temperature, Eh, pH and oxygenation as water travels between surface and subsurface environments. Lastly, sites with significant groundwater–surface water interactions may require additional consideration regarding exposure risk quantification. If contaminants are transported from the subsurface into surface water, or vice versa, assessing exposure risk pathways must include potential effects on aquatic ecosystems and human exposure risks. With these general themes in mind, the examination and quantification of groundwater–surface water interactions should not be overlooked when assessing any site where the determination of fate and transport is an important outcome. To preserve client confidentiality, only publicly available site information is discussed in the following to demonstrate the applicability of groundwater–surface water interactions to site investigations.

7.3.1 *Grindsted Å, Jutland, Denmark: A brief example*

The *Grindsted Å* site, located in Jutland, Denmark, is a well-studied stream impacted by discharges from a nearby factory, which is currently inactive (for example, Rassmussen *et al.*, 2016; Balbarini *et al.*, 2017; Sonne *et al.*, 2017; Cremeans *et al.*, 2019) (Figure 7.4). Beginning in 1914, the factory handled over 1,000 chemicals and produced explosives, pharmaceuticals and enzymes (NIRAS, 2009). The geology in the 200 km^2 catchment is mainly sand and sandy clay (Heron *et al.*, 1998), and the stream reach in this study has been shown to receive most of its input from groundwater, with a small amount coming from tributaries (Rasmussen *et al.*, 2016). Though the inactive factory did not have a point-source discharge into the stream, samples from the *Grindsted Å* have consistently contained chlorinated solvents and pharmaceutical compounds. These compounds originated at the factory and, according to Nielsen *et al.* (2014), were transported from the factory to the stream via groundwater. As such, groundwater–surface water exchange is a crucial part of the fate and transport assessment for this site.

One study assessed groundwater–surface water exchange on one reach of the river using several tools, including mini-piezometers, SBPVPs

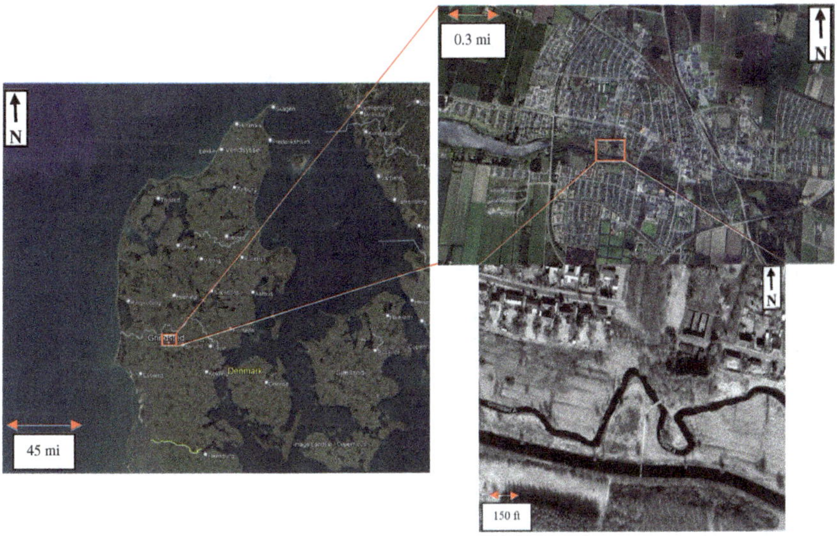

Figure 7.4. Left: imagery showing Denmark and surrounding area, with Grindsted shown at the centre of Jutland. Right top: an image of the town of Grindsted, with a box showing, right bottom, the studied reach of the Grindsted *Å*.

and temperature spears across 26 transects with 3–5 measurement points per transect (Cremeans *et al*., 2018). Across a single stream meander, all methods agreed: The discharge of chlorinated ethene was shown to be dominated by 2–3 transects, with little to no flux occurring elsewhere throughout the reach. This assessment demonstrated, as expected, that mass flux across the groundwater–surface water interface is as variable as the streambed. Further, the evaluation with three tools showed that the groundwater–surface water exchange could be quantified in absolute and relative terms on a sufficiently detailed scale to potentially inform remediation design.

One significant hurdle in this process is the dynamic systems related to rivers and streams. As water moves through the channel, the geometry, depth and, sometimes, direction of that channel change (Figure 7.5). As such, the data presented in the study, which were collected over a short period in June 2016, should not be considered representative of the reach today. Rather, these data represent a snapshot of the system that has undergone (and will continue to undergo) significant change. As such, while

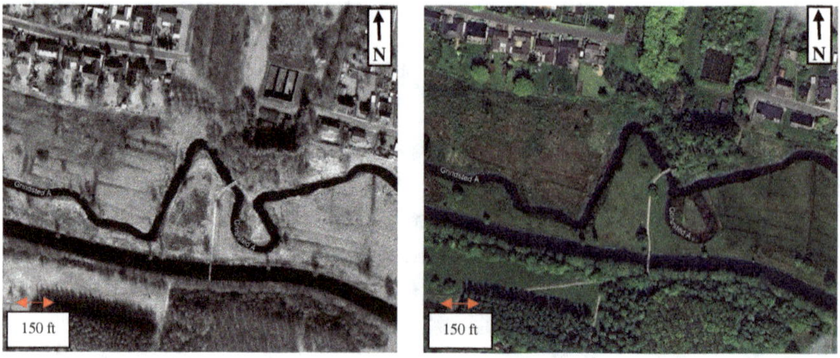

Figure 7.5. Google Earth images depicting large-scale changes in the river between 2014 (left) and 2024 (right) highlight the dynamic nature of such systems.

this study determined that mass flux occurred in the streambed in specific "hot spots" of flow, those hots pots will undoubtedly change over time. Any remedial solution would have to account for the changes within the system by shortening treatment time or stabilising change throughout the period of treatment. However, if these problems can be overcome (or if they are not a factor, such as in a low-energy surface water environment), this type of characterisation of the groundwater–surface water interface could be a valuable way to tailor and increase the effectiveness of remediation solutions.

7.3.2 *Tar Creek, Oklahoma, US: A brief example*

The Tar Creek Superfund site, located in northeastern Oklahoma, US, represents a significant environmental challenge stemming from decades of lead and zinc mining activities. The site's historical mining operations have left a legacy of environmental contamination, including heavy metals such as lead, zinc and cadmium, as well as acid rock drainage. Understanding the interactions between groundwater and surface water is crucial for assessing and remediating the environmental impacts of contamination at the Tar Creek site.

Situated in the Tri-State Mining District, the Tar Creek site encompasses approximately 40 square miles in Ottawa County, Oklahoma. The area is characterised by karst topography and is underlain by a complex geological formation consisting of fractured limestone and shale formations. The presence of extensive underground mine workings, tailings piles and waste disposal sites has contributed to the contamination of soil, groundwater and surface water throughout the region. The hydrogeology of the Tar Creek site is influenced by the fractured nature of the bedrock, which allows for the movement of groundwater through interconnected fractures and fissures. Groundwater flow paths are highly dynamic, with contaminants migrating from abandoned mine shafts, tailings piles and waste disposal sites into surrounding aquifers. Groundwater–surface water interactions occur through seeps, springs and direct discharge points, leading to the contamination of surface water bodies such as Tar Creek and its tributaries.

Constituents at the Tar Creek site include heavy metals such as lead, zinc and cadmium, as well as acid rock drainage resulting from the oxidation of sulphide minerals leached from the abandoned mine workings. These contaminants pose risks to human health and the environment, with elevated levels of heavy metals observed in soil, groundwater and surface water samples throughout the site. The interaction between contaminated groundwater and surface water exacerbates the spread of pollutants downstream, impacting aquatic ecosystems and potentially exposing downstream communities to hazardous substances. Remediation activities at the Tar Creek site have focused on containment, treatment and monitoring to address the legacy of contamination. Strategies include the cleanup and stabilisation of mine waste, the installation of groundwater monitoring wells to track contaminant migration and the implementation of containment and treatment measures to mitigate the transport of metals. Additionally, habitat restoration projects aim to restore the ecological health of the Tar Creek watershed and promote the recovery of native species and habitats.

One remediation effort in the Tar Creek area uses groundwater–surface water interactions as the energy source for a passive bioreactor system (Nairn *et al.*, 2009). Specifically, the system is designed to exploit groundwater that arrives at the surface as a spring to fill the first pond in the bioreactor. This accumulated groundwater is then treated in a sequence of other ponds, each designed to chemically remove certain contaminants, before being returned to Tar Creek. In this case, the remediation solution relies on the groundwater–surface water interaction to transport the water to the surface and then uses simple gravity drainage to direct water between ponds. In a twist on groundwater–surface water interaction research, Cremeans *et al.* (2019) assessed flux within two of the manmade ponds to develop

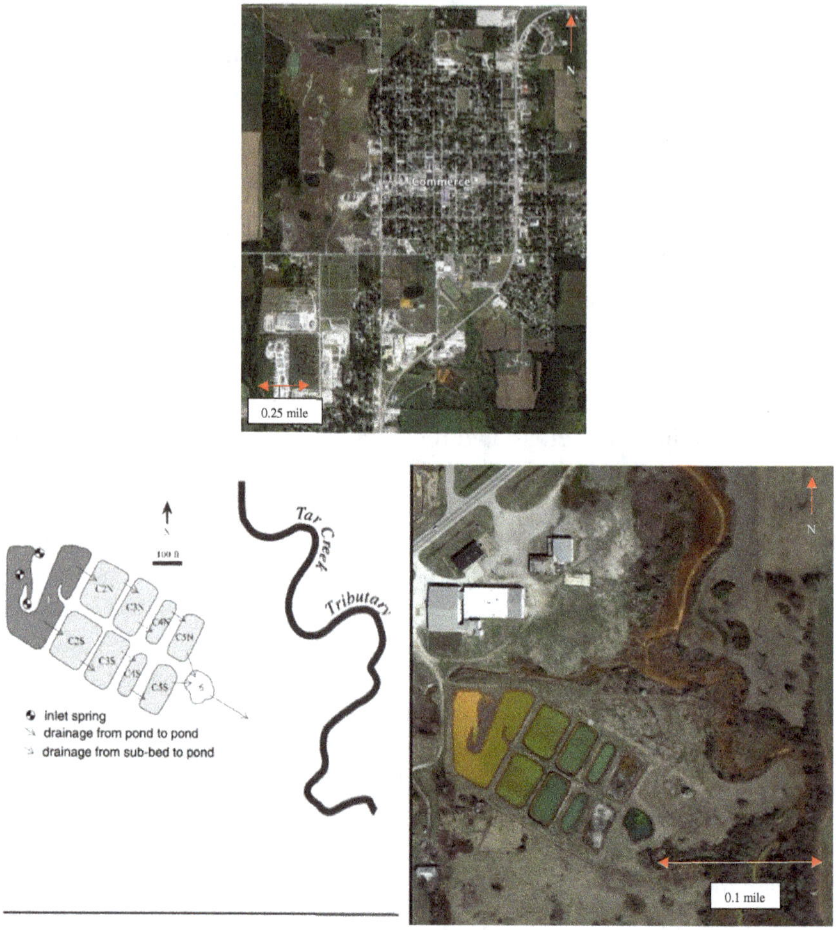

Figure 7.6. From left to right: The Tar Creek Superfund site includes the town of Commerce, OK. Southeast of town, a passive remediation system, relying on a double chain of treatment ponds, has been installed to treat artesian mine water (centre figure adapted from Cremeans *et al.*, 2019). Those ponds have been successfully pilot tested, with assessment of groundwater–surface water interactions.

operational and maintenance guidelines for that portion of the system (US EPA, n.d.c).[1] These cells are noted as C3N and C3S in Figure 7.6. Knowledge of the head differences between the ponds and output flow

[1]For more information, please see the USEPA landing page for this site: https://cumulis.epa.gov/supercpad/cursites/csitinfo.cfm?id=0601269.

allowed for an assessment of overall flow through the system. However, this method could not assess heterogeneities in the flow that affected the removal rates within ponds containing a mushroom compost substrate.

To this end, the ponds were assessed with greater resolution using the SBPVP. To ensure that coverage of measurements was adequate, mass balance was calculated using geochemical data paired with measured outflow values and SBPVP outflows. While the mass fluxes were compared and found to be similar, the SBPVP data indicated that some areas of the substrate were passing water much more quickly than others (due to degradation processes occurring preferentially near the inflow pipes). Ultimately, knowledge of groundwater–surface water interactions enabled a detailed understanding of flow through the system which allowed for the development of effective and efficient maintenance of the remediation structure at this site.

7.4 Conclusions

In a limited and dynamic system, such as the Earth, nothing can increase forever without consequences. Evaluation of groundwater–surface water interactions can provide valuable insights into contaminant and nutrient loads, as well as water supply. As noted in the examples above, together with the background information, many researchers are already quantifying groundwater–surface water interactions to inform decision-making. Specifically, those studies indicate the following:

- While often overlooked in assessments of contaminant fate and transport, assessing and remediating the water involved in groundwater–surface water interactions can be a vital part of assessing, and remediating, contaminated sites. To capture as much contaminant mass as possible, thereby ensuring that the total burden of contaminants in the environment does not increase with time, we must evaluate this aspect of contaminant fate and transport to design the most effective solutions.
- Knowledge and methods pertaining to groundwater–surface water interactions continue to be a major part of water and nutrient flux studies, aiming to more accurately describe the movement and transformation of water and its constituents through the highly reactive hyporheic zone. Movement of water in this area of the hydrogeologic system affects water quality, nutrient cycling, temperature regulation

in natural waters, contaminant fate and transport and water supply, among other phenomena that are of great interest to environmental scientists from various disciplines. Looking forward, we may be able to leverage these natural phenomena to remediate contaminated groundwater and/or surface water.

As discussed throughout this chapter, the tools and methods for quantifying groundwater–surface water interactions are not perfect; however, many studies have benefitted and will continue to benefit from fate and transport assessments that include this component of flow. Concurrently, more methodological development is needed to make the application of groundwater–surface water methods more approachable and practical in the field. As a starting point, Table 7.1 provides a summary of the pros and cons of the methods discussed in this chapter (as described in their cited papers above).

Related to the methods discussed above, more work is needed to iron out the operating procedures for standardising the design and implementation of the tools and data analysis. Much work remains in applying these methods to non-sandy, highly heterogeneous, deep and fast-flowing environments. In addition to method-specific questions, many fundamental hydrogeologic questions remain that are pertinent to quantifying groundwater–surface water interactions. For example, issues related to scale and estimations of soil and aquifer properties (such as hydraulic conductivity and porosity) remain significant sources of error and uncertainty across the discipline. Lack of ability to conduct extensive testing on many sites, due to client or budgetary restrictions, can exacerbate these issues.

The future of groundwater–surface water interactions research lies in interdisciplinary, collaborative research to advance our understanding of the complex hydrogeological, geochemical and ecological processes that govern these interactions. Key research priorities include advances in design and implementation that reduce time and monetary costs of assessing groundwater–surface water interactions, integrating modelling with high-resolution monitoring, further exploration of the biogeochemical mechanisms occurring in the hyporheic zone, development of remote-sensing technologies for quantifying groundwater–surface water interactions and further refinement of tools and methods for assessing groundwater–surface water interactions. Interdisciplinary collaboration and knowledge exchange will continue to play a vital role in addressing emerging challenges and fostering innovation in groundwater–surface water interaction research.

Table 7.1. Summary of methods.

Method	Pros	Cons
Mini-Piezometers	Prolifically used and accepted across the discipline	Reliant on estimations of K and, sometimes, n
Data type: *point measurement or time series*	Inexpensive to make and install Can be used at a variety of spatial and temporal resolutions, for single-time point measurements or for time-series measurements Easy to deploy in *most* shallow environments	Estimating gradients over small distances is associated with significant errors Not especially suitable for deep or fast-flowing water Some guesswork required on the number of piezometers needed to evaluate spatial and temporal variability Generally, cannot be installed in hard substrates
Large-Scale Tracer Tests	Can be used to gain an average picture of groundwater–surface water interactions at the reach scale	Does not indicate variability in flux or hydraulic conditions throughout test
Data type: *temporal and spatial average*	Due to lack of tool installation, deep and fast-flowing waters are generally not a concern When designed well, an incredibly effective method to determine large-scale patterns of groundwater–surface water interactions	Requires some knowledge of gaining versus losing reaches prior to designing an efficient study May require expensive installation of wells, in addition to the expense of tracer dyes and measurement packets
Temperature Profiling Data type: *point measurement or time series*	Fast, easy measurement process in *most* environments Fairly inexpensive and reusable—temperature spear	Cannot be installed in hard substrates Flux is estimated using temperatures, which results in a relative value (highly dependent on contrast, which may vary seasonally)

(*Continued*)

Table 7.1. (*Continued*)

Method	Pros	Cons
	Can collect data at high resolution (spatial and temporal) if using DTS cabling	Must abide by the assumptions of the analytical method, which eliminates many environments
	Analysis can easily be coded into an Excel sheet	Can be expensive—DTS cabling
Streambed Point Velocity Probes	Eliminates errors associated with Darcy's law through use of mini-tracer test	Cannot collect continuous data Cannot be installed in hard substrates
Data type: *point measurement or time series*	Can be permanently or temporarily installed	Will not function properly in environments with large pore spaces
	Capable of capturing temporal and spatial changes with enough probes and tests over time	Not well suited to very shallow, deep or fast-moving water Injection process requires finesse
	Can capture a wide range of groundwater–surface water velocities (observed 18–2,700 cm/day)	
Seepage Meters	Provides a time-average flux measurement	Cannot be installed in hard substrates
Data type: *temporal average*	Widely used and accepted Simple deployment and data collection	Not well suited to deep or fast-moving water Some designs tend to underestimate flow Slow testing process

References

Acuña, V. and Tockner, K. (2009). Surface-subsurface water exchange rates along alluvial river reaches control the thermal patterns in an Alpine river network. *Freshwater Biology*, 54, 306–320. Available at: https://www.doi.org/10.1111/j.1365-2427.2008.02109.x.

Aley, T. (2019). Ozark underground laboratory's groundwater tracing handbook. *KIP Articles*. 3918. Available at: https://digitalcommons.usf.edu/kip_articles/3918.

Anderson, M. P. (2005). Heat as a ground water tracer. *Ground Water*, 43(6), 951–968.

Balbarini, N., Boon, W. M., Nicolajsen, E., Nordbotten, J. M., Bjerg, P. L., and Binning, P. J. (2017). A 3-D numerical model of the influence of meanders on groundwater discharge to a gaining stream in an unconfined sandy aquifer. *Journal of Hydrology*, 552, 168–181. Available at: https://doi.org/10.1016/j.jhydrol.2017.06.042.

Baxter, C. and Hauer, F.R. (2000). Geomorphology, hyporheic exchange, and selection of spawning habitat by bull trout (Salvelinus confluentus). *Canadian Journal of Fisheries and Aquatic Sciences*, 57(7), 1470–1481. https://doi.org/10.1139/f00-056.

Baxter, C., Hauer, F. R., Woessner, W. W. (2003). Measuring groundwater-stream water exchange: New techniques for installing minipiezometers and estimating hydraulic conductivity. *Transactions of the American Fisheries Society*, 132, 493–502. Available at: https://www.doi.org/10.1577/1548-8659(2003)132<0493:MGWENT>2.0.CO;2.

Bhaskar, A. S., Harvey, J. W., and Henry, E. J. (2012). Resolving hyporheic and groundwater components of streambed water flux using heat as a tracer. *Water Resources Research*, 48(8). Available at: https://doi.org/10.1029/2011WR011784.

Blanchfield, P. J. and Ridgeway, M. S. (1996). Use of seepage meters to measure ground water flow at brook trout redds. *Transactions of the American Fisheries Society*, 125, 813–818.

Bourke, S. A., Cook, P. G., Shanafield, M., Dogramaci, S., and Clark, J. F. (2014). Characterisation of hyporheic exchange in a losing stream using radon-222. *Journal of Hydrology*, 519(A), 94 105. Available at: https://www.doi.org/10.1016/j.jhydrol.2014.1006.1057.

Boussinesq, J. (1877). Essa sur latheories des eaux courantes. *Mem. Presentes par divers savants a l'Academic des Sciences de l'Institut National de France*, 23(1).

Bredehoeft, J. D. and Papadopulos, I. S. (1965). Rates of vertical groundwater movement estimated from the Earth's thermal profile. *Water Resources Research*, 1(2), 325–328.

Brock, T. D., Lee, D. L., Janes, D., and Winek, D. (1982). Groundwater seepage as a nutrient source to a drainage lake, Lake Mendota, Wisconsin. *Water Resources*, 16, 1255 1263.

Brodie, R. S., Sundaram, B., Tottenham, R., Hostetler, S., and Ransley, T. (2007). An Overview of Tools for Assessing Groundwater-Surface Water Connectivity. Canberra, Australia: Bureau of Rural Sciences. Available at:

http://www.southwestnrm.org.au/sites/default/files/uploads/ihub/brodie-r-et-al-2007-overview-tools-assessing-groundwater-surface-water.pdf.

Butler Jr, J. J. (2019). *The Design, Performance, and Analysis of Slug Tests*. 2nd ed. Boca Raton, Florida: CRC Press.

Carslaw, H. S. and Jaeger, J. C. 1959. *Conduction of Heat in Solids*. New York: Oxford University Press.

Caschetto, M., Barbieri, M., Galassi, D. M. P., Mastrorillo, L., Rusi, S., Stoch, F., Di Cioccio, A., and Petitta, M. (2014). Human alteration of groundwater-surface water interactions (Sagittario River, Central Italy): Implication for flow regime, contaminant fate and invertebrate response. *Environmental Earth Sciences*, 71, 1791–1807. Available at: https://doi.org/10.1007/s12665-013-2584-8.

Cey, E. E., Rudolph, D. L., Parkin, G. W., and Aravena, R. (1998). Quantifying groundwater discharge to a small perennial stream in southern Ontario, Canada. *Journal of Hydrology*, 210(1–4), 21–37. Available at: https://www.doi/org/10.1016/s0022-1694(98)00172-3.

Chow, R., Frind, M. E., Frind, E., Jones, J. P., Sousa, M. R., Rudolph, D. L., Molson, J. W., and Nowak, W. (2016). Delineating baseflow contribution areas for streams — A model and methods comparison. *Journal of Contaminant Hydrology*, 195, 11–22. Available at: https://doi.org/10.1016/j.jconhyd.2016.11.001.

Coluccio, K. (2018). A comparison of methods for estimating groundwater-surface water interactions in braided rivers. MS Thesis, University of Canterbury.

Constantz, J. (1998). Interaction between stream temperature, streamflow, and groundwater exchanges in Alpine streams. *Water Resources Research*, 34(7), 1609–1615.

Constantz, J. and Stonestrom, D. (2003). Heat as a tracer of water movement near streams. In Stonestrom, D. and Constantz, J. (Eds.), *Heat as a Tool for Studying the Movement of Ground Water Near Streams*. U.S. Geological Survey Circular 1260.

Cook, P. G., Lamontagne, S., Berhane, D., and Clark, J. F. (2006). Quantifying groundwater discharge to Cockburn River, southeastern Australia using dissolved gas tracers ^{222}Rn and SF_6. *Water Resources Research*, 42, W10411. Available at: https://www.doi.org/10.1029/2006WR004921.

Cook, P. G., Rodellas, V., Andrisoa, A., and Stieglitz, T. C. (2018). Exchange across the sediment-water interface quantified from porewater tracer profiles. *Journal of Hydrology*, 559, 873–883. Available at: https://doi.org/10.1016/j.jhydrol.2018.02.070.

Cremeans, M. M., Devlin, J. F., Osorno, T. C., McKnight, U. S., and Bjerg, P. L. (2019a). A comparison of tools and methods for estimating groundwater-surface water exchange. *Groundwater Monitoring & Remediation*, 40(1), 24–34. Available at: https://doi.org/10.1111/gwmr.12362.

Cremeans, M. M., Devlin, J. F., Osorno, T. C., and Nairn, R. W. (2019b). Assessment of bed hydraulics and metal loadings in a passive vertical flow bioreactor in commerce, Oklahoma. *Groundwater Monitoring & Remediation*, 39(3), 40–47. Available at: https://doi.org/10.1111/gwmr.12337.

Deming, D. (2019). The eighth wonder of the world. *Groundwater*, 58(1), 152–161. Available at: https://doi.org/10.1111/gwat.12958.

Devlin, J. F. and McElwee, C. D. (2007). Effects of measurement error on horizontal hydraulic gradient estimates. *Groundwater*, 45(1), 62–73. Available at: https://doi.org/10.1111/j.1745-6584.2006.00249.x.

Dimova, N. T., Burnett, W. C., Chanton, J. P., and Corbett, J. E. (2013). Application of radon-222 to investigate groundwater discharge into small shallow lakes. *Journal of Hydrology*, 486, 112–122. Available at: https://doi.org/10.1016/j.jhydrol.2013.01.043.

DiStefano, R. J., Magoulick, D. D., Imhoff, E. M., and Larson E. R. (2009). Imperiled crayfishes use hyporheic zone during seasonal drying of an intermittent stream. *Journal of the North American Benthological Society*, 28(1), 142–152. Available at: https://doi.org/10.1899/08-072.1.

Domenico, P. A. and Schwartz, F. W. (1998). *Physical and Chemical Hydrogeology*. 2nd ed. New York: John Wiley & Sons Inc.

Erickson, D. R. (1981). A study of littoral groundwater seepage at Williams Lake, Minnesota using seepage meters and wells. M. S. Thesis, University of Minnesota, Minneapolis, 135 p.

Fleckenstein, J. H., Krause, S., Hannah, D. M., and Boano, F. (2010). Groundwater-surface water interactions: New methods and models to improve understanding of processes and dynamics. *Advances in Water Resources*, 33(11), 1291–1295. Available at: https://doi.org/10.1016/j.advwatres.2010.09.011.

Frei, S. and Gilfedder, B. S. (2015). FINIFLUX: An implicit finite element model for quantification of groundwater fluxes and hyporheic exchange in streams and rivers using radon. *Water Resources Research*, 51, 6776–6786. Available at: https://www.doi.org/10.1002/2015WR017212.

Garcia-Solsona, E., Garcia-Orellana, J., Masque, P., Rodellas, V., Mejıas, M., and Ballesteros, B. (2010). Groundwater and nutrient discharge through karstic coastal springs (Castello, Spain). *Biogeosciences*, 7(9), 2625–2638. Available at: https://doi.org/10.5194/bg-7-2625-2010.

González-Pinzón, R., Ward, A. S., Hatch, C. E., Wlostowski, A. N., Singha, K., Gooseff, M. N., Haggerty, R., Harvey, J. W., Cirpka, O. A., and Brock, J. T. (2015). A field comparison of multiple techniques to quantify groundwater–surface-water interactions. *Freshwater Science*, 34(1). Available at: https://doi.org/10.1086/679738.

Gooseff, M. N., McKnight, D. M., Carr, M., and Baeseman, J. (2010). Antarctic McMurdo Dry Valley stream ecosystems as analogue to fluvial systems on Mars. In Doran, P., Lyons, W. B., and McKnight, D. M. (Eds.), *Life in*

Antarctic Deserts and Other Cold Dry Environments: Astrobiological Analogues. Cambridge, UK: Cambridge University Press, pp. 139–160.

Halder, S., Roy, M. B., and Roy, P. K. (2023). Understanding the exchange process between ground and surface water using mini drive point piezometer and mathematical models to identify suitable managed aquifer recharge sites. *Environmental Science and Pollution Research*, 30, 92736–92767. Available at: https://doi.org/10.1007/s11356-023-28885-z.

Harvey, J. W., Böhlke, J. K., Voytek, M. A., Scott, D., and Tobias, C. R. (2013). Hyporheic zone denitrification: Controls on effective reaction depth and contribution to whole stream mass balance. *Water Resources Research*, 49(10), 6298–6316. Available at: https://doi.org/10.1002/wrcr.20492.

Harvey, J. W. and Wagner, B. J. (2000). Quantifying hydrologic interactions between streams and their subsurface hyporheic zones. In Jones, J. A. and Mulholland, P. J. (Eds.), *Streams and Ground Waters*. San Diego: Academic Press, pp. 3–43.

Hatch, C. E., Fisher, A. T., Revenaugh, J. S., Constantz, J., and Ruehl, C. (2006). Quantifying surface water–groundwater interactions using time series analysis of streambed thermal records: Method development. *Water Resources Research*, 42(10). Available at: https://doi.org/10.1029/2005WR004787.

Hausner, M. B., Suárez, F., Glander, K. E., van de Giesen, N., Selker, J. S., and Tyler, S. W. (2011). Calibrating single-ended fiber-optic Raman spectra distributed temperature sensing data. *Sensors*, 11(11), 10859–10879. Available at: https://doi.org/10.3390/s111110859.

Heron, G., Bjerg, P. L., Gravesen, P., Ludvigsen, L., and Christensen, T. H. (1998). Geology and sediment geochemistry of a landfill leachate contaminated aquifer (Grindsted, Denmark). *Journal of Contaminant Hydrology*, 29, 301–317. Available at: https://doi.org/10.1016/S0169-7722(97)00028-4.

Hughes, B. (2006). Streambed Conductance Survey (SKM Report). Available at: http://www.marlborough.govt.nz/Environment/Groundwater/~/media/Files/MDC/Home/Environment/Groundwater/Pre%202008%20Reports/Stream_Depletion_Report_Stage_1_December_2006.ashx.

Irvine, D. J., Briggs, M. A., Lautz, L. K., Gordon, R. P., McKenzie, J. M., and Cartwright, I. (2017a). Using diurnal temperature signals to infer vertical groundwater-surface water exchange. *Groundwater*, 55(1), 10–26. Available at: https://doi.org/10.1111/gwat.12459.

Irvine, D. J., Cartwright, I., Post, V. E. A., Simmons, C. T., and Banks, E. W. (2016). Uncertainties in vertical groundwater fluxes from 1-D steady state heat transport analyses caused by heterogeneity, multidimensional flow, and climate change. *Water Resources Research*, 52(2), 813–826. Available at: https://doi.org/10.1002/2015WR017702.

Irvine, D. J., Kurylyk, B. L., Cartwright, I., Bonhamd, M., Post, V. E. A., Banks, E. W., and Simmons, C. T. (2017b). Groundwater flow estimation using temperature-depth profiles in a complex environment and a changing climate. *Science of the Total Environment*, 574, 272–281.

Israelsen, O. W. and Reeve, R. C. (1944). Canal lining experiments in the Delta Area. Utah Technical Bulletin 313: Utah Agricultural Experiment Station, Logan.

Jensen, J. K. and Engesgaard, P. (2011). Nonuniform groundwater discharge across a streambed: heat as a tracer. *Vadose Zone Journal*, 10, 98–109. Available at: https://doi.org/10.2136/vzj2010.0005.

Kalbus, E., Reinsdorf, F., and Schirmer, M. (2006). Measuring methods for groundwater-surface water interactions: A review. *Hydrology and Earth System Sciences*, 10, 873–887. Available at: www.hydrol-earth-syst-sci.net/10/873/2006/.

Kaminsky, E., Englisch, C., Griebler, C., Steiner, C., Götzl, G., Knoeller, K., Sandén, H., Laaha, G., and Stumpp, C. (2023). What are the driving factors affecting urban groundwater quality? A multi-tracer approach for the assessment of Vienna's shallow aquifers. In *EGU General Assembly 2023*, Vienna, Austria, 24–28 April 2023, EGU23-5588. Available at: https://doi.org/10.5194/egusphere-egu23-5588.

Keery, J., Binley, A., Crook, N., and Smith, J. W. N. (2007). Temporal and spatial variability of groundwater–surface water fluxes: development and application of an analytical method using temperature time series. *Journal of Hydrology*, 336(1–2), 1–16. Available at: https://doi.org/10.1016/j.jhydrol.2006.12.003.

Kennedy, C. D., Murdoch, L. C., Genereux, D. P., Corbett, D. R., Stone, K., Pham, P., and Mitasova, H. (2010). Comparison of Darcian flux calculations and seepage meter measurements in a sandy streambed in North Carolina, United States. *Water Resources Research*, 46(9). Available at: https://doi.org/10.1029/2009WR008342.

Krabbenhoft, D. P., Bowser, C. J., Anderson, M. P., and Valley, J. W. (1990). Estimating groundwater exchange with lakes: 1. The stable isotope mass balance method. *Water Resources Research*, 26(10), 2445–2453. Available at: https://doi.org/10.1029/WR026i010p02445.

Krause, S., Blume, T., and Cassidy, N. J. (2012). Investigating patterns and controls of groundwater up welling in a lowland river by combining fibre-optic distributed temperature sensing with observations of vertical hydraulic gradients. *Hydrology and Earth System Sciences*, 16(6), 1775–1792. Available at: https://doi.org/10.5194/hess-16-1775-2012.

Larned, S. T., Gooseff, M. N., Packman, A. I., Rugel, K., and Wondzell, S. M. (2015). Groundwater-surface-water interactions: current research directions. *Freshwater Science*, 34(1). Available at: https://doi.org/10.1086/679491.

Lautz, L. K. (2010). Impacts of nonideal field conditions on vertical water velocity estimates from streambed temperature time series. *Water Resources Research*, 46(1). Available at: https://doi.org/10.1029/2009WR007917.

Lee, D. R. (1977). A device for measuring seepage flux in lakes and estuaries. *Limnology and Oceanography*, 22, 140–147.

Lee, D. R. and Cherry, J. A. (1979). A field exercise on groundwater flow using seepage meters and mini-piezometers. *Journal of Geological Education*, 27(1), 6–10. Available at: https://www.doi.org/10.5408/0022-1368-27.1.6.

Lewandowski, J., Arnon, S., Banks, E., Batelaan, O., Betterle, A., Broecker, T., Coll, C., Drummond, J. D., Garcia, J. G., Galloway, J., Gomez-Valez, J., Grabowski, R. C., Herzog, S. P., Hinkelmann, R., Höhne, A., Hollender, J., Horn, M. A., Jaeger, A., Krause, S., Prats, A. L., Magliozzi, C., Meinikmann, K., Mojarrad, B. B., Mueller, B. M., Peralta-Maraver, I., Popp, A. L., Posselt, M., Putschew, A., Radke, M., Raza, M., Riml, J., Robertson, A., Rutere, C., Schaper, J. L., Schirmer, M., Schulz, H., Shanafield, M., Singh, T., Ward, A. S., Wolke, P., Wörman, A., and Wu, L. (2019). Is the hyporheic zone relevant beyond the scientific community? *Water*, 11(11), 2230. Available at: https://doi.org/10.3390/w11112230.

Loheide III, S. P. (2018). Groundwater Research: A brief history of the past 100 years. *AGU (American Geophysical Union) Fall Meeting Abstracts H22C-07*. Available at: https://ui.adsabs.harvard.edu/abs/2018AGUFM.H22C..07L.

Malard, F., Mangin, A., Uehlinger, U., and Ward, J. V. (2001). Thermal heterogeneity in the hyporheic zone of a glacial floodplain. *Canadian Journal of Fisheries and Aquatic Sciences*, 58(7), 1319–1335. Available at: https://www.doi.org/10.1139/cjfas-58-7-1319.

Moore, W. S. (1999). The subterranean estuary: A reaction zone of ground water and sea water. *Marine Chemistry*, 65(1–2), 111–125. Available at: https://doi.org/10.1016/S0304-4203(99)00014-6.

Murdoch, L. C. and Kelly, S. E. (2003). Factors affecting the performance of conventional seepage meters. *Water Resources Research*, 39(6). Available at: https://doi.org/10.1029/2002WR001347.

Nairn, R., Beisel, T., Thomas, R. C., LaBar, J. A., Strevett, K. A., Fuller, D., Strosnider, W. H., Andrews, W. J., Bays, J., and Knox, R. C. (2009). Challenges in design and construction of a large multi-cell passive treatment system for ferruginous lead-zinc mine waters. In Barnhisel, R. I. (Ed.), *Conference Proceeding: 2009 National Meeting of the American Society of Mining and Reclamation, Billings, MT, Revitalizing the Environment: Proven Solutions and Innovative Approaches*, 30 May–5 June, 2009. Lexington: ASMR.

Nielsen, S. S., Tuxen, N., Frimodt, O., Sonne, A. T., Binning, P. J., Fjordbøge, A. S., and Aabling, J. (2014). Risk assessment of surface water impacted by point source related contaminated groundwater [in Danish]. *Danish EPA, Copenhagen*.

NIRAS. (2009). A historical overview of the contamination in Grindsted [in Danish], Technical Report No 04/350, Ribe County.

Null, K. A., Dimova, N. T., Knee, K. T., Esser, B. K., Swarzenski, P. W., Singleton, M. J., Stacey, M., and Paytan, A. (2012). Submarine groundwater discharge-derived nutrient loads to San Francisco Bay: Implications to future ecosystem changes. *Estuaries and Coasts*, 35(5). Available at: https://doi.org/10.1007/s12237-012-9526-7.

Osorno, T. C. (2021). Point Velocity Probe Handbook. Ozark Underground Laboratory, pp. 12–13.

Paulsen, R. J., Smith, C. F., O'Rourke, D., and Wong, T. F. (2001). Developed and evaluation of an ultrasonic ground water seepage meter. *Ground Water*, 39, 904–911.

Post, V. E. A. and von Asmuth, J. R. (2013). Review: Hydraulic head measurements—New technologies, classic pitfalls. *Hydrogeology Journal*, 21(4), 737–750. https://doi.org/10.1007/s10040-013-0969-0.

Rasmussen J. J., McKnight, U. S., Sonne, A. T., Wiberg-Larsen, P., and Bjerg, P. L. (2016). Legacy of a chemical factory site: contaminated groundwater impacts stream macroinvertebrates. *Archive of Environmental Contamination & Toxicology*, 70(2), 219–230. Available at: https://doi.org/10.1007/s00244-015-0211-2.

Ren, J. and Packman, A. I. (2004). Stream-subsurface exchange of zinc in the presence of silica and kaolinite colloids. *Environmental Science and Technology*, 38(24), 6571–6581. Available at: https://doi.org/10.1021/es035090x.

Rodgers, P., Soulsby, C., Petry, J., Malcolm, I., Gibbins, C., and Dunn, S. (2004). Groundwater-surface water interactions in a braided river: A tracer-based assessment. *Hydrological Processes*, 18(7), 1315–1332. Available at: https://www.doi.org/10.1002/hyp.1404.

Rosenberry, D. O. (2005). Integrating seepage heterogeneity with the use of ganged seepage meters. *Limnology and Oceanography: Methods*, 3(2), 131–142. Available at: https://doi.org/10.4319/lom.2005.3.131.

Rosenberry, D. O. (2008). A seepage meter designed for use in flowing water. *Journal of Hydrology*, 359(1–2), 118–130. Available at: https://doi.org/10.1016/j.jhydrol.2008.06.029.

Rosenberry, D. O. and LaBaugh, J. W. (2008). Field techniques for estimating water fluxes between surface water and ground water. *United States Geological Survey Techniques and Methods* 4-D2.

Rosenberry, D. O. and Morin, R. H. (2004). Use of an electromagnetic seepage meter to investigate temporal variability in lake seepage. *Ground Water*, 42(1), 68–77.

Schillig, P. C. (2012). VelProbePE: An automated spreadsheet program for interpreting point velocity probe breakthrough curves. *Computers & Geosciences*, 39, 161–170. Available at: https://www.doi.org/10.1016/j.cageo.2011.06.007.

Schillig, P., Devlin, J. F., McElwee, C. D., Walter, K., and Gibson, B. (2014). Assessment of density-induced tracer movement in groundwater velocity measurements with point velocity probes (PVPs). *Ground Water Monitoring and Remediation*, 34(4), 44–50.

Schmadel, N. M., Neilson, B. T., and Kasahara, T. (2013). Deducing the spatial variability of exchange within a longitudinal channel water balance. *Hydrological Processes*, 28, 3088–3103.

Schmidt, C., Bayer-Raich, M., and Schirmer, M. (2006). Characterization of spatial heterogeneity of groundwater-stream water interactions using multiple depth streambed temperature measurements at the reach scale. *Hydrology and Earth System Sciences*, 10, 849–859. Available at: https://doi.org/10.5194/hess-10-849-2006.

Schmidt, C., Conant, B., Bayer-Raich, M., and Schirmer, M. (2007). Evaluation and field-scale application of an analytical method to quantify groundwater discharge using mapped streambed temperatures. *Journal of Hydrology*, 347(3–4), 292–307. https://doi.org/10.1016/j.jhydrol.2007.08.022.

Schulz, H. (2017). Application of the streambed-Point Velocity Probe for investigating contaminant discharge to streams. Master's Thesis, Technical University of Denmark Department of Environmental Engineering, 125 p.

Selker, J. S., van de Giesen, N., Westhoff, M., Luxemburg, W., and Parlange, M. B. (2006a). Fiber optics opens window on stream dynamics. *Geophysical Research Letters*, 33(24). Available at: https://doi.org/10.1029/2006GL027979.

Selker, J. S., Thévenaz, L., Huwald, H., Mallet, A., Luxemburg, W., van de Giesen, N., Stejskal, M., Zeman, J., Westhoff, M., and Parlange, M. B. (2006b). Distributed fiber-optic temperature sensing for hydrologic systems. *Water Resources Research*, 42(12). Available at: https://doi.org/10.1029/2006WR005326.

Shanafield, M., Hatch, C. E., and Pohll, G. (2011). Uncertainty in thermal time series analysis estimates of streambed water flux. *Water Resources Research*, 47, W03504. Available at: https://www.doi.org/10.1029/2010WR009574.

Silliman, S. E., Ramirez, J., and McCabe, R. L. (1995). Quantifying downflow through creek sediments using temperature time series: One-dimensional solution incorporating measured surface temperature. *Journal of Hydrology*, 167, 99–119.

Solder, J. E., Gilmore, T. E., Genereux, D. P., and Solomon, D. K. (2016). A tube seepage meter for in situ measurement of seepage rate and groundwater sampling. *Groundwater*, 54(4), 588–595. Available at: https://doi.org/10.1111/gwat.12388.

Sonne, A. T., McKnight, U. S., Rønde, V., and Bjerg, P. L. (2017). Assessing the chemical contamination dynamics in a mixed land use stream system. *Water*

Research, 125, 141–151. Available at: https://doi.org/10.1016/j.watres.2017.08.031.

Stallman, S. (1965). Steady one-dimensional fluid flow in a semi-infinite porous medium with sinusoidal surface temperature. *Journal of Geophysics Research*, 70(12), 2821–2827.

Suzuki, S. (1960). Percolation measurements based on heat flow through soil with special reference to paddy fields. *Journal of Geophysics Research*, 65(9), 2883–2885.

Taniguchi, M. and Fukuo, Y. (1993). Continuous measurements of ground water seepage using a automatic seepage meter. *Ground Water*, 31(7), 675–679.

Turcotte, D. L. and Schubert, G. (1982). *Geodynamics: Applications of Continuum Physics to Geological Problems*. New York: John Wiley & Sons.

Tyler, S. W., Selker, J. S., Hausner, M. B., Hatch, C. E., Torgersen, T., Thodal, C. E., and Schladow, S. G. (2009). Environmental temperature sensing using Raman spectra DTS fiber-optic methods. *Water Resources Research*, 45(4). Available at: https://doi.org/10.1029/2008WR007052.

US EPA. (n.d.a). OAK RIDGE RESERVATION (USDOE) OAK RIDGE, TN. Cumulis.epa. https://cumulis.epa.gov/supercpad/SiteProfiles/index.cfm?fuseaction=second.cleanup&id=0404152 (Accessed 4 February 2024).

US EPA (n.d.b). HANFORD 100-AREA (USDOE) BENTON COUNTY, WA. Cumulis.epa. https://cumulis.epa.gov/supercpad/cursites/csitinfo.cfm?id=1001114 (Accessed 4 February 2024).

US EPA (n.d.c). *TAR CREEK (OTTAWA COUNTY) OTTAWA COUNTY, OK*. Cumulis.epa. https://cumulis.epa.gov/supercpad/cursites/csitinfo.cfm?id=0601269 (Accessed 4 February 2024).

Van Damme, K., Brancelj, A., and Dumont, H. J. (2009). Adaptations to the hyporheic in Aloninae (Crustacea: Cladocera): allocation of *Alona protzi* Hartwig, 1900 and related species to *Phreatalona* gen. nov. *Hydrobiologia*, 618(1), 1–34. Available at: https://www.doi.org/10.1007/s10750-008-9607-6.

van de Giesen, N., Steele-Dunne, S. C., Jansen, J., Hoes, O., Hausner, M. B., Tyler, S., and Selker, J. S. (2012). Double-ended calibration of fiber-optic Raman spectra distributed temperature sensing data. *Sensors*, 12(5), 5471–5485. Available at: https://doi.org/10.3390/s120505471.

Ward, A. S., Gooseff, M. N., Voltz, T. J., Fitzgerald, M., Singha, K., and Zarnetske, J. P. (2013). How does rapidly changing discharge during storm events affect transient storage and channel water balance in a headwater mountain stream? *Water Resources Research*, 49, 5473–5486.

Wilson, S. J., Moody, A., McKenzie, T., Bayani Cardenas, M., Luijendijk, E., Sawyer, A. H., Wilson, A., Michael, H. A., Xu, B., Knee, K. L., Cho, H. M., Weinstein, Y., Paytan, A., Moosdorf, N., Chen, C. T. A., Beck, M., Lopez, C.,

Murgulet, D., Kim, G., Charette, M. A., Waska, H., Ibanhez, J. S. P., Chaillou, G., Oehler, T., Onodera, S., Saito, M., Rodellas, V., Dimova, N., Montiel, D., Dulai, H., Richardson, C., Du, J., Petermann, E., Chen, X., Davis, K. L., Lamontagne, S., Sugimoto, R., Wang, G., Li, H., Torres, A. I., Demir, C., Bristol, E., Connolly, C. T., McClelland, J. W., Silva, B. J., Tait, D., Kumar, B. S. K., Viswanadham, R., Sarma, V. V. S. S., Silva-Filho, E., Shiller, A., Lecher, A., Tamborski, J., Bokuniewicz, H., Rocha, C., Reckhardt, A., Bottcher, M. E., Jiang, S., Stieglitz, T., Gbewezoun, H. G. V., Charbonnier, C., Anschutz, P., Hernandez-Terrones, L. M., Babu, S., Szymczycha, B., Sadat-Noori, M., Niencheski, F., Null, K., Tobias, C., Song, B., Anderson, I. C., and Santos, I. R. (2024). Global subterranean estuaries modify groundwater nutrient loading to the ocean. *Limnology and Oceanography Letters*, 9(4), 411–422. Available at: https://doi.org/10.1002/lol2.10390.

Winter, T. C. (2007). The Rold of ground water in generating streamflow in headwater areas and in maintaining base flow. *JAWRA Journal of the American Water Resources Association*, 43(1), 15–25. Available at: https://doi.org/10.1111/j.1752-1688.2007.00003.x.

Winter, T. C., Harvey, J. W., Franke, O. L., and Alley, W. M. (1998). Ground Water and Surface Water: A Single Resource. U.S. Geological Survey Circular 1139.

Woessner, W. W. and Sullivan, K. E. (1984). Results of seepage meter and mini-piezometer study, Lake Mead, Nevada. *Ground Water*, 22, 561–568.

Wondzell, S. M. (2015). Groundwater-surface-water interactions: Perspectives on the development of the science over the last 20 years. *Freshwater Science*, 34(1). Available at: https://doi.org/10.1086/679665.

Younger, P. L. (2012). Crouching enemy, hidden ally: The decisive role of groundwater discharge features in two major British battles, Flodden 1513 and Prestonpans 1745. *Geological Society, London, Special Publications*, 362(1), 19–33. Available at: https://doi.org/10.1144/SP362.2.

Zamora, C. (2006). Estimating rates of exchange across the sediment/water interface in the lower Merced River, CA. MS Thesis. California State University, Sacramento, 102p.

https://doi.org/10.1142/9781800617841_0008

Chapter 8

Arsenic Contamination in the Ganges–Brahmaputra–Meghna River Delta Aquifer: Understanding Water Security Challenges

Madhumita Chakraborty[*,|], **Md Ilias Mahmud**[†,‡], **Sarmin Sultana**[§], **Nguyen Pham Minh Ngoc**[*], **and Abhijit Mukherjee**[¶]

[*]*Department of Earth and Environmental Geoscience, Washington and Lee University, Lexington, VA 24450, USA*

[†]*Department of Geology and Geological Engineering, University of Mississippi, Oxford, Oxford, MS 38677, USA*

[‡]*Department of Geology and Mining, University of Barishal, Barishal, 8254, Bangladesh*

[§]*Department of Geology, University of Dhaka, Dhaka, 1000, Bangladesh*

[¶]*Department of Geology and Geophysics, Indian Institute of Technology Kharagpur, Kharagpur, WB, 721302, India*

[|]*mchakraborty@wlu.edu*

Abstract

The Ganges–Brahmaputra–Meghna River (GBM) delta, shared between Bangladesh and India, confronts a critical environmental and public health challenge due to widespread arsenic (As) contamination in its groundwater. The source of As within the delta is geogenic and non-point, primarily sourced from the unconsolidated sediments from the Himalayas that constitute the delta's subsurface. The shift from surface to groundwater use in the 1970s, originally intended to protect the delta population from waterborne diseases and enhance agricultural productivity, inadvertently exposed up to an estimated 77 million people to toxic levels of As. Chronic exposure to As is associated with severe health outcomes, including cancers, cardiovascular diseases and impaired cognitive development. This chapter provides a brief description of the origin, occurrence and distribution of groundwater As in the GBM delta and discusses the challenges of management and mitigation of As contamination under the prevailing plans and policies in the region.

8.1 Introduction

Arsenic (As) stands out as one of the most concerning heavy metal contaminants in soil and water, posing a severe threat to millions of people across the globe. Although As contamination can be anthropogenically induced, the more widespread and severe instances of As contamination, which significantly impact larger geological regions and give rise to serious public health concerns, are geogenic in nature. Naturally occurring As concentrations in groundwater exceeding the World Health Organization (WHO) drinking-water guideline of 10 μg/L have been reported from many parts of the world (Ravenscroft *et al.*, 2009). The problem is particularly severe in South and Southeast Asia, putting millions of people at risk of adverse As-related health effects. Of these, the most severe case of As contamination is reported in the Ganges–Brahmaputra–Meghna (GBM) River delta, shared between Bangladesh and India, where an estimated 77 million individuals are exposed to the risk of As contamination in their drinking water (Edmunds *et al.*, 2015), marking it "the largest mass poisoning of a population in history" (Smith *et al.*, 2000). This large-scale enrichment of As in groundwater from the alluvial and deltaic aquifers of the GBM delta has attracted widespread global interest from scientific, research and non-governmental organisations, focusing on comprehending and addressing As toxicity in the delta's groundwater.

8.1.1 *Geological and sedimentological setting*

The GBM delta is the largest fluvio-deltaic system in the world. The GBM River system drains vast areas of the Himalayas and its surroundings. Throughout the delta's history, particularly since the mid-Miocene, unlithified sediment up to 22 km thick has accumulated (Alam *et al.*, 2003). These sediments form highly productive aquifers that constitute the GBM delta aquifer system (Chakraborty *et al.*, 2021). Figure 8.1 shows the

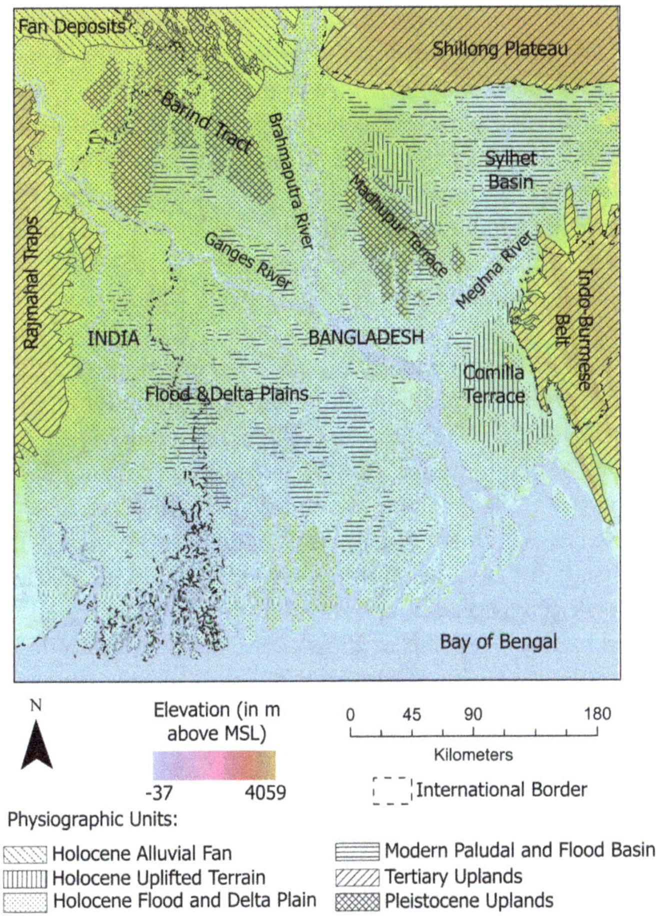

Figure 8.1. The SRTM 90-m digital elevation map of the GBM delta and the surrounding area showing the major physiographic features.

geological setting of the GBM delta, along with the adjacent physiographic provinces. Geomorphologically, the GBM delta constitutes extensive low-lying, gently sloping Holocene flood and delta plains, which are locally disrupted by the uplifted blocks of Pleistocene units (notably the Madhupur Terrace and Barind Tract), which are remnants of the otherwise eroded Pleistocene landscape (Mukherjee *et al*., 2009). Spatially heterogeneous influences of sediment supply, eustatic cycles and active tectonism on the late Quaternary evolution of the delta have resulted in varied sedimentation patterns in the delta subsurface (Goodbred and Kuehl, 2000).

Drawing from Goodbred and Kuehl (2000), the sedimentary sequences of the delta can be classified into the following: (a) the western-central inland delta exhibits an unconfined, continuous aquifer unit dominated by channel sand deposits, with a near absence of fine-grained units; (b) the lower delta in the southern and eastern extents is characterised by alternating fine and coarse sediments, resulting in a semiconfined to confined, vertically segregated and multilayered aquifer system with limited hydraulic connectivity; (c) the northeastern part of the delta, along the present and old Brahmaputra channels and the Sylhet Basin, exhibits a mud-dominated sequence.

The alluvial sediments underlying the Holocene flood and delta plains can be classified into overlying Holocene and underlying Pleistocene aquifers. The Holocene aquifers are characterised by grey sediments and abundant organic matter (OM), indicating their reduced nature (Chakraborty *et al*., 2015). These are underlain by discontinuous oxidised Pleistocene sands, which are orange-brown to yellow-brown, with calcareous and ferruginous nodules and low OM (BGS/DPHE, 2001). In contrast, the Pleistocene uplands consist of Plio-Pleistocene yellow-brown sands, topped by thick deposits of Pleistocene red-brown mottled clay and silt (Ravenscroft *et al*., 2005). As enrichment is mainly observed in the groundwater beneath the shallow Holocene aquifers, while Pleistocene aquifers generally have low levels of dissolved As.

8.1.2 *The arsenic problem*

In the 1970s, the delta experienced a large-scale shift to groundwater usage for drinking and irrigation as an alternative to polluted surface waters, leading to the installation of millions of tube wells. Unbeknownst at the time, this inadvertently exposed millions of people to toxic levels of

As, causing widespread poisoning and long-term public health challenges in both India and Bangladesh. High As levels in groundwater are pervasive across the delta (Figure 8.2), with studies indicating that 42–56% of wells are contaminated with As above the WHO guideline of 10 μg/L for drinking water, and 25–37% exceed 50 μg/L (BGS/DPHE, 2001; Rahman *et al.*, 2001). Human exposure to As primarily occurs through drinking water and food irrigated with high-As groundwater (Chakraborty *et al.*, 2015). Arsenic poisoning, also called arsenicosis, develops insidiously

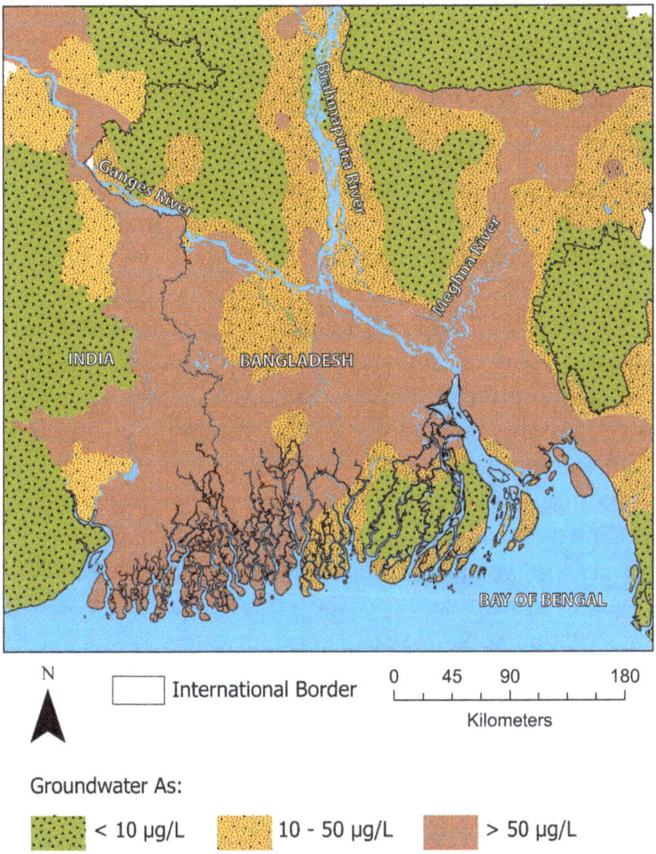

Figure 8.2. Map depicting the distribution and severity of groundwater contamination in the GBM delta. The map was generated using disjunctive kriging on the BGS/DPHE (2001) As data in Bangladesh and identified As-contaminated districts in West Bengal according to Mukherjee *et al.* (2009).

after months or years of exposure. Symptoms commonly manifest as dermatological issues, such as melanosis, leucomelanosis and keratosis. As a Class-I carcinogen, As also causes severe health effects, including multiple-organ malfunctions and cancers (Chakraborty *et al.*, 2015). Although it is challenging to estimate the exact number of people at risk, approximately 30–35 million individuals in Bangladesh are exposed to As levels exceeding 50 μg/L, with around 77 million exposed to levels above 10 μg/L (Chakraborty *et al.*, 2015). In West Bengal, about 15 million people are exposed to As concentrations over 50 μg/L, and 50 million are exposed to levels above 10 μg/L in their drinking water.

8.2 Sources of Arsenic

The primary origin of As in the groundwater of the GBM delta is geogenic via release from soils and sediments. The unconsolidated delta sediments typically contain As within the range of <1–15 mg/kg (Smedley and Kinniburgh, 2002), which is somewhat higher than the crustal average of 1–2 mg/kg (Chakraborty *et al.*, 2015). However, specific hotspots can have concentrations as high as 490 mg/kg (Mukherjee and Bhattacharya, 2001). These relatively moderate concentrations of solid-phase As observed in delta sediments contrast with the exceptional levels found in groundwater, implying that elevated As concentrations in groundwater do not necessitate exceptionally high As content in sediments and are instead driven by mobilisation under favourable biogeochemical and hydrological conditions (Smedley and Kinniburgh, 2002). Further, an alternative argument contends that sediments derived from the Himalayan convergent margin orogenic system serve as the primary source of As within the GBM delta (Mukherjee *et al.*, 2019). This hypothesis is grounded in the finding that, globally, As-enriched basins are often situated adjacent to orogenic systems, where magmatic processes transport As-enriched minerals, which, once exposed at the surface and weathered, deposit As-laden sediments in foreland basins and downstream areas, where As can be released within aquifers under favourable conditions.

The sandy aquifers of the GBM delta have As concentrations typically below 10 mg/kg, while the fine-grained layers of clays and muds, often interbedded with OM-rich horizons, typically exhibit As contents as high as 100 mg/kg (Harvey *et al.*, 2005). In the sands, As is not held within the quartzo-feldspathic grains but is instead sorbed onto the iron (oxyhydr) oxide coatings around the grains (Davis *et al.*, 1990). As can also be retained by other oxides, silicates, phosphates and carbonates; however, their contributions to the total As load are limited due to their low abundance in the delta sediments. Studies suggest that pyrite and other sulphide minerals host the largest share of the total As load (60–70%) within the delta sediments (Polizzotto *et al.*, 2005; Lowers *et al.*, 2007), although most sulphide minerals are believed to be sinks for As rather than being sources.

Apart from natural factors, As in soils and shallow sediments can be contributed by anthropogenic activities. Traditionally, dry-season rice cultivation in the delta uses irrigation involving prolonged flooding of fields with groundwater, which leads to the sequestration of As in soils and sediments under oxic conditions (Chakraborty *et al.*, 2015). The As load from the standing water on the fields, combined with As-laden young sediments at the surface, results in soil As concentrations as high as 40 mg/kg, often exceeding those in aquifer sediments (Polizzotto *et al.*, 2006). Such soils can subsequently act as a potential source of groundwater As in later circulation cycles.

8.3 Processes for Arsenic Release into the Groundwater

Most studies indicate that the principal process for As mobilisation in the delta is the microbially mediated reductive dissolution of iron (oxyhydr) oxide coatings on aquifer sand grains (Figure 8.3). In the strongly reducing conditions of the grey, OM-rich Holocene delta aquifers, microbial metabolism reduces iron (oxyhydr)oxides from insoluble ferric to soluble ferrous states, driving iron dissolution and liberating the adsorbed As (as reduced As (III) species) into the groundwater (Bhattacharya *et al.*, 1997; Harvey *et al.*, 2002; Mukherjee *et al.*, 2008). Alternatively, in locally oxic environments within the aquifers, such as shallow aquifers exposed to the atmosphere due to pumping-induced lowering of the water table or enhanced recharge carrying dissolved oxygen, the oxidation of sulphide minerals (e.g., pyrite and arsenopyrite) can release soluble sulphate and As (as oxidised As(V) species) into the groundwater (Mallick and Rajgopal, 1995). Various chemical species in solution, such as phosphates, silica, carbonates and bicarbonates, can compete with and displace sorbed As, thereby mobilising it into groundwater through competitive ion

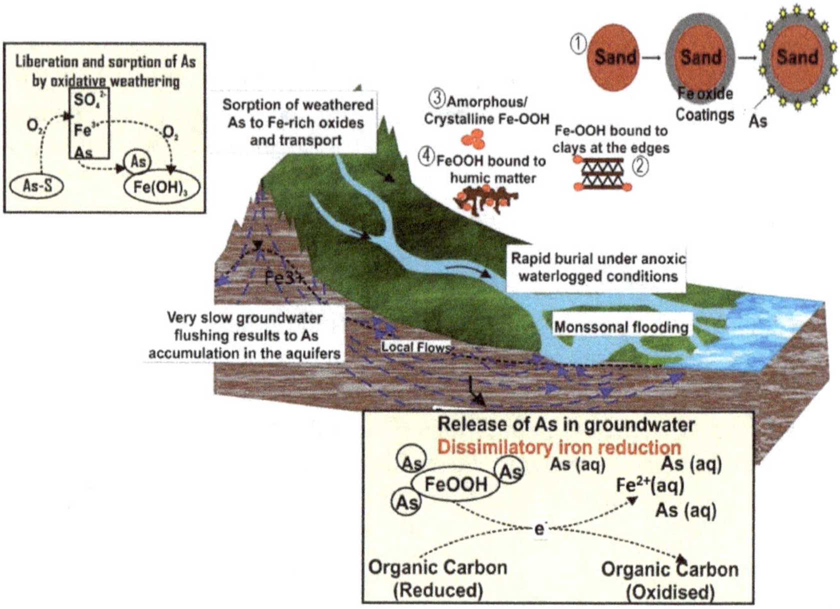

Figure 8.3. Schematic diagram demonstrating the key processes influencing As release into groundwater from its source to its sink.

Source: Adapted from Chakraborty *et al.* (2021).

exchange. Thus, anthropogenic inputs, such as phosphate from fertilisers (Stollenwerk *et al.*, 2007), may cause As liberation into groundwater, although the large-scale impact of these inputs is debated.

None of the aforementioned processes can singularly account for the spatial occurrence and distribution of As within the delta's groundwaters. Instead, the phenomenon is truly governed by the simultaneous operation of multiple processes. This, coupled with low flushing rates within the delta, has led to widespread As concentrations within the region.

8.4 Geologic and Anthropogenic Controls on Groundwater Arsenic Distribution

In a recent study, Chakraborty *et al.* (2022) demonstrated the influence of aquifer geometry on As distribution within the Sonar Bangla Aquifer (SBA) system, located west of the Ganges–Padma channel. They identified three subsystems based on the distribution of silt and clay confining units: (a) the Type I subsystem to the northwest is characterised by a largely unconfined continuous sand aquifer with no intercalating confining units of significant extent or thickness; (b) the Type II subsystem exhibits a transitional character, featuring a semiconfined aquifer system with more than one vertically segregated aquifer unit intercalated with discontinuous (<30 km in length) confining layers; (c) the Type III aquifer subsystem along the delta front to the south and the Padma estuary is characterised by a nearly confined multilayered aquifer system with more than three horizontally stratified aquifer units separated by extensive yet discontinuous (>30 km in length) aquitard units. Results show that the Type II and Type III subsystems exhibit a decline in the percentage of As-contaminated wells with depth, with 68% and 98% of tube wells, respectively, yielding As-safe water (<10 μg/L), while no such depth decline was observed within the Type I aquifer, with only 34% of wells found to be As-safe. This decline in Type II and Type III aquifers, as opposed to Type I, can be attributed to their layered aquifer architecture being intercalated by the confining units, which act as natural barriers against invasion of shallow As or As-mobilising species into the deeper aquifers (Burgess *et al.*, 2010; Mahmud *et al.*, 2017; Chakraborty *et al.*, 2022) (Figure 8.4). This control of aquifer architecture on As contamination in the deep aquifers is expected to hold true across the delta, where

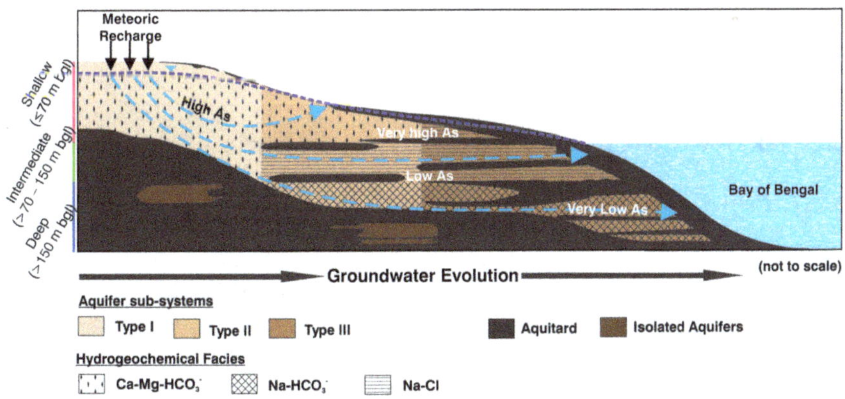

Figure 8.4. Schematic model illustrating the aquifer subsystems of the GBM delta, highlighting the associated hydrogeochemical profiles and dissolved As concentrations across various aquifer types and depths.

the deep aquifers in the multilayered systems remain protected by the overlying impermeable caps across the delta and have a higher probability of yielding As-safe waters, while the deep aquifers within the thick continuous sandy systems would be more vulnerable to As-contamination.

However, researchers have reported high As contamination from parts of the deep Pleistocene aquifers, especially from the western delta (Khan *et al.*, 2019). This, on most occasions, has been attributed to either *in situ* mobilisation of As in the deep aquifers (McArthur *et al.*, 2016) and/or the lateral or downward migration of As or As-mobilising species from contaminated Holocene paleochannel-fill aquifers. The latter can be caused by the following factors: (a) naturally driven deep groundwater inflows in areas dominated by a higher vertical hydraulic conductivity resulting from the absence of interlayered confining units (Khan *et al.*, 2019); (b) enhanced vertical gradients caused by extensive deep irrigation or municipal pumping, resulting in a rapid downward flow of OM-rich waters from surficial sources (such as irrigation return flow and ponds) into deeper aquifers, thereby contaminating them (Mukherjee *et al.*, 2011). Additionally, under conditions of heavy pumping stress, the confining clay units are known to expel As or dissolved organic carbon (DOC) into the aquifers below in response to depressurisation (due to a drop in groundwater pressure heads) and/or compaction (due to land subsidence) (Erban *et al.*, 2013; Mihajlov *et al.*, 2020). Additional anthropogenic influences on As include fertiliser use, pit-latrines and septic tanks, which can leak As-mobilising (e.g., OM) and/or As-retarding species (e.g., nitrates, sulphates and phosphates) (McArthur *et al.*, 2012). However, their impact is believed to be limited to local scale.

8.5 Management and Mitigation Efforts

8.5.1 *Plans and policy interventions*

Over the years, the WHO has progressively tightened the acceptable levels of As in drinking water, justified by global epidemiological effects and detection capabilities. Initially set at 2000 μg/L in 1958, it was reduced to 50 μg/L by 1963 (Yamamura, 2003). By 1993, the WHO proposed a provisional limit of 10 μg/L. The Bureau of Indian Standards (BIS) lowered its As threshold to 10 μg/L in 2004, from 50 μg/L established in 1991, though the permissible level in the absence of alternative water sources remains at 50 μg/L (BIS, 2003). Meanwhile, Bangladesh maintains a national standard of 50 μg/L (BBS/UNICEF, 2009).

Following the detection of As in the groundwaters of the GBM delta, one of the first regional-scale hydrogeochemical surveys to map As contamination across Bangladesh was jointly conducted by the British Geological Survey and the Directorate of Public Health Engineering of Bangladesh in 1998–1999 (BGS/DPHE, 2001). This initiative spurred a comprehensive evaluation of over 5 million tube wells from 2000 to 2006 as part of the Bangladesh Arsenic Mitigation Water Supply Project (Johnston and Sarker, 2007). Concurrently, similar screening efforts were executed in India, where the Public Health Engineering Department (PHED) of West Bengal, in partnership with UNICEF, screened numerous wells in Assam and West Bengal during 2004–2005 to evaluate As concentrations.

Both the Indian and Bangladeshi governments have implemented policies to regulate As contamination in groundwater. The Ministry of Drinking Water & Sanitation (MDWS), Government of India, has launched the National Rural Drinking Water Programme (NRDWP) to support state-level water infrastructure improvements, focusing on areas affected by water scarcity and poor water quality. Priority in fund allocation and installation of community water purification plants has been given to fluoride- and As-affected habitations in the country. This programme also institutionalised community involvement in regular water quality monitoring through the National Rural Drinking Water Quality Monitoring & Surveillance Programme (NRDWQM&SP) initiated in 2006 (Shrivastava, 2016). In addition, the government of West Bengal created the Arsenic Task Force in 2005 to develop a master plan for As mitigation, using findings from the UNICEF/PHED As database (Bhavan *et al.*, 2007). Similarly, in 2004, Bangladesh established a National Policy of Arsenic Mitigation (NPAM) and Implementation Plan for Arsenic Mitigation (IPAM), aiming to provide sustainable water solutions nationwide (GoB, 2004). The intervention strategies have been based on contamination levels exceeding 50 μg/L: (a) emergency response in villages where over 80% of tube wells are As-contaminated; (b) medium-term response targeting areas where 40–80% of tube wells are As-contaminated; (c) long-term response for the entire nation, focusing on establishing sustainable water supply solutions. Supporting laws such as the Safe Drinking Water Act (2005), Bangladesh Water Act (2013) and Bangladesh Water

Rules (2018) bolster these efforts by setting standards and protecting water quality. All of these initiatives share similar target areas of concern and strategic interventions. They encompass the following:

(a) Regular monitoring of tubewells (domestic, community and irrigation wells) using field test kits and laboratory support: Wells are colour-coded to indicate contamination levels.
(b) Development of region-specific safe water alternatives, such as improved dug wells, deep wells, surface water treatment, rainwater harvesting (RWH) and As-removal technologies: To maintain safety standards, all As-removal technologies must be pre-approved by the Bangladesh Council for Scientific and Industrial Research (BCSIR) before they are marketed and used publicly. Section 8.5.2 provides a comprehensive list of the commonly used As-mitigation options deployed within the delta. The policy advocated the use of surface water over groundwater to minimise As exposure and preferred the installation of piped water systems where practical.
(c) Agricultural interventions aided by research on the impacts of As on soil fertility and crop health: The NRDWQM&SP policy aims to set national standards for permissible As levels in irrigation water and agricultural outputs to safeguard food supplies.
(d) Healthcare initiatives through regular community screenings to detect affected individuals, development of standard protocols for diagnosis and treatment of arsenicosis, training healthcare providers and supporting patient rehabilitation.
(e) Raising public awareness about the dangers of As, clarifying that arsenicosis is not contagious, and educating about alternative safe water sources.
(f) Institutional and capacity building of governmental, community-level and private entities in As management, including building grass-roots workers and well-equipped district and sub-district laboratories for As and water quality analysis.
(g) Enhancing coordination and cooperation among all stakeholders involved, including national and local governmental agencies, NGOs, private agencies (such as donor organisations) and civil society in research, planning and service delivery.
(h) Addressing broader social issues such as gender equity, rights of the economically disadvantaged and the integration of sanitation with groundwater management in As-mitigation strategies.

8.5.2 *Arsenic-safe water sources*

The following section details common mitigation strategies implemented by the governments of India and/or Bangladesh to provide As-safe water for the residents of the delta.

(A) Identifying As-Safe Wells: Several studies in Bangladesh have identified As-free tube wells as the most socially acceptable and infrastructurally sustainable mitigation option (Hossain *et al.*, 2015; Kumar *et al.*, 2021). This widespread acceptance of tube wells as a potable water source is rooted in three key advantages: (a) tube wells provide an unrestricted supply of water at any time of the day (Hossain *et al.*, 2015); (b) given women's primary role in fetching water, households exhibit preference for nearby water sources, such as tube wells, which are extensively distributed across the delta, ensuring close proximity to homes; (c) tube wells are a cost-effective and low-maintenance alternative to other available options. Within the delta, groundwater As concentrations exhibit wide local-scale variations (influenced by varying well depths, sediment facies and ages, aquifer geometry, etc.), commonly resulting in the coexistence of As-safe wells within areas containing numerous As-affected wells. This enables the practice of "well switching"—shifting from using an As-contaminated well to an As-safe well—which is a strategy recognised as a highly effective and readily available mitigation measure (Hossain *et al.*, 2015). Efforts to mark and identify As-contaminated wells have been undertaken in both West Bengal and Bangladesh. Conventionally, As-contaminated well heads are painted red, while As-free wells are painted green for easy identification. In a performance survey conducted by Hossain *et al.* (2015) to evaluate different As mitigation measures—including As-removal filters (ARF), surface water filters, pond sand filters (PSF), RWH and As-safe tube wells—deployed in the Matlab upazila (sub-district) of Bangladesh, it was found that 89% of households opted for As-safe tube wells as their primary source of domestic water supply, marking a notable increase from the 58% recorded during the prior three years. However, well switching can impose unsustainable stress on a specific well, leading to the well drying up or changes in As levels due to enhanced inflow of contaminated groundwater from the surroundings. Additionally, temporal or seasonal variations in As concentrations in well water may render initially safe wells unsafe over time, highlighting the need for regular monitoring of these identified safe wells (Kumar *et al.*, 2021). Surveys from the region also highlight challenges in promoting well switching, with 36% of the population using unsafe wells due to unawareness of As hazards and 29% due to the unavailability of nearby As-safe tube wells (Hossain *et al.*, 2015).

(B) Installing As-Safe Tube Wells: Due to the prevalence of unconsolidated sediments and shallow water levels across the delta, drilling hand-pumped tube wells is labour- and cost-efficient. Consequently, installing tube wells in As-safe aquifers can serve as an efficient mitigation measure. However, the spatial distribution of As-safe aquifers within the delta is influenced by a complex interplay of various (hydro)geological and anthropogenic factors (see Section 8.4), posing a challenge in identifying them. Here, we discuss some of the potentially safe aquifer sources identified by previous research across regions within the delta.

- **Deep tube wells:** Building upon the finding of an overall decline in groundwater As concentrations with depth, numerous researchers have advocated for deep tube wells as a safe and viable source of potable water (Burgess *et al.*, 2010). In response to these findings and subsequent mitigation policies in both nations, many deep (>150 m) tube wells have been installed in As-affected regions as a reliable water option, which often incurs high costs due to the additional depth of drilling and pipeline installation. Consequently, these wells can rarely be installed and maintained by individual households and need organisational or community resources (Kumar *et al.*, 2021).

However, the findings of Khan *et al.* (2019) and Chakraborty *et al.* (2022) challenge the ubiquitous suitability of deep tube wells as a reliable source of As-free drinking water. Furthermore, extensive and unsustainable pumping from the deep aquifer is anticipated to trigger a rapid inflow from shallow depths, which frequently harbour high As, thereby leading to cross-contamination of the deeper aquifers (Chakraborty *et al.*, 2022). Michael and Voss (2008) proposed restricting the use of deep aquifers for domestic water and shallow aquifers for irrigation to establish a hydraulic barrier between these depths, safeguarding against such cross-contamination. Also, contamination from shallow to deeper aquifers may often seep through the annular space surrounding the borehole. Therefore, for the effective and successful installation of deep tube wells, considerations include additional costs, local-scale availability and identification of

As-safe deeper aquifers (preferably separated from shallower ones by an impermeable layer), sustainable pumping rates and structural integrity through annular sealing of boreholes to prevent leakage from shallow aquifers. Various national and state agencies have installed thousands of deep tube wells, mostly in the southeastern part of the delta, where the multilayered aquifer geometry provides natural protection to the deeper aquifers against As invasion from shallow depths (Ali *et al.*, 2006). Testing conducted by both the Bangladesh Arsenic Mitigation Water Supply Project (BAMWSP) in Haziganj upazila, Chandpur District, Bangladesh and the Bangladesh Government's Arsenic Policy Support Unit (APSU), primarily from the southern and southeastern regions of the country, revealed all sampled deep wells to meet the WHO guideline and the Bangladesh standard for As. However, prevalent issues were reported concerning elevated levels of iron, ammonia, manganese and the colour of the water, sometimes also accompanied by the presence of thermotolerant coliforms (TTCs), likely resulting from compromised wellbore integrity.

- **Other well depths:** Besides deep aquifers, many regions across the delta often possess one or more shallow or intermediate aquifers that yield As-safe groundwater. Rahman and Muyeed (2009) state that in numerous regions, As-safe water can be found at surficial and shallow depths; however, at these depths, the aquifers are typically composed of fine-grained materials, making them unsuitable for standard tube-well installation. Consequently, during installation, an artificial packing with coarser sands (known as shrouding) is constructed around the tube well screen to effectively draw water from these aquifers. Also, these wells face a notable risk of microbial contamination if they are not adequately distanced from contaminant sources, such as latrines. A recent study by Hossain *et al.* (2023) identified a predominantly As-safe aquifer at a depth of 120 m in the Matlab upazila, a hotspot for As contamination in Bangladesh. Notably, 91.4% of tube wells installed at this depth met the WHO standard for As, while 89.3% met the standard for manganese. The authors suggested that installing tube wells within this depth range elsewhere could serve as a promising strategy for As mitigation, provided similar suitable sand layers exist in the subsurface at that depth.

Another study conducted by Hossain *et al.* (2014) identified sediment colour as an indicator of groundwater As contamination in the Matlab upazila of Bangladesh. They found that reddish sands (potentially oxidised) are typically As-safe, with all tested wells producing water with As levels below the Bangladesh national drinking-water standard and 62% meeting the WHO standard. Conversely, blackish sands (potentially reduced) are typically As-contaminated, with 83% of tested wells exceeding the Bangladesh standard and 91% exceeding the WHO standard. Utilising this correlation and leveraging local drillers' expertise, a sediment colour tool was developed. This user-friendly tool, which has also been digitised into a mobile phone application named "ASMITAS" (O'Kelly, 2020), educates local drillers on targeting safe aquifers based on sediment colour characteristics. This tool holds the promise of serving as a bridge between local and scientific knowledge, empowering local drillers—who play a pivotal role in tube-well installation—to effectively locate As-free aquifers not only across the delta but also in similar deltaic or alluvial environments worldwide. One advantage of drilling shallow and intermediate wells over deep tube wells is significant cost reduction, potentially empowering communities to drill their own wells instead of relying on larger stakeholders. However, a thorough investigation is necessary to identify the presence of these aquifers and assess their productivity and sustainability over time before installing tube wells to exploit them.

(C) Dug Wells: Dug wells, constructed by digging shallow holes in the ground to access the water table, are an ancient method of groundwater extraction that has been traditionally used within the delta region (Ali, 2006). Despite the widespread adoption of tube wells in the late 1970s, approximately 1.3 million people in Bangladesh rely on dug wells (GoB, 2002), particularly in areas where geological constraints make deeper drilling challenging. Typically, these dug wells are low in As due to immobilisation under oxic environments, rendering them a cost-effective As-safe water supply option (DPHE, 2004). Consequently, many governmental, national and international non-governmental organisations have installed several thousand of these wells across the delta (APSU, 2005a). Reports indicate that only 3% of dug wells tested by APSU (2004), 2% by DASCOH (2004) and 3% by DCH (2003) exceed the Bangladesh standard of 50 μg/L for As (Ali, 2006). However, additional testing by NIPSOM (2003) and JICA and AAN (2004) reported that 15% and 43% of sampled dug wells exceeded the 50 μg/L threshold.

More commonly, dug wells are highly susceptible to bacteriological contamination, particularly during the monsoons (APSU, 2005b). Various water testing reports indicate the prevalent presence of TTCs in the well waters (Ali, 2006). Consequently, the Bangladesh NPAM and IPAM plans for As mitigation recommend "improved dug wells" (GoB, 2004), designed to offer better sanitary protection, featuring a top covering to prevent the entry of dust and debris and a small hole for aeration and sunlight to deter unpleasant odours. Despite this, issues such as foul taste, discolouration, high turbidity and elevated levels of iron, manganese and ammonia have also been observed in these wells (Ali, 2006), necessitating treatment before use. *In situ* disinfection has been proven to be ineffective in treating dug well water (Majed, 2005). Further, dug wells are not a suitable alternative in regions covered with a thick impermeable layer (e.g., clay or fine silt) or in areas with very deep water tables.

(D) Pond or River Sand Filters: Surface water is abundantly available across the GBM delta in ponds, lakes and rivers and is typically low in As due to oxic conditions. Consequently, the Bangladesh NPAM and IPAM plans recommend the utilisation of surface water over groundwater as an As-safe alternative water supply (Ali, 2006). While easily accessible for water supply, most of these sources have microbiological contamination, necessitating treatment before consumption. One of the most common community-based surface water treatments within the delta is pond and river sand filters (PSFs and RSFs), widely in use in the southwest coastal regions of Bangladesh, where shallow groundwater is often brackish (Hossain *et al.*, 2015). Organisations such as BAMWSP, DPHE-UNICEF, DPHE-Danida, DPHE-GoB-IV and AAN have installed hundreds of pond filters as an alternative water supply in As-affected areas (APSU, 2005a). These filters are sand- or gravel-packed (often size-graded) slow filtration units installed on the banks of ponds or rivers and are effective for meeting domestic needs. Raw water is manually pumped into the filter using hand pumps, and the treated water is collected through a tap. Often, pre-treatment units for turbidity reduction, such as roughing filters, are required to prevent clogging of the sand filters. The thickness of the sand filter bed must be sufficient to effectively remove all pathogen loads, and the efficiency of these filters is often limited by inadequate filter depths (Ali, 2006). Therefore, treated water often requires post-treatment with chlorination for further disinfection. These sand filters require regular cleaning, typically every two months. Also, the PSFs/RSFs have slow rates of filtration and discharge, which restrict their usability on a large scale.

Surveys indicate that a substantial portion of the installed PSFs often malfunction within a few years due to the following reasons: drying up of the corresponding pond during dry seasons; degradation of pond water quality from activities such as fish culture, washing and bathing; poor water quality due to frequent presence of bacteriological contamination, unpleasant odours and other impurities; and maintenance issues, including damage to the pipe that links the pond and PSF, breaking and theft of tube well components and PSF tape and a lack of cooperation from users (Hossain *et al.*, 2015). Therefore, ensuring the year-round sustainability of these units requires identifying perennial surface water systems as the raw water source, protecting them against external contamination and implementing regular maintenance and cleaning of the filters. These measures entail significant expenses and necessitate community involvement in the maintenance and operation of these filters, which often pose a major challenge to the functionality of these units.

(E) Rainwater Harvesting: RWH is the practice of collecting, storing and utilising rainwater and has long been observed in the coastal areas of the delta (in the presence of saline surface waters and shallow groundwater) to fulfil domestic water needs, predating the recognition of widespread As contamination in groundwater (Kumar *et al.*, 2021). A typical RWH system comprises a catchment surface for collecting rainwater, a storage tank and a pipeline or gutter system to transport water from the catchment area to the storage tank (Ali, 2006). Organisations such as DPHE-UNICEF and DPHE-Danida have installed several thousands of these units in Bangladesh (APSU, 2005b). Rainwater is typically of high potable quality, with negligible As levels. Nonetheless, these units require periodic maintenance, including flushing out the first rain at the onset of the monsoon to remove dirt, leaves and other impurities. In this method, the availability of water is ensured within the homestead, eliminating the need for people to walk long distances to fetch water.

However, in the GBM delta, rainfall distribution is highly skewed, primarily occurring during the monsoonal months from June to September with annual variations. Moreover, the delta region is densely populated and characterised by prevalent joint-family structures (Hossain *et al.*, 2015), where limited space and financial resources pose constraints on constructing large storage tanks, leading to a failure to meet the needs of these sizable households. These challenges limit the ability to rely on RWH as the primary year-round water source, although it can serve as a supplementary water source. Other commonly reported issues include high construction and maintenance costs; frequent infrastructural issues, such as dirty catchment surfaces, accumulation of debris and decaying leaves in the tank, damage to pipes or gutters, catchment surfaces and filtration nets as well as cracking in the storage tank; lack of taste due to rainwater typically being devoid of essential minerals; and the development of bacteriological contamination in stored rainwater, most commonly during the monsoon (Hossain *et al.*, 2015). Hossain *et al.* (2015) found that RWH units in their study area in Matlab, Bangladesh, were operational for only six months per year and about 40% of these units malfunctioned within just 3–4 years.

(F) Arsenic Removal Filters: The existing technologies for As removal can be grouped into four main categories as given in the following. However, many of these processes, on most occasions, fail to meet the WHO guidelines for As and have been suggested to be used in conjunction with other methods.

i. **Coprecipitation-based filters:** Also called coagulation-based filters, they typically utilise chemical coagulants that react with solutes and suspended matter to form cationic colloids, which then agglomerate to form larger flocs, settling out of the water as insoluble precipitate while concurrently inducing co-precipitation of the As as the adsorbed species onto it. These As-laden flocs are then filtered out of the water through gravity separation or sand filtration (Singh *et al.*, 2021). Rapid mixing followed by continued moderate stirring is necessary for inducing coagulation and agglomeration of the flocs. The advantages of this process lie in its use of commonly available chemicals, low installation costs and the ability to treat As across a wide range of natural water chemistries (Singh *et al.*, 2021). However, this process generates large quantities of toxic metal-rich sludge, posing challenges for disposal and raising environmental concerns. Also, the sand filter beds used for entrapment of the flocs are prone to clogging and thus require frequent cleaning. While the process has reasonably high removal efficiency, it often fails to meet the WHO As guideline of 10 μg/L (Kabir and Chowdhury, 2017). Lime softening, which is commonly used in municipal systems where lime is used as a coagulant for treating hard water, can also be used for As removal, but it faces limited adoption as the primary treatment due to low removal efficiency (Singh *et al.*, 2021). Some of the commonly deployed coprecipitation-based As filters within the delta are as follows:

- *Bucket treatment unit (BTU)*: The BTUs are among the most widely used coagulation-based As filters in the delta. These household-level filtration units consist of two buckets, each with a capacity of 20 L, arranged one over the other and connected by a pipe (Ahmed, 2001; Singh *et al.*, 2021) (Figure 8.5(a)). In the BTUs developed by the DPHE-Danida Arsenic Mitigation Pilot Project (AMPP), a dosage of 200 mg/L aluminium sulphate (alum) and 2 mg/L potassium permanganate powder is used as the coagulant and oxidiser. Raw water is introduced into the upper bucket, where coagulants are added and vigorously stirred for 30–60 s to initiate coagulation, followed by gentle stirring for the next 90 s to facilitate floc formation. Following this, the water is left undisturbed for 1–2 h for the As-rich flocs to settle. Subsequently, the water is directed to the lower bucket through a connecting pipe, positioned slightly above the bottom of the upper bucket to prevent the escape of precipitated sludge. The lower bucket has a sand filter, which ensures that any escaped flocs or particulate matter from the upper bucket are trapped, and the clean water is filtered out of the lower bucket.

  DPHE-Danida supplied several thousand BTUs to rural households in Bangladesh. While performance testing in both field and lab settings demonstrated significant removal efficiency (Sarkar *et al.*, 2000; Kohnhorst and Paul, 2000), testing under the BAMWSP, DFID and WaterAid (2001) rapid assessment programme revealed instances where these units failed to meet Bangladesh's threshold in drinking water. This poor performance was attributed to inadequate mixing and the variability of water quality, particularly groundwater pH (Ahmed, 2001). The Bangladesh University of Engineering and Technology (BUET) improved the BTU design and operations using 100 mg/L ferric chloride and 1.4 mg/L potassium permanganate as coagulants, achieving improved removal efficiency and meeting the Bangladesh guidelines.
- *Stevens Technology for Arsenic Removal (STAR) bucket treatment*: A BTU modification by the Stevens Institute of Technology involves a dual-bucket system, where the first bucket is designated

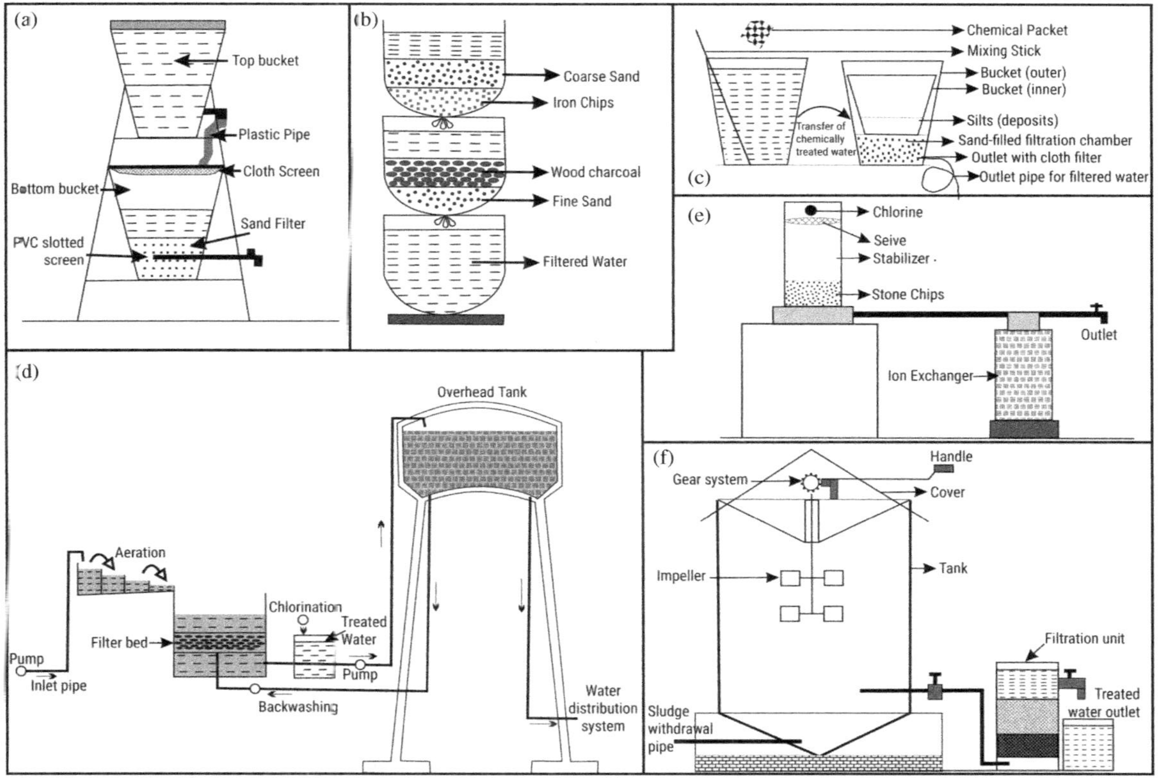

Figure 8.5. Schematic diagrams of various common As filtration systems: (a) bucket treatment unit, (b) three-pitcher filter, (c) Stevens Technology for Arsenic Removal (STAR) bucket treatment, (d) arsenic and iron removal plant (AIRP), (e) tetrahedron ion exchange resin filter and (f) fill and draw units.

Sources: Adapted from Rahman and Muyeed (2009) and Ivy *et al.* (2023).

for chemical mixing and the second bucket, featuring an additional inner bucket placed over a sand filter bed and with slits on the sidewall, facilitates floc separation through sedimentation and filtration processes (Raju, 2022; Ahmed *et al.*, 2001) (Figure 8.5(c)). A premade and packaged mix of iron sulphate and calcium hypochlorite is used as the coagulant. Results indicate efficient removal of As to meet the Bangladesh standard (50 μg/L) in approximately 80–95% of cases, yet it falls short in reducing As concentrations below the WHO guideline (10 μg/L) (Kabir and Chowdhury, 2017). Furthermore, the filter is ineffective for treating waters exceeding 500 μg/L As (Ivy, 2023).

- *Fill and draw units*: This community-level treatment, piloted by the DPHE-Danida AMPP, comprises a 600 L tank with slight tapering at the bottom for sludge accumulation and removal (Ahmed *et al.*, 2001) (Figure 8.5(f)). The tank features a manually operated stirrer with flatbed impellers, rotated for 30 s at 60 rpm after the addition of coagulant and oxidiser to the raw water to accelerate floc production. Following overnight floc sedimentation, clean water is extracted through a pipe positioned a few inches above the tank's bottom to restrict sludge. The water then passes through a sand filter before being collected via a tap.
- *Filter units attached to handpumps*: These units, widely used across the delta and attached to the mouths of tubewells, are based on coagulation using alum (and sodium hypochlorite), followed by sedimentation and filtration. While the unit has been found to have good removal efficiency, it does not meet the WHO limit (Singh *et al.*, 2021).

Apart from these chemical coagulation-based filters, electrocoagulation is another process that involves the application of electric current to oxidise metal anodes (such as Al, Fe and Mg), releasing metal ions that then react with the hydroxyl ions in the water to form metal hydroxides, which subsequently adsorb As onto them (Singh *et al.*, 2021).

- *ECAR filter*: The electrochemical As remediation (ECAR) technology, a highly efficient electrocoagulation method for As removal that meets international drinking-water standards, was developed and patented by the Gadgil Laboratory at the University of California, Berkeley, and Lawrence Berkeley National Laboratory and was piloted in West Bengal, India. ECAR employs a low-voltage electric current to initiate the dissolution of an iron electrode, leading to the formation of iron (oxyhydr)oxide (rust) (Shrivastava, 2016). The As in the raw water is oxidised and binds to the rust particles, resulting in the creation of larger particles that can be subsequently filtered out of the solution. This process is recognised for its improved As-removal efficiency, with the potential to meet the WHO guideline of 10 μg/L, while remaining cost-effective (Kabir and Chowdhury, 2017), although its applicability is often impeded by the need for continuous electricity, which can be challenging in rural delta regions (Singh *et al.*, 2021).

ii. **Adsorption-based filters:** These filtration systems remove As by adsorption onto the filter media. Commonly used media include activated alumina, activated carbon and iron (oxyhydr)oxides, such as ferric hydroxide, ferrihydrite and hydrated ferric oxide (Ahmed, 2001). This filtration method yields high removal rates, although the efficiency varies depending on groundwater pH, As content and oxidation state, the presence of competing ions and contact time (Kumar *et al.*, 2021; Singh *et al.*, 2021). Ferrate ions (Fe(VI)) are found to be highly effective oxidising agents and coagulants across a wide range of natural-water pH levels, facilitating the oxidation of arsenite to arsenate and subsequent coagulation with ferric ions (derived from ferrate ions), resulting in the removal of As from water while also being environmentally benign. Some of the common adsorption-based As filters deployed within the delta are listed in the following:

- *MAGC-Alcan filter*: Developed by MAGC Technologies and Alcan, this household-level two-bucket unit utilises the AAFS50 filter, a globally patented porous iron-enhanced activated alumina adsorptive media, in both buckets to adsorb As (Practical Action, n.d.). The unit can filter between 11,000 and 80,000 L of water (Jadhav *et al.*, 2015), after which the adsorption media become clogged, necessitating periodical regeneration and eventual replacement after 3–4 regenerations (Ahmed *et al.*, 2001). These filters are highly effective, although the removal efficiency is pH-dependent, being optimal at slightly acidic pH levels (5.5–6) and declining at pH above 7. Also, these filters demonstrate improved results following peroxidation of the raw water. A major challenge hindering the wide-scale acceptability of these filters is their high cost (Jadhav *et al.*, 2015). While the Alcan filter is sold in Bangladesh, an alternative to this called the Nirmal filter is marketed in India. The Nirmal includes a second stage of filtration through a ceramic candle. This filter is less expensive, although it requires regeneration every six months (Practical Action, n.d.).

- *Sidko filter*: Installed in India and Bangladesh by M/S Pal Trockner(P) Ltd, India, and Sidko Limited, Bangladesh, these units utilise AdsorpAs®, a granular ferric hydroxide-based As-selective absorbent developed by Technical University, Berlin, Germany, which is known for its high removal efficiency (Ahmed, 2001). Pre-treatment for iron removal is necessary to prevent clogging of the filter bed.
- *Read-F arsenic removal unit*: These filtration units developed by Shin Nihon Salt Co. Ltd., Japan, are available for both community and household levels and use cerium oxide as the adsorbent held in ethylene-vinyl alcohol copolymer (EVOH) (Ahmed, 2001). These units demonstrate high selectivity for As under diverse physicochemical conditions and a broad pH range, achieving high removal efficiencies for both arsenate and arsenite forms without necessitating pre-oxidation processes. In the GBM delta's high-iron groundwater, these units require iron removal from the raw water before filtration along with periodic regeneration of the filter.
- *BUET iron sand filter*: These small-scale units utilise iron-coated sand as the adsorbent, with an iron content of 25 mg/g of sand. The unit demonstrates good efficiency in treating As in water (both arsenate and arsenite) up to 300 μg/L and can filter through 350 bed volumes to comply with the Bangladesh standard of 50 μg/L (Ahmed, 2001).
- *Kanchan™ arsenic filter (KAF)*: Co-developed by the Massachusetts Institute of Technology (MIT), the Environment and Public Health Organization (ENPHO) of Nepal and Nepal's Rural Water Supply and Sanitation Support Program (RWSSSP), the KAF is another common household-level slow sand filter that uses iron oxide for As removal (Shrivastava, 2016). KAF consists of a plastic or concrete vessel filled with fine sand with a diffuser basin on top, containing iron nails exposed to the air–water interface. The iron nails undergo rapid rusting to form ferric (oxyhydr)oxide on their surfaces, which effectively adsorbs As from the water poured into the system. Subsequently, upon flushing, the As-laden

iron particles are trapped within the fine sand layer below, due to its relatively small pores, and the As-free water flows down and out of the filter. As the ferric (oxyhydr)oxide particles detach from the iron nails, fresh iron surfaces are exposed, sustaining the adsorption capacity for As. Field research by MIT and ENPHO, along with independent studies by Tribhuvan University, Kathmandu University, and the United States Peace Corps in Nepal, demonstrated the KAF's As removal efficiency to range from 85 to over 95%. Notably, this method can also effectively remove microbial contamination by forming a biofilm known as *schmutzdecke* on the sand particles when the sand layers remain saturated with water for an extended period.

- *Sono three-pitcher filter*: These units are made up of vertically stacked three clay pitchers (or *kolshis* in the local language) and are an improved version of the two-pitcher unit that has traditionally been used in Bangladesh to eliminate suspended particles and iron from water (Ivy *et al.*, 2023). The topmost bucket contains coarse sand and iron (zero-valent) chips, while the middle pitcher holds fine sand and wood charcoal or coke media for water filtration (Ahmed, 2001) (Figure 8.5(b)). The bottom-most bucket collects filtered water (Ivy *et al.*, 2023). These units have removal efficiencies of 90–95% but need media replacement every 3–5 years due to clogging (Kabir and Chowdhury, 2017). Also, these filters are prone to bacteriological contamination when not properly maintained and monitored (Ivy *et al.*, 2023).
- *Safi filter*: Named after its Bangladeshi developer, Prof. Sayeed Safi-ullah, this filter is made of two vertically stacked buckets, each of 15–20 L capacity. Raw water filters down from the top to the bottom bucket through a "permeable candle" that has a potential for As adsorption and is composed of ferric oxide, aluminium hydroxide, manganese dioxide, laterite soil and porous silica (Ivy *et al.*, 2023). These units can filter only 40 L/day (Hossain *et al.*, 2015). The candle needs replacement every two years (Kabir and Chowdhury, 2017).
- *Shapla filter*: This household-level filtration unit developed by International Development Enterprises (IDE), Bangladesh, uses iron-coated brick chips as the matrix for As adsorption. The iron coating on the bricks is built by treating them in a ferrous sulphate solution (Ahmed, 2005). These units can filter 4,000 L of water at

the rate of 30 L/day (Practical Action, n.d.). The use of locally available inexpensive materials has made it popular in rural areas (Ivy *et al.*, 2023). However, these units have questionable As removal capacity and are prone to bacterial contamination, necessitating regular cleaning.

- Some filters, like the *Granet Home-made filter* and *Chari filter*, use indigenous materials such as brick chips and sands as the filtering media, without the use of any chemicals (Ahmed, 2001). These units do not yield acceptable water quality.

iii. **Oxidation filters:** In this process, As-bearing water undergoes oxidation, transforming arsenite into arsenate, which precipitates due to its low solubility in water. Although atmospheric oxygen can initiate oxidation, this is a slow process that can take several days. The process can be accelerated with oxidising agents such as chlorine, permanganate, ozone, hypochlorite, chlorine dioxide, hydrogen peroxide, sand-coated manganese oxide and Fenton's reagent (Singh *et al.*, 2021). These agents cause the high iron content in delta groundwaters to oxidise and precipitate, with As co-precipitating as adsorbed species on the iron (oxyhydr)oxides. The precipitates are then removed by filtration. Some examples of oxidation-based filters from the delta are given as follows:

- *In situ chemical oxidation*: The DPHE-Danida AMPP has investigated *in situ* chemical oxidation as a method for As mitigation, which involves aerating groundwater with atmospheric oxygen or oxidising agents, then reintroducing it to aquifers (Raju, 2022). This facilitates *in situ* oxidisation of ferrous iron to ferric iron and arsenite to arsenate within the aquifers and subsequent precipitation, thereby mitigating As contamination directly within the aquifer.
- *Arsenic and iron removal plant (AIRP)*: This is another commonly used technique for As and iron removal through oxidative sequestration. The unit comprises three or four chambers connected in an array (Figure 8.5(d)). Groundwater is pumped into the first chamber through a perforated screen, which increases the oxygen levels, thus promoting rapid oxidation and precipitation of iron and As. The water is then passed through the next one or two chambers which contain varied filter media (such as sand, gravel, graded

brick or stone chips,) that trap As and iron particulates. These units are designed for both household and municipal use and have the capacity to produce 1200 L/day of filtered water (Ivy *et al.*, 2023). Despite good removal efficiency, their efficacy diminishes at As concentrations above 200 ppb. Also, a study by Rahman *et al.* (2021) reported that close to half of these units failed to meet the WHO drinking-water standard for As.

- *SORAS technology*: The Solar Oxidation and Removal of Arsenic (SORAS) technology, developed by the Swiss Federal Institute of Aqua Partnership Project in Bangladesh, is a household-level technology (Wegelin *et al.*, 2000). The water is exposed to sunlight within polyethylene terephthalate (PET) or other UV-transparent containers, which leads to UV-catalysed photochemical oxidation of arsenite and ferrous ions, succeeded by the adsorption of arsenates onto ferrous (oxyhydr)oxides and their precipitation (Choong *et al.*, 2007). The filtration procedure developed for SORAS begins with partial (90%) filling of a plastic container with pumped groundwater (Wegelin *et al.*, 2000). Approximately six drops of fresh lemon juice per litre of water are added as an easily available alternative to citrates, which accelerates the photochemical oxidation and results in faster and more complete precipitation of As and iron (oxyhydr)oxides (Wegelin *et al.*, 2000). The mixture is vigorously shaken for 30 s, and the container is then placed horizontally in sunlight for a day before being turned vertically, allowing solids to settle overnight. The next day, the clarified water is carefully decanted or filtered through a tight mesh. Field trials conducted in Bangladesh demonstrated removal efficiencies of 45–78%, averaging 67% (Shrivastava, 2016). Notably, SORAS is particularly effective in treating raw water containing As concentrations below 100–150 μg/L. This low-cost, small-scale method is suitable for household use but is influenced by water chemistry and operational conditions.

iv. **Ion-exchange filters:** Ion exchange involves the transfer of ions between a solution and the charged surface of a solid medium. Typically, for As removal, the medium comprises a weak base resin imbued with ionic functional groups such as chloride, which exchange with As in aqueous environments (Raju, 2022). The process exhibits minimal pH dependence, yet its efficacy can be adversely impacted by the presence of sulphate and nitrate, which are preferentially exchanged before As (Ahmed, 2001), or by the presence of Fe (III) ions, which form complexes with As. The efficiency of As removal is markedly improved by the pre-oxidation of As(III) to As(V), although any residual oxidants must be neutralised to prevent resin damage (Ahmed, 2001).

- *Tetrahedron ion-exchange resin filter*: In a rapid assessment programme in Bangladesh, a tetrahedron ion exchange resin filter demonstrated promising As removal capabilities, providing 624 L of water daily, sufficient for multiple households (Ivy *et al.*, 2023). The operation involves pre-oxidising arsenite with sodium hypochlorite, with residual chlorine serving to limit microbial proliferation in the system (Ahmed, 2001; Singh *et al.*, 2021) (Figure 8.5(e)). Periodic resin regeneration with sodium chloride is necessary, and the management of As-enriched waste resin remains an environmental challenge.

  Despite the exploration of various resins in Bangladesh, extensive field evaluations are needed. The development of resins specifically tailored for As could significantly improve the viability of ion exchange as a strategy for As reduction in drinking water.

v. **Membrane separation:** In this As removal process, raw water is filtered through specialised membranes that block the passage of solutes based on their molecular size and charge. High-pressure-driven membrane filtration techniques, such as nanofiltration (NF) and reverse osmosis (RO), demonstrate promising As removal efficiencies of up to 95%, adhering to WHO standards by effectively excluding As ions due to the extremely fine filter pore sizes of 0.001 and 0.0005 μm, respectively (Singh *et al.*, 2021; Ivy *et al.*, 2023; Ahmed, 2001; Raju, 2022; Kabir and Chowdhury, 2017). These methods require pressures exceeding 20 bars for RO and 10–20 bars for NF, leading to high operational costs (Kabir and Chowdhury, 2017). Innovations by Oh *et al.* (2000) and others have introduced low-pressure adaptations, employing bicycle pumps for operation (Ahmed, 2001). The efficiency of NF is influenced by factors such as pH (with an optimal range around 8.0), temperature and As concentration and necessitates the pre-oxidation of As(III) to As(V) for effective removal. The effectiveness of RO is not dependent on pH, although arsenite removal is more efficient at higher pH levels due to its charged nature (Kabir and Chowdhury, 2017).

Commercial implementations, such as the MRT-1000 household RO dispenser by B&T Science Co. Limited and broad-spectrum RO filters by Reid System Limited, have been launched in Bangladesh, achieving arsenite removal efficiencies of over 80% (Ahmed, 2001). Moreover, the AMRIT system developed by the Indian Institute of Technology Madras, which utilises nanoscale iron oxyhydroxide for As filtration, is being considered by India's Department of Drinking Water and Sanitation. Despite the high efficiency of these filtration methods, the inherent low filtration rates of these fine-pored systems, coupled with high costs, maintenance requirements and the availability of more cost-effective alternatives, limit their application in the delta region (Singh *et al.*, 2021). Although research into nanomaterials for As removal in Bangladesh is currently limited, they show significant promise for future applications at both household and community levels. In contrast, lower-pressure methods such as microfiltration and ultrafiltration have achieved limited success due to their larger pore sizes, and other techniques such as electrofiltration, electrodialysis and membrane distillation remain underexplored for practical As removal from water (Singh *et al.*, 2021).

8.6 Closing Remarks

Both Bangladesh and India have made significant progress in developing As-mitigation solutions over the past few decades. Within the GBM delta, groundwater sources are widely trusted for drinking, leading to the extensive implementation of As-removal filters and deep wells as effective solutions (Rahman and Muyeed, 2009). Recent advancements in As-removal technologies have been notable. Governments have evaluated various As-removal technologies, mandating prior testing and validation by authorities such as BCSIR in Bangladesh for any As filter to enter the market. However, studies report that most As-removal filters malfunction within a few years of installation due to inadequate maintenance. In contrast, As-safe deep tubewells have been found to remain functional, making them the most widely accepted and utilised alternative drinking water source. Despite this, As concentrations in tube wells can vary over time, necessitating periodic testing. Additionally, studies show that not all deep tubewells are As-safe (Khan *et al.*, 2019; Chakraborty *et al.*, 2022), emphasising the need to identify As-safe aquifers before allocating additional resources to deep drilling. Other safe drinking water options, such as PSFs and RWH, often experience complicated and expensive maintenance, also leading to a decline in usage and eventual malfunctioning in most cases.

Despite advancements in remediation technologies capable of reducing As levels in groundwater to below WHO guidelines, the high costs often exceed the financial capacity of many affected populations (World Bank, 2017). Thus, there is a critical need for low-cost, simple technologies, especially targeted towards the poor and disadvantaged communities. Effective sludge management and media regeneration could enhance the potential of these technologies. While several As-remediation techniques show promise, no single solution can address the diverse social and economic needs of the entire As-affected delta population. Thus, the integration of multiple region-specific technologies is believed to be necessary.

A survey by Inauen *et al.* (2013) conducted in 2009 found that despite all respondents having access to at least one safe water alternative, only 62% of them used a safe water option. The study attributed the lack of usage of safe water sources to a lack of any pursuit of or planning for better living standards. Therefore, the challenges are not solely infrastructural but also cultural. Suitable As-mitigation measures must be region-specific and consider institutional, psychological and technical aspects (Johnston *et al.*, 2014). Addressing these multifaceted challenges requires a holistic approach that integrates technological innovation, community engagement and policy support to ensure sustainable and effective As mitigation.

References

Ahmed, M. F. (2001). An overview of arsenic removal technologies in Bangladesh and India. In *Proceedings of BUET-UNU International Workshop on Technologies for Arsenic Removal from Drinking Water*, Dhaka, pp. 5–7.

Alam, M., Alam, M. M., Curray, J. R., Chowdhury, M. L. R., and Gani, M. R. (2003). An overview of the sedimentary geology of the Bengal Basin in relation to the regional tectonic framework and basin-fill history. *Sedimentary Geology*, 155(3–4), 179–208.

Ali, M. A. (2006). Arsenic contamination of groundwater in Bangladesh. *International Reviews for Environmental Strategies*, 6(2), 329–360.

Arsenic Policy Support Unit (APSU). (2004). Risk Assessment of Arsenic Mitigation Options, Phase-I (RAMMO-I). Dhaka: Arsenic Mitigation Policy Support Unit and ITB-Bangladesh Centre for Water Supply and Waste Management.

Arsenic Policy Support Unit (APSU). (2005a). Progress with Provision of Arsenic Mitigation Options to the End of December 2004. Dhaka: Arsenic Mitigation Policy Support Unit and ITB-Bangladesh Centre for Water Supply and Waste Management.

Arsenic Policy Support Unit (APSU). (2005b). Risk Assessment of Arsenic Mitigation Options, Phase-II (RAMMO-II). Dhaka: Arsenic Mitigation Policy Support Unit and ITB-Bangladesh Centre for Water Supply and Waste Management.

BAMWSP, DFID and WaterAid Bangladesh. (2001). Rapid Assessment of Household Level Arsenic Removal Technologies, Phase-I and Phase-II, Final Report, WS Atkins International Limited.

BBS/UNICEF (Bangladesh Bureau of Statistics/United Nations Children's Fund). (2011). Bangladesh National Drinking Water Quality Survey of 2009. UNICEF, Dhaka, Bangladesh, p. 192.

BGS/DPHE (British Geological Survey/Department of Public Health Engineering [Bangladesh]). (2001). Arsenic contamination of groundwater in Bangladesh. In Kinniburgh, D. G. and Smedley, P. L. (Eds.), Technical Report WC/00/19. Keyworth: British Geological Survey.

Bhattacharya, P., Chatterjee, D., and Jacks, G. (1997). Occurrence of arsenic-contaminated groundwater in alluvial aquifers from delta plains, Eastern India: options for safe drinking water supply. *International Journal of Water Resources Development*, 13(1), 79–92.

Bhavan, Y. (2007). Report of the Task Force on Formulating Action Plan for Removal of Arsenic Contamination in West Bengal. New Delhi: Government of India Planning Commission.

Bureau of Indian Standards (BIS). (2003). Indian Standard Drinking Water—Specification (First Revision) (Incorporating Amendment No. 1) UDC 628.1.033, IS 10500: 1991 Edition 2.1 (1993-01).

Burgess, W. G., Hoque, M. A., Michael, H. A., Voss, C. I., Breit, G. N., and Ahmed, K. M. (2010). Vulnerability of deep groundwater in the Bengal Aquifer System to contamination by arsenic. *Nature Geoscience*, 3(2), 83–87.

Chakraborty, M., Mukherjee, A., and Ahmed, K. M. (2015). A review of groundwater arsenic in the Bengal Basin, Bangladesh and India: From source to sink. *Current Pollution Reports*, 1, 220–247.

Chakraborty, M., Mukherjee, A., and Ahmed, K. M. (2021). Transboundary groundwater of the Ganges–Brahmaputra–Meghna River delta system. In *Global Groundwater*, pp. 129–141. Amsterdam: Elsevier.

Chakraborty, M., Mukherjee, A., Ahmed, K. M., Fryar, A. E., Bhattacharya, A., Zahid, A., ... and Chattopadhyay, S. (2022). Influence of hydrostratigraphy on the distribution of groundwater arsenic in the transboundary Ganges River delta aquifer system, India and Bangladesh. *GSA Bulletin*, 134(9–10), 2680–2692.

Choong, T. S., Chuah, T. G., Robiah, Y., Koay, F. G., and Azni, I. (2007). Arsenic toxicity, health hazards and removal techniques from water: An overview. *Desalination*, 217(1–3), 139–166.

Davis, J. A. and Kent, D. B. (1990). Surface complexation modeling in aqueous geochemistry. *Reviews in Mineralogy and Geochemistry*, 23(1), 177–260.

Development Association for Self-reliance, Communication and Health (DASCOH). (2004). Evaluation of Dug Well as a Safe Water Alternative Option. Dhaka: DASCOH.

Dhaka Community Hospital (DCH). (2003). Water Quality Test: Community Based Arsenic Mitigation Project. Dhaka: DCH.

DPHE, National Policy for Arsenic Mitigation. (2004). Available at: http://www.dphe.gov.bd/index.php?option=com_content&view=article&id=80&Itemid=85.

Edmunds, W. M., Ahmed, K. M., and Whitehead, P. G. (2015). A review of arsenic and its impacts in groundwater of the Ganges–Brahmaputra–Meghna delta, Bangladesh. *Environmental Science: Processes & Impacts*, 17(6), 1032–1046.

Goodbred Jr, S. L. and Kuehl, S. A. (2000). The significance of large sediment supply, active tectonism, and eustasy on margin sequence development: Late Quaternary stratigraphy and evolution of the Ganges–Brahmaputra delta. *Sedimentary Geology*, 133(3–4), 227–248.

Government of Bangladesh (GoB). (2002). Arsenic Mitigation in Bangladesh. Dhaka: Local Government Division, Ministry of LGRD & Cooperatives. Government of People's Republic of Bangladesh.

Government of Bangladesh (GoB). (2004). National Policy for Arsenic Mitigation and Implementation Plan for Arsenic Mitigation in Bangladesh. Dhaka: Local Government Division, Ministry of Local Government Rural Development and Cooperatives.

Harvey, C. F., Swartz, C. H., Badruzzaman, A. B. M., Keon-Blute, N., Yu, W., Ali, M. A., ... and Ahmed, M. F. (2002). Arsenic mobility and groundwater extraction in Bangladesh. *Science*, 298(5598), 1602–1606.

Harvey, C. F., Swartz, C. H., Badruzzaman, A. B. M., Keon-Blute, N., Yu, W., Ali, M. A., ... and Ahmed, M. F. (2005). Groundwater arsenic contamination on the Ganges Delta: biogeochemistry, hydrology, human perturbations, and human suffering on a large scale. *Comptes Rendus Geoscience*, 337(1–2), 285–296.

Hossain, M. I., Bukhari, A., Almujibah, H., Alam, M. M., Islam, M. N., Chowdhury, T. A., ... and Hasnat, M. A. (2023). Validation of the efficiency of arsenic mitigation strategies in southwestern region of Bangladesh and development of a cost-effective adsorbent to mitigate arsenic levels. *Journal of Environmental Management*, 348, 119381.

Hossain, M., Rahman, S. N., Bhattacharya, P., Jacks, G., Saha, R., and Rahman, M. (2015). Sustainability of arsenic mitigation interventions—An evaluation of different alternative safe drinking water options provided in Matlab, an arsenic hot spot in Bangladesh. *Frontiers in Environmental Science*, 3, 30.

Hossain, M., Bhattacharya, P., Frape, S. K., Jacks, G., Islam, M. M., Rahman, M. M., ... and Ahmed, K. M. (2014). Sediment color tool for targeting arsenic-safe aquifers for the installation of shallow drinking water tubewells. *Science of the Total Environment*, 493, 615–625.

Ivy, N., Mukherjee, T., Bhattacharya, S., Ghosh, A., and Sharma, P. (2023). Arsenic contamination in groundwater and food chain with mitigation options in Bengal delta with special reference to Bangladesh. *Environmental Geochemistry and Health*, 45(5), 1261–1287.

Jadhav, S. V., Bringas, E., Yadav, G. D., Rathod, V. K., Ortiz, I., and Marathe, K. V. (2015). Arsenic and fluoride contaminated groundwaters: A review of current technologies for contaminants removal. *Journal of Environmental Management*, 162, 306–325.

JICA & AAN. (2004). Water Quality and Follow-up Survey on Arsenic Contamination of Dugwells in Sharsha Upazila. Japan International Cooperation Agency and Asian Arsenic Network.

Johnston, R. and Sarker, M. H. (2007). Arsenic mitigation in Bangladesh: National screening data and case studies in three upazilas. *Journal of Environmental Science and Health*, 42(12), 1889–1896.

Kabir, F. and Chowdhury, S. (2017). Arsenic removal methods for drinking water in the developing countries: Technological developments and research needs. *Environmental Science and Pollution Research*, 24, 24102–24120.

Khan, M. R., Michael, H. A., Nath, B., Huhmann, B. L., Harvey, C. F., Mukherjee, A., ... and van Geen, A. (2019). High-arsenic groundwater in the southwestern Bengal Basin caused by a lithologically controlled deep flow system. *Geophysical Research Letters*, 46(22), 13062–13071.

Kohnhorst, A. and Paul, P. (2000). Testing simple arsenic removal methods. In *Water, Sanitation and Hygiene: Challenges of the Millennium, Pre-prints of the 26 WEDC Conference*, Dhaka, Bangladesh, pp. 177–181.

Kumar, R., Kumar, S., and Ghosh, A. (2021). Current scenario of groundwater arsenic contamination in West Bengal and its mitigation approach. In Kumar, N. (Ed.), *Arsenic Toxicity: Challenges and Solutions*. Singapore: Springer, pp. 193–216. https://doi.org/10.1007/978-981-33-6068-6_8.

Lowers, H. A., Breit, G. N., Foster, A. L., Whitney, J., Yount, J., Uddin, M. N., and Muneem, A. A. (2007). Arsenic incorporation into authigenic pyrite, Bengal Basin sediment, Bangladesh. *Geochimica et Cosmochimica Acta*, 71(11), 2699–2717.

Mahmud, M. I., Sultana, S., Hasan, M. A., and Ahmed, K. M. (2017). Variations in hydrostratigraphy and groundwater quality between major geomorphic units of the Western Ganges Delta plain, SW Bangladesh. *Applied Water Science*, 7, 2919–2932.

Majed, N. (2005). Contamination of dug well water and its control. M.Sc. Engineering Thesis, Department of Civil Engineering, Bangladesh University of Engineering and Technology, Dhaka, Bangladesh.

Mallick, S. and Rajagopal, N. R. (1995). Groundwater development in the arsenic-affected alluvial belt of West Bengal—Some questions. *Current Science*, 70(11), 956–958.

McArthur, J. M., Ghosal, U., Sikdar, P. K., and Ball, J. D. (2016). Arsenic in groundwater: The deep late pleistocene aquifers of the Western Bengal Basin. *Environmental Science & Technology*, 50(7), 3469–3476.

McArthur, J. M., Sikdar, P. K., Hoque, M., and Ghosal, U. (2012). Waste-water impacts on groundwater: Cl/Br ratios and implications for arsenic pollution of groundwater in the Bengal Basin and Red River Basin, Vietnam. *Science of the Total Environment*, 437, 390–402.

Michael, H. A. and Voss, C. I. (2008). Evaluation of the sustainability of deep groundwater as an arsenic-safe resource in the Bengal Basin. *Proceedings of the National Academy of Sciences*, 105(25), 8531–8536.

Mihajlov, I., Mozumder, M. R. H., Bostick, B. C., Stute, M., Mailloux, B. J., Knappett, P. S., ... and van Geen, A. (2020). Arsenic contamination of Bangladesh aquifers exacerbated by clay layers. *Nature Communications*, 11(1), 2244.

Mukherjee, A. B. and Bhattacharya, P. (2001). Arsenic in groundwater in the Bengal Delta Plain: Slow poisoning in Bangladesh. *Environmental Reviews*, 9(3), 189–220.

Mukherjee, A. and Fryar, A. E. (2008). Deeper groundwater chemistry and geochemical modeling of the arsenic affected western Bengal basin, West Bengal, India. *Applied Geochemistry*, 23(4), 863–894.

Mukherjee, A., Fryar, A. E., and Thomas, W. A. (2009). Geologic, geomorphic and hydrologic framework and evolution of the Bengal basin, India and Bangladesh. *Journal of Asian Earth Sciences*, 34(3), 227–244.

Mukherjee, A., Gupta, S., Coomar, P., Fryar, A. E., Guillot, S., Verma, S., ... and Charlet, L. (2019). Plate tectonics influence on geogenic arsenic cycling: From primary sources to global groundwater enrichment. *Science of the Total Environment*, 683, 793–807.

National Institute of Preventive and Social Medicine (NIPSOM). (2003). Modified Dug Well as a Source of Alternative Safe Water: Quality and Social Acceptability. Dhaka: Department of Occupational and Environmental Health, NIPSOM.

O'Kelly, E. (2020). A Preliminary Assessment of the Novel Application ASMITAS Using Sediments from Matlab, Bangladesh. Degree Project in Environmental Engineering, Kth Royal Institute of Technology School of Architecture and the Built Environment, Stockholm, Sweden.

Oh, J. I., Yamamoto, K., Kitawaki, H., Nakao, S., Sugawara, T., Rahaman, M. M., and Rahaman, M. H. (2000). Application of low-pressure nanofiltration coupled with a bicycle pump for the treatment of arsenic-contaminated groundwater. *Desalination*, 132, 307–314.

Polizzotto, M. L., Harvey, C. F., Li, G., Badruzzman, B., Ali, A., Newville, M., ... and Fendorf, S. (2006). Solid-phases and desorption processes of arsenic within Bangladesh sediments. *Chemical Geology*, 228(1–3), 97–111.

Polizzotto, M. L., Harvey, C. F., Sutton, S. R., and Fendorf, S. (2005). Processes conducive to the release and transport of arsenic into aquifers of Bangladesh. *Proceedings of the National Academy of Sciences*, 102(52), 18819–18823.

Practical Actions. (n.d.). Removing Arsenic from Water. Available at: https://www.pseau.org/outils/ouvrages/practical_action_arsenic_removal_2014.pdf.

Rahman, M. H. and Al-Muyeed, A. (2009). Arsenic crisis of Bangladesh and mitigation measures. *Journal of Water Supply: Research and Technology—AQUA*, 58(3), 228–245.

Rahman, M. H., Kumar, S., Bari, A., Sharma, A., and Rahman, M. (2021). Efficiency of arsenic and iron removal plants (AIRPs) for groundwater treatment in rural areas of southwest Bangladesh. *Water*, 13, 354.

Rahman, M. M., Chowdhury, U. K., Mukherjee, S. C., Mondal, B. K., Paul, K., Lodh, D., ... and Chakraborti, D. (2001). Chronic arsenic toxicity in Bangladesh and West Bengal, India—A review and commentary. *Journal of Toxicology: Clinical Toxicology*, 39(7), 683–700.

Raju, N. J. (2022). Arsenic in the geo-environment: A review of sources, geochemical processes, toxicity and removal technologies. *Environmental Research*, 203, 111782.

Ravenscroft, P., Brammer, H., and Richards, K. (2011). *Arsenic Pollution: A Global Synthesis*. Chichester, UK: John Wiley & Sons.

Ravenscroft, P., Burgess, W. G., Ahmed, K. M., Burren, M., and Perrin, J. (2005). Arsenic in groundwater of the Bengal Basin, Bangladesh: Distribution, field relations, and hydrogeological setting. *Hydrogeology Journal*, 13, 727–751.

Sarkar, A., Thogersen, C., Rahaman, A., and Choudhury. (2000). Bucket treatment unit for arsenic removal, In *Water, Sanitation and Hygiene: Challenges of the Millennium, Pre-prints of the 26 WEDC Conference*, Dhaka, Bangladesh, pp. 308–310.

Shrivastava, B. K. (2016). Policy intervention for arsenic mitigation in drinking water in rural habitations in India: Achievements and challenges. *Journal of Water and Health*, 14(5), 827–838.

Singh, P., Borthakur, A., Singh, R., Bhadouria, R., Singh, V. K., and Devi, P. (2021). A critical review on the research trends and emerging technologies for arsenic decontamination from water. *Groundwater for Sustainable Development*, 14, 100607.

Smedley, P. L. and Kinniburgh, D. G. (2002). A review of the source, behaviour and distribution of arsenic in natural waters. *Applied Geochemistry*, 17(5), 517–568.

Smith, A. H., Lingas, E. O., and Rahman, M. (2000). Contamination of drinking-water by arsenic in Bangladesh: A public health emergency. *Bulletin of the World Health Organization*, 78(9), 1093–1103.

Stollenwerk, K. G., Breit, G. N., Welch, A. H., Yount, J. C., Whitney, J. W., Foster, A. L., ... and Ahmed, N. (2007). Arsenic attenuation by oxidized aquifer sediments in Bangladesh. *Science of the Total Environment*, 379(2–3), 133–150.

Wegelin, M., Gechter, D., Hug, S., Mahmud, A., and Motaleb, A. (2000). SORAS—A simple arsenic removal process. Available at: http://phys4.harvard.edu/ wilson/mitigation/SORAS_ Paper.html.

Yamamura, S. (2003). Chapter 5: Drinking water guidelines and standards. In Hashizume, H. and Yamamura, S. (Eds.), *Arsenic, Water and Health: The State of the Art*. Geneva: World Health Organization.

https://doi.org/10.1142/9781800617841_0009

Chapter 9

Practical Considerations of Linear- and Branched-PFAS Isomers for Developing Conceptual Site Models and Site Remediation

Sheau-Yun (Dora) Chiang

Jacobs Solutions, Dallas, TX, USA

dora.chiang@jacobs.com

Abstract

Per- and polyfluoroalkyl substances (PFASs) have been used in many industrial and consumer applications since the 1950s. In the 1970s, municipalities, hydrocarbon-processing industries and the US military began using PFAS-based aqueous film-forming foam (AFFF) to efficiently extinguish hydrocarbon-based fires. Repeated use of AFFF for firefighting activities resulted in PFAS contamination in surface water, groundwater, sediments and soils. In this case study, a former fire training area (FTA) was contaminated with PFAS from the repeated use of AFFF. This FTA has an ongoing full-scale pump-and-treat system for removing contaminants, including PFOS and PFOA, from groundwater. Since the operation of the full-scale granular activated carbon (GAC) system, frequent GAC changeouts have been an issue, resulting in higher operational and maintenance costs. This provided a need to fully understand the site-specific conceptual site model (CSM) with respect to contaminant loading capabilities. *In situ* remediation efforts, including soil vapour extraction, air sparging and *in situ* anaerobic biodegradation, had been implemented at the FTA; thus, we hypothesised that *in situ* remedial activities can have a potential impact on the transformation of precursors associated with AFFF.

Due to the mixed use of electrochemical fluorination (ECF) and fluorotelomer-based (FT) AFFF at the site, precursors are abundant, and *in situ* transformation is also expected there. The methodologies of analysing precursors are still evolving; however, a CSM in relation to PFAS can be improved based on the uses of fingerprinting tools provided in this study, including an expanded PFAS analyte list, perfluoroalkyl carboxylates (PFCA)/perfluoroalkyl sulphonates (PFSA) ratios, branched (Br) / linear (L) ratios and total oxidisable precursor (TOP) assay data.

This study assessed the Br and L isomers of 15 individual PFASs: C4–C8 PFCAs, C4–C8 PFSAs and C6 and C8 perfluoroalkyl sulphonamides (FASAs) along with 4:2, 6:2 and 8:2 fluorotelomer sulphonates (FtS). In the influent of a GAC system, most of the PFCAs, PFSAs and FASAs comprised both Br and L compositions, which indicated an ECF-based AFFF. However, PFCAs can be derived from either ECF or FT AFFF; therefore, the Br/L ratio for PFCAs can vary when L-FtS are present and converted into L-PFCAs. The detection of both Br- and L-PFAS and the predominance of L-PFCA isomers over Br-PFCA isomers indicate mixed use of both ECF and FT-based AFFF. Additionally, the larger concentrations of L-PFCA isomers indicate that the AFFF products used were *in situ* transformed by biological and/or remedial activities at this demonstration site. The TOP assay was used to estimate the mass loading of PFCA precursors in the influent, in addition to that of PFOS and PFOA. Combining fingerprinting of the influent data and the site investigation data provides insights into the precursor transformation. We recommend that a CSM study should include expanded PFAS analyte characterisation, along with the evaluation of fingerprinting data, to better design treatment systems and optimise operational costs.

9.1 Introduction

9.1.1 *PFAS isomers and their distribution in the environment*

Per- and polyfluoroalkyl substances (PFAS) were discovered in the 1930s, have been produced since the late 1940s and have been commonly used in industrial and consumer applications since the 1950s (ITRC, 2021; Wang *et al.*, 2017). Since the 1960s, municipalities, hydrocarbon-processing industries and the US Military have been using PFAS-based aqueous film-forming foam (AFFF) to efficiently extinguish hydrocarbon-based fires. Repeated use of products containing PFAS has led to widespread contamination in surface water, groundwater, sediment and soils worldwide. The distribution, fate and transport of PFAS in the environment are influenced by various factors, such as geographic location, environmental conditions and PFAS composition and their characteristics at the time of production, use and release. Electrochemical fluorination (ECF) and fluorotelomerisation (FT) were the two primary manufacturing processes for AFFF; they resulted in high concentrations of PFOA and PFOS, which are among the most commonly studied PFAS. Studies on PFAS have highlighted the distinction between linear and branched isomers, particularly relevant to regions where PFAS were produced via the ECF process (Schulz *et al.*, 2020). The ECF process produced linear and branched perfluorinated isomers, and this process continued from the 1950s to 2002. Research conducted on PFOA produced by 3M Co. across 18 production lots over a span of 20 years revealed an approximate composition of 78% linear and 22% branched isomers. Similarly, for PFOS, 3M Co. manufactured it from the 1950s to 2002, with a distribution of 70% linear and 30% branched isomers (Benski *et al.*, 2010; Buck *et al.*, 2011; Londhe *et al.*, 2022). Following the 2002 phase-out, linear PFOA (L-PFOA) and other linear perfluoroalkyl carboxylic acids (L-PFCAs) have predominantly been produced by the telomerisation process, which produces isomerically pure PFCAs but can also contain chain length impurities.

In addition to PFOA and PFOS, it is important to note that various other PFAS can also have both linear and branched isomers, including PFBA, PFHxA, PFHpA, PFNA, PFUnA, PFPeS, PFHxS and FOSA (Londhe *et al.*, 2022). Novel sources of branched PFOS (Br-PFOS) isomers have been identified, expanding beyond the generally known source of impurities in the manufacture via ECF (Lehmler, 2005). In a study investigating the different isomers of perfluoroethylcyclohexanesulphonate (PFECHS), two PFOS isomers were found within the PFECHS mixture (Stefanac *et al.*, 2018). Linear PFOS (L-PFOS) was found in the mixture as well. This shows that while ECF production of PFOS is most likely the primary contributor to its branched isomers, these compounds may also arise from the manufacture of other PFAS. Further, isomer analysis of perfluorinated alkyl substances beyond PFOS itself can provide valuable insights into the origins of Br-PFOS. In a separate study examining the release of endocrine-disrupting hormones from various medical devices, branched, but not linear, PFAS were identified (Gosetti *et al.*, 2018).

As PFAS contamination has been detected globally (Stock *et al.*, 2007; Grunfeld *et al.*, 2024), the analysis of PFAS isomers has emerged as a potential approach to differentiate PFAS releases from the two major manufacturing processes: ECF and telomerisation. An isomer profiling study was conducted to quantitatively assess the contributions from ECF and telomerisation processes in regions across North America, Asia and Europe (Benskin *et al.*, 2010). The study revealed varying proportions of PFOA from different sources in Tokyo Bay, with 33% originating from ECF, 53% from telomerisation and the remaining 14% coming from isopropyl-telomer. Additionally, Br-PFOS was found to be enriched in the Mississippi River, consistently between sampling locations. Other PFAS isomer profiles were also reported. Overall, the study found that ECF manufacturing, which produces both linear and branched isomers, has contributed to the bulk of PFAS contamination around these source regions, while other sources were also significant. A study evaluating PFAS composition around a major fluorochemical manufacturing park in China documented that the branched PFBS isomer accounted for 15–27% of the total PFBS in water samples. Additionally, branched isopropyl PFOA was found (28%) in the influent of a wastewater treatment plant (WWTP), suggesting the involvement of isopropyl telomerisation (Jin *et al.*, 2015)

Figure 9.1 shows PFAS isomer data for PFOS, PFOA and PFHxS in different environmental matrices (plot generated based on data presented by Schultz *et al.*, 2020). The distribution data based on a limited number of field studies suggest enriched linear PFOS, PFOA and PFHxS in sediment and soil. This phenomenon can be attributed to the higher hydrophobic sorption of linear PFAAs compared to branched PFAAs. L-PFOS composition ratios in surface water exhibit wide variability, with averages lower than the approximately 70% starting point from manufacture. These lower ratios represent the potential enrichment of Br-PFOS in surface water, as L-PFOS are preferentially sorbed onto soil/sediment through natural processes, leaving branched isomers to be enriched in the water phase. On the other hand, Br-PFOS precursors are more susceptible to transformation into Br-PFOS in soil microcosms (Liu *et al.*, 2019). Therefore, the preferential transformation of branched isomers can also lead to an enrichment of Br-PFOS in soil and water phases. Overall, the distribution of PFAS data reveals significant variations in the concentrations of branched and linear PFAS in the environment.

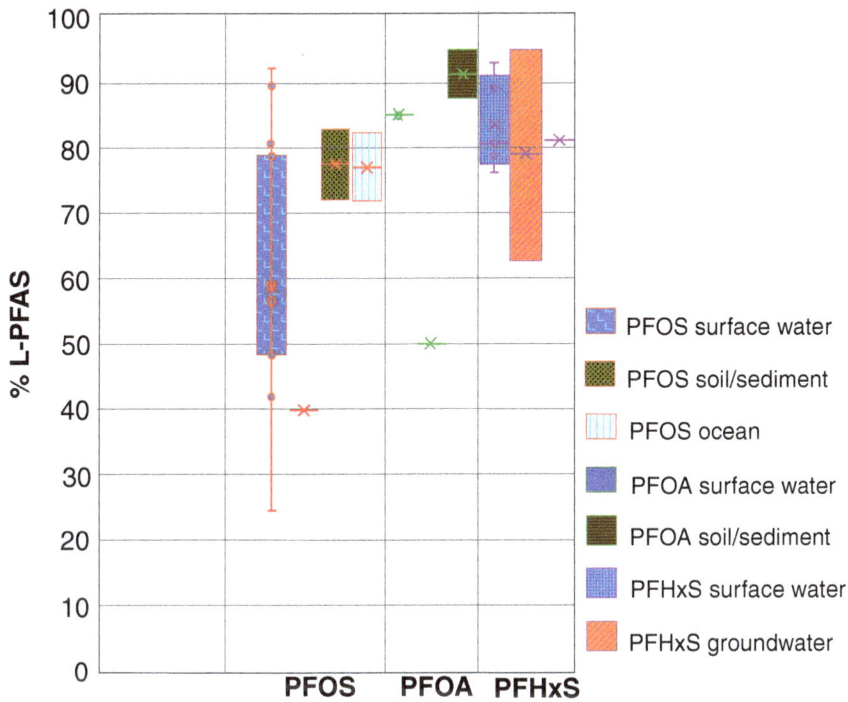

Figure 9.1. PFOS, PFOA and PFHxS isomer distribution in environmental media. *Source*: Schultz *et al.* (2020).

9.1.2 *Site investigation of PFAS isomers*

While PFAS isomers are often considered a means to differentiate PFAS manufacturing processes and possibly track PFAS sources, the environmental data reveal wide variations in the ratio of L-PFAS to Br-PFAS in water matrices. The distribution of PFAS isomers over short-to-long transport ranges can be influenced by several factors, in addition to the initial isomeric distribution from the manufacturing process. The environmental factors contributing to these variations include the following: interactions with hydrogeological and geochemical processes *in situ*, interactions with remediation activities at the site and isomer-specific precursor transformation. After PFAS isomers are released, both natural and anthropogenic processes can modify the ratio between linear and branched isomers. For instance, natural processes may lead to the preferential sorption of L-PFAS to suspended particles, sediments, biomass and natural organic matter during the scavenging of L-PFAS and enrichment of L-PFAS in abiotic solids. Consequently, L-PFAS sorption on aquifer materials would lead to higher Br-PFAS ratios (enriched Br-PFAS) in associated water matrices.

PFAS precursor isomer transformation has not been well studied. In a study by Liu *et al.* (2019), *N*-ethylperfluorooctane sulphonamide ethanol (EtFOSE), a precursor of PFOS, was incubated in soil microcosms for 105 days to examine isomer-specific transformation processes. Experimental data combined with a mathematical model suggest new biotransformation pathways leading to PFOS, including a direct pathway involving the hydrolysis of the sulphonamide bond. A similar rate of biotransformation was observed for EtFOSE, with estimated half-lives of 8.7 and 9.6 days for the branched and linear isomers, respectively, without statistical difference. Furthermore, two transformation intermediates, *N*-ethylperfluorooctanoic acid (EtFOSAA) and perfluorooctane sulphonamide (FOSA), exhibited preferential biotransformation of branched isomers. On the contrary, one intermediate *N*-ethyl perfluorooctane sulphonamide (EtFOSA) exhibited the preferred transformation of the linear isomer, with estimated half-lives of 80.8 and 11.2 days for the branched and linear isomers, respectively. As PFOS is likely to be generated through more than one pathway or one precursor, its final isomer ratio is collectively determined by several upstream reactions, each exhibiting isomer-specific transformation kinetics. Though the soil showed enrichment of branched PFOS isomers during the four-month incubation, compared to PFOS standards, the study indicates that some uncertainty arises in concluding the preferential generation of branched PFOS from its precursors due to the lack of standards for branched isomers. The complexity of isomer-specific biotransformation also suggests applying the PFOS isomer ratio for source tracking in environmental microbial systems is challenging. A large-scale field study suggests that the composition of linear isomers of PFOS, PFOA and PFHxS implies precursor degradation might be an important source of PFAS in addition to the 3M ECF (Xu *et al.*, 2024).

Current site investigations typically do not include PFAS isomers in the conceptual site model, leading to a limited understanding of their importance in environmental processes. The evaluation of PFAS isomers based on different natural and remediation-derived processes, such as sorption, biotransformation and oxidation, remains largely unknown. However, some PFAS treatment studies reveal differences in treatment performance between L-PFAS and Br-PFAS. In some cases, treated water may exhibit changes in the ratio of L-PFAS to Br-PFAS compared to the influent. For treatment facilities, the discharges of PFAS isomers that are more difficult to be adsorbed/removed can return to the environment.

9.1.3 *Significance of PFAS isomers for PFAS remediation systems*

Due to the absence of any differentiation in the analysis and regulation of PFAS isomers, the removal of linear versus branched isomers by PFAS treatment systems has not been well evaluated. While USEPA sets the maximum concentration level (MCL) for PFOA and PFOS as 4 ng/L individually (USEPA, 2024), more PFAS treatment systems are expected to be designed and built for PFAS removal. The difference between the PFAS isomers in treatment efficacy may have long-term impacts on treatment system operation and maintenance (Table 9.1).

Certain treatment technologies can change PFAS isomeric ratios, leading to preferential treatment of specific PFAS isomers, and conversely specific isomers may exhibit rapid breakthrough. Although the differences in physicochemical properties seem to be small, because of the low regulatory limits, these differences could influence the overall and long-term treatment efficiency. Adsorption techniques such as granular activated carbon (GAC), where the dominant mechanism is hydrophobic interaction with contaminants, L-PFAS tend to show better removal than Br-PFAS. This has been reported in several studies (Belkouteb *et al.*, 2020; Eschauzier *et al.*, 2012; McCleaf *et al.*, 2017) using GACs as well as materials such as Geothite (Uwayezu, 2019). McCleaf *et al.* (2017) found that the branched isomers of PFHxS, PFOS and FOSA showed an 8–13% lower removal efficiency than the linear isomers during pilot-scale GAC treatment. Another pilot-scale GAC system indicated that the Br-PFAS showed an earlier breakthrough than the linear counterpart, attributed to more robust hydrophobic interactions between L-PFAS and GAC (Rodowa *et al.*, 2020). Similarly, a PFAS breakthrough study at a full-scale drinking-water treatment facility in Sweden concluded the PFAS removal was influenced by the total operation time of GAC filters, GAC type and surface loading rate; the study also reported the different removal efficiencies for linear versus branched isomers (Belkouteb *et al.*, 2020). The removal efficiency for L-PFHxS was higher compared to that of Br-PFHxS across

Table 9.1. Conceptual significance of PFAS isomers for site investigation and remediation.

Processes/ Treatment	Mechanism	Enriched Isomers in Water	Significance	Reference
Natural processes	Sorption in soils	Branched	Preferential sorption of L isomers leads to enriched Br isomers in water bodies	Liu *et al.* (2019) and Schultz *et al.* (2020)
Natural processes	Biotransformation	Branched	Preferential degradation of Br precursors resulting in the generation of branched terminal products	Liu *et al.* (2019)
GAC	Sorption	Branched	Preferential removal of L isomers, rapid breakthrough of Br isomers	Belkouteb *et al.* (2020), Eschauzier *et al.* (2012), McCleaf *et al.* (2017) and Rodowa *et al.* (2020)
Anion exchange resin	Electrostatic interactions	None	Br and L have similar electrostatic interactions	McCleaf *et al.* (2017) and Park *et al.* (2020)
Reductive defluorination of PFOS	Reductive defluorination	Linear	Greater Br isomer degradation than L isomers	Ochoa-Herrera *et al.* (2008)
Electron beam	Electron affinity	Linear	Greater Br isomer degradation than L isomers	Trojanowicz *et al.* (2020)

Table 9.1. (*Continued*)

Processes/ Treatment	Mechanism	Enriched Isomers in Water	Significance	Reference
UV-Sulphite	Degradation	Linear	Tertiary — CF3 in Br isomers is more susceptible to degradation	Yamamoto *et al.* (2007), Gu *et al.* (2016) and Liu *et al.* (2020)
Electrochemical oxidation	PFAS sorption followed by direct electron transfer and mineralisation by hydroxyl radicals	Branched	Br-PFAS enriched in treated water because sorbed L-PFAS are more susceptible to electrochemical oxidation	Chaplin (2014), Radjenovic and Sedlak (2015), Uwayezu *et al.* (2021) and Wang *et al.* (2020)

various operation durations and volumes of treated water. For instance, L-PFHxS removal was 66% while that of Br-PFHxS was 37% for a GAC system operated for 360 days with 29,300 bed volume (BV) of treated water. This result is in agreement with previous studies which found that the removal of Br-PFAS through GAC is less effective than for L-PFAS due to different adsorption strengths (Eschauzier *et al.*, 2012; McCleaf *et al.*, 2017).

Another study involving two-stage carbon filters indicated that the Br-PFOS/L-PFOS ratio increased in treated waters as water passed through the filters (Eschauzier *et al.*, 2012). However, in adsorption processes where charged interactions (e.g., ionic) are the dominant mechanism, there would be minimal effect on the final ratio of L-PFAS to Br-PFAS after treatment (McCleaf *et al.*, 2017; Park *et al.*, 2020). In a previous study that utilised anion exchange resins (AIX) and GAC to remove PFCAs, PFSAs and FOSA, similar removals were observed for PFOS, PFHxS and FOSA isomers using AIX, but branched isomers showed lower removals using GAC (McCleaf *et al.*,

2017). In another study aimed at removing PFCAs and PFSAs utilising magnetic AIX, identical uptake was observed for both Br-PFAS and L-PFAS isomers (Park *et al.*, 2020). Given that both Br-PFAS and L-PFAS will likely exhibit similar electrostatic interactions, it is hypothesised that treatment techniques that rely on charged interactions with PFAS will not significantly alter the isomeric distribution of PFAS in treated water.

In a study conducted to assess the reductive defluorination of PFOS, Br-PFOS showed more susceptibility to reductive defluorination compared to L-PFOS (Ochoa-Herrera *et al.*, 2008). Additionally, when PFOS degradation was performed by an electron beam, Br-PFOS was found to preferentially degrade over L-PFOS, attributed to the higher electron affinity of branched isomers (Trojanowicz *et al.*, 2020). Similar results were observed for PFOS using a UV-sulphite system and using photodegradation, where the Br-PFOS degraded faster than L-PFOS due to the tertiary CF3 group being more susceptible to degradation (Yamamoto *et al.*, 2007; Gu *et al.*, 2016). In another study utilising UV-sulphite to degrade PFAS of different chain lengths and functional groups, rate constants for the degradation of branched isomers for PFOS, PFHpA, PFHpS, PFHxS and PFOA ($>2\ h^{-1}$) were an order of magnitude higher than the corresponding linear isomers ($0.018–0.440\ h^{-1}$) (Liu *et al.*, 2020). Thus, it can be concluded that unlike adsorption techniques, destructive treatment methods will likely enrich the ratio of L-PFAS to Br-PFAS due to the preferential degradation of branched isomers. However, it is worth noting that this may not be valid for certain destructive techniques such as electrochemical oxidation processes (eAOPs), which employ a two-step mechanism. The first step is the adsorption of PFAS on the surface of the electrodes, which is followed by a direct electron transfer (DET) reaction and mineralisation of the PFAS radical by hydroxyl radicals (Chaplin, 2014; Radjenovic and Sedlak, 2015; Uwayezu *et al.*, 2021; Wang *et al.*, 2020). Since the first step of this technique involves adsorption, the linear isomers would likely be preferentially adsorbed and participate more in the DET reaction, ultimately undergoing degradation. Consequently, this may result in a scenario where more L-PFAS are degraded than Br-PFAS, potentially leading to a possible enrichment of Br-PFAS in the treated water. Therefore, with more and more destructive techniques under development, it becomes essential to understand the degradation mechanisms and kinetics of PFAS isomers and monitor for concentrations of PFAS isomers in the influent and treated effluent.

9.1.4 *PFAS isomer analysis*

The structural differences among PFAS isomers dictate their distinct physical and chemical properties, such as hydrophobicity, thereby influencing their fate and transport mechanisms in the environment. In some cases, the branched isomers may predominate over linear isomers in environmental samples due to the higher hydrophobic sorption of linear isomers to soil, aquifer materials and PFAS treatment sorbents. Consequently, branched isomers may be enriched in water (e.g., groundwater, surface water and treated water effluent). Moreover, due to their higher hydrophobic sorption, linear isomers tend to be more bioaccumulative and exhibit a preference for binding to human serum albumin (Schulz *et al.*, 2020). However, there are only a few studies that measure or model the physicochemical properties of PFAS isomers (Chen *et al.*, 2015; Beesoon and Martin, 2015); therefore, the significance of separating PFAS isomers in the data reporting has not received much attention (Table 9.2).

Table 9.2. Reported and predicted physicochemical properties of PFAS isomers in the literature.

PFAS Analyte	Parameter	Reported Values	% Difference for Br-PFAS Relative to L-PFAS	Reference
L-PFOS	Sediment derived log *Koc* (cm^3/g)	~3.3	~6.2–34	Chen *et al.* (2015)
Br-PFOS		~2.2–3.1		
L-PFOA		~3.1	~4.8–11	
Br-PFOA		~2.7–2.9		
L-PFOSA		~4.4	~18	
Br-PFOSA		~3.6		
L-PFOS	*Kd* for human serum albumin	$8(\pm4) \times 10^{-8}$	~10^5 to 5×10^5	Beesoon *et al.* (2015)
3m-PFOS		$4(\pm2) \times 10^{-4}$		
4m-PFOS		$8(\pm1) \times 10^{-5}$		
5m-PFOS		$8(\pm5) \times 10^{-5}$		
L-PFOA		$1(\pm9) \times 10^{-4}$	~200–300	
3m-PFOA		$4(\pm2) \times 10^{-4}$		
4m-PFOA		$3(\pm2) \times 10^{-4}$		

Source: Londhe *et al.* (2022).

PFAS isomers solely vary in the perfluoroalkyl chain structure; they have proven to be difficult to separate and identify individually through established analytical methods. The trace levels of individual isomers present in environmental matrices have posed challenges in achieving sufficient sensitivity when developing a method to quantify individual isomers (Chu and Letcher, 2009). Although numerous attempts have been made in the past to determine individual isomers, it has been more typical to crudely separate L-PFAS from Br-PFAS and report their relative contributions to total PFAS. While there is an expectation that L-PFOS/Br-PFOS ratios can offer insights and be used for tracking PFAS sources, it is crucial to acknowledge that PFAS precursor transformation and transport mechanisms can be highly complex. For instance, different branched isomers exhibit varying chemical and physical properties, consequently leading to distinct toxicological, persistence and bioaccumulation potentials (Fang *et al.*, 2016; Houde *et al.*, 2008). It should be noted that this simplified branched versus linear isomer separation may result in the loss of significant information regarding the specific contributions of individual isomers to the overall isomer profile.

USEPA Methods 537.1, 533 and 1633 are common analytical methods used by commercial and research laboratories to analyse target PFAS in a water or environmental sample. For contaminated site investigation, USEPA Method 1633 is now frequently employed to quantify target PFAS concentrations in non-drinking water samples. US commercial laboratories also offer the modified USEPA Method 537.1 for challenging environmental matrices with significant sample matrix interferences. Modified USEPA Method 537.1 has been utilised not only for traditional PFAS analysis but also for TOP assays and additional target PFAS not covered in the above-mentioned USEPA methods. Although liquid chromatography–tandem mass spectrometry (LC-MS/MS) has the capability to differentiate branched and linear PFAS, the USEPA methods currently mandate users to integrate both peaks together and report total concentrations rather than isomer-specific concentrations. This approach can potentially introduce bias in the quantification of total PFAS levels due to differences in instrument sensitivity, collision energy and the ion transitions of each isomer (Londhe *et al.*, 2022).

Despite USEPA Method 1633 being finalised, which has not changed the requirement for reporting linear and branched isomers separately when both are present in the sample, the method still necessitates PFAS to be reported as a single result calculated from the combined responses of the linear and branched isomers when either qualitative or quantitative standards containing both types of isomers are used.

9.2 Case Study Assessing PFAS Isomers in a Groundwater Treatment System

This case study collected data from a PFAS treatment system to evaluate the treatment system's performance and highlight the necessity of investigating PFAS isomers at a PFAS contaminated site. The site description and study data are largely published (Rodowa *et al.*, 2020). The selected demonstration site (fire training area, FT-02) is a fire training area contaminated with PFAS from AFFF releases. The site has a full-scale pump and treat (P&T) system to remove PFOS and PFOA from groundwater using GAC. However, frequent GAC changeout has posed challenges during operation of the full-scale GAC system, leading to higher operation and maintenance costs. One of the contributing factors to the increased frequency of GAC changeout was the higher PFAS mass loading onto the GAC system than the mass data collected from PFOS and PFOA. The pre-design study for this full-scale GAC system only assessed PFOS and PFOA breakthroughs, neglecting other PFAS. As a result, the PFAS mass loading was underestimated, leading to more rapid PFAS breakthroughs than anticipated. To address this issue, the project intends to use a pilot-scale system to characterise the presence and abundance of all quantifiable PFAS (e.g., PFAAs, PFAS isomers and PFAS precursors) in the influent and to evaluate the breakthrough of those compounds.

9.2.1 *Site description*

Wurtsmith Air Force Base (WAFB) is located in Oscoda, Michigan, in the northern part of Michigan's Lower Peninsula, approximately one mile west of Lake Huron. The former base is bounded to the northeast by Van Etten Lake, to the east by the community of Oscoda, to the south by the Au Sable River and to the west by the Au Sable State Forest. Covering 5,221 acres, the base is approximately 170 miles north of Detroit, Michigan.

A former fire training area (FT-02) was active from 1958 to 1991. During training activities, waste fuels were ignited after they were left to float on a water lens that surrounded a simulated steel aeroplane. The fires were extinguished with AFFF, Halon 1211 (bromochloromethane) and/or multipurpose dry chemicals (potassium bicarbonate or sodium bicarbonate). During a typical exercise, 2,000 gallons of waste fuel were consumed. The volume of AFFF applied by firefighters was unknown. Prior to PFAS plume discovery, several non-PFAS remedial actions had been completed at the FT-02 site. Such remedial actions were conducted after the 1990s and continued after PFAS were identified. These completed remedial actions included soil vapour extraction (SVE) for contaminated soil, an air-sparging (AS) system to treat contaminants in groundwater, an *in situ* anaerobic biodegradation pilot study for nitrate reduction proximal to the FT-02 source, and an AS curtain installed downgradient from the FT-02 source area to address the non-PFAS plume (Figure 9.2). Like most other contaminated sites, it is highly possible that original PFAS compositions from AFFF releases have been altered due to various redox transformations and physical treatments of other non-PFAS contaminants.

PFAS were first discovered at WAFB in the groundwater collected in 1998 and 1999, up to 1,650 ft downgradient of FT-02. The full-scale groundwater pump and treat system at FT-02 began to operate on 25 April 2015. Two 20,000-pound (lb) GAC vessels were operated in a series (lead-lag) configuration using Calgon DSR-A carbon at an operating design flow rate of 241 gallons per minute (gpm) using seven recovery wells and with an empty bed contact time (EBCT) of approximately 17.5 minutes per vessel (Figure 9.3). The treated effluent from the FT-02 PTS was discharged to a series of six lateral infiltration galleries (Figure 9.2). The flow was distributed evenly through the infiltration gallery, which is situated downgradient of the air sparging curtain. Since April 2015, the FT-02 PTS has operated within the design parameters; however, the change-out of the carbon was more frequent than expected. On average, the changeout is completed every 3–4 months instead of every 7–8 months as expected. The frequent changeout has resulted in higher O&M costs for the PTS, indicating the need to investigate the causes of frequent media changeout and search for a more efficient adsorbent or different ways to improve the GAC performance.

A pilot-scale system was built to evaluate the PFAS fate and transport in the GAC P&T system and investigate opportunities to improve treatment performance. Funded by the US Air Force, the pilot system includes three parallel treatment trains. Treatment train #1 investigated the PFAS removal using GAC simulating the full-scale operational conditions; treatment train #2 investigated the PFAS removal using GAC coupled

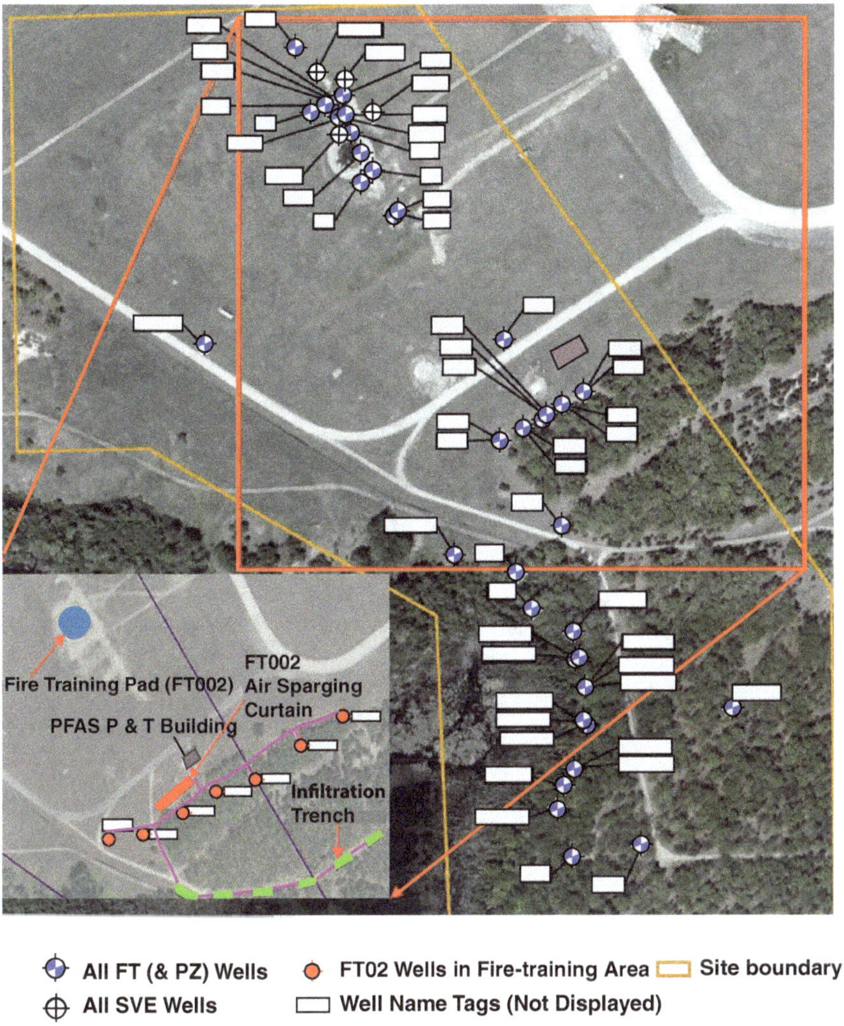

Figure 9.2. Case study site layout and FT-02 location.

with enzymatic destruction of PFAS; and treatment train #3 investigated a different carbon material (RemBind) for PFAS removal. The PFAS isomer assessment is limited to treatment train #1. Figure 9.4 presents the sampling locations associated with treatment train #1.

Treatment train #1 incorporated two sequential 55-gallon GAC canisters containing 200 lbs of DSR-A GAC (Figure 9.5). To replicate the full-scale treatment system, the hydraulic loading rate was approximately 2.1

Figure 9.3. Full-scale GAC P&T system for PFAS treatment.

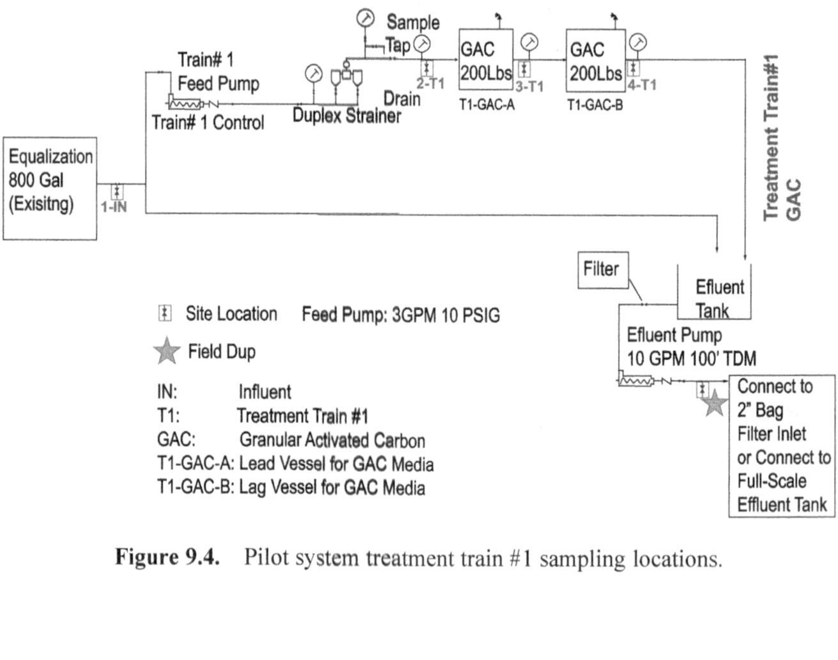

Figure 9.4. Pilot system treatment train #1 sampling locations.

Figure 9.5. Pilot system treatment trains.

gpm (Table 9.3). No changeout event was proposed for the pilot study to observe the complete breakthroughs of PFAS. UV light was not included in the original design. Based on biomass buildup potential, UV light was added to the system. A bag filter was operated to capture particulate matter that could plug the GAC vessels. The pressure drop across the bag filter was monitored daily, and the bag was changed when the pressure drop exceeded an acceptable value (maximum of 5 psi) based on the pilot system equipment design. The pressure buildup from the GAC vessels was also monitored daily, and the feed pump control valve was adjusted to maintain the desired flow rate through pilot system.

Table 9.3. Pilot treatment train #1 design parameters.

Design Parameter	Units	Value
GAC, type	–	DSR-A
Number of GAC vessels	–	2
GAC quantity per vessel	pounds	200
Volume per vessel	gallons	55
Hydraulic loading rate	gallons/minute	2.1 (approx.)
Empty bed contact time	minutes	26

9.2.2 *Analytical methods*

During the pilot system operation, weekly samples were collected from influent, from intermediate sampling points between lead and lag GAC vessels and from treated effluent after lag GAC vessel. All the samples were analysed for PFAS, TOPs and PFAS isomers. The laboratory analytical methods are described by Rodowa *et al.* (2020) and summarised as follows.

On the day of analysis, samples were brought to room temperature and inverted several times for homogenisation. A 3 mL aliquot was taken from each sample, originally stored in 30 mL HDPE containers. To this aliquot, 0.97–1.0 g of NaCl, 10 μL hydrochloric acid and 0.72 ng of all mass labelled internal standards were added to a 15 mL centrifuge tube. Micro liquid–liquid extraction was then carried out under the conditions described by Backe *et al.* (2013). Briefly, the samples were extracted in triplicate with a 10:90 mixture of 2,2,2-trifluoroethanol in ethyl acetate, resulting in a total extract volume of 1 mL. Subsequently, 500 μL of methanol was added to achieve a final volume of 1.5 mL. Finally, a 0.72 ng aliquot of the M2-PFOA isotopically labelled standard was added to the extract to calculate recovery (%) during extraction. Each sample batch typically consisted of multiple weeks of samples, with a single week of samples comprised of influent (n = 4) and lead effluent (n = 1) and lag effluent (n = 1).

LC-QToF-MS: Preliminary data (not shown) collected for the creation of an LC-MS/MS method were generated as part of the study by Barzen-Hanson *et al.* (2017) using a Shimadzu LC-30AD (Columbia, MO) coupled to a SCIEX Triple TOF 5600 qTOF-MS (Framingham, MA) with electrospray ionisation (ESI) in both positive and negative ionisation modes. The LC-QToF-MS system was fitted with two zirconium-modified diol (Zr-diol) guard columns (Agilent, Santa Clara, CA) and a Zorbax Eclipse Plus C18 (Agilent). Operational parameters were the same as those specified by Barzen-Hansen *et al.* To add a PFAS analyte to the subsequent LC-MS/MS method, a peak needed to meet the criteria given by Barzen-Hanson *et al.* and to meet the criteria for quantification on the LC-MS/MS method, as described in the following, totalling 52 individual PFAS.

LC-MS/MS: An Agilent 1100 series HPLC (Agilent, Santa Clara, CA) was connected to a Waters Triple Quadrupole Detector MS/MS system (Waters Corporation, Milford, MA). The LC-MS/MS was equipped with two zirconium-modified diol (Zr-diol) guard columns (Agilent) and a Zorbax Eclipse Plus C18 analytical column (Agilent). Operational parameters for both HPLC and the mass spectrometer can be found in the work by Backe *et al.* (2013), with a modified analyte list of PFAS (Table 9.4). Data quality tiers of confidence can be referenced from Robel *et al.* (2017) Quantitative (Qn) analytes contain both authentic analytical and isotopically labelled internal standards; semi-quantitative (Sq) analytes have reference materials but unmatched internal standards; and qualitative (Ql) analytes have neither authentic nor internal standards. Note that N-EtFOSAA and N-MeFOSAA are in Method 537 and will be acquired for this project but have not been yet observed in Wurtsmith groundwater. Additionally, longer-chained compounds than PFNA (e.g., C10–C14) in the PFCA series and longer-chained compounds than PFOS (e.g., PFNS and PFDA) in the PFSA series have not yet been detected, but the project acquired data for these longer-chain PFCAs and PFSAs regardless.

A total of five treatment train #1 samples were selected over the 11-month experiment, and the influent, lead vessel influent and lead vessel effluent (Figure 9.4) were subjected to the TOP assay process (Houtz and Sedlak, 2012). Each vial received 1.5 mL of sample added to 1.5 mL of fixed composition reaction solution (60 mM potassium persulphate and 125 mM sodium hydroxide in water). Samples were placed in a water bath at 85–90°C for six hours and then cooled in 4°C until analysis. On the day of analysis, samples were brought to room temperature, 0.72 ng of all isotopically labelled internal standards were added and the sample was then extracted and analysed as described above.

Table 9.4. List of PFAS for analyses.

Class #	Class	Chain Length Distribution Observed	In Method 537?	Structure	Notes (Type of AFFF)
1	Perfluoroalkyl carboxylates (PFCAs)	C3–C9 (n=7)	C6–C14 (9 total)	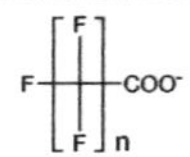	3M (branched and linear) Telomer (linear only)
2	Perfluoroalkyl sulphonates (PFSAs)	C3–C8 (n=6)	C4, C6, C8 (3 total)		3M (branched and linear)
3	Fluorotelomer sulphonates (FTSs)	4:2, 6:2 and 8:2 (n=3)	none	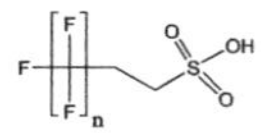	Telomer (linear only), stable biodegradation product of Ansul active ingredient under anaerobic conditions
4	N-Sulphopropyl perfluoroalkanesulphonamide (N-SP-FASA)	C4–C6 (n=3)	none	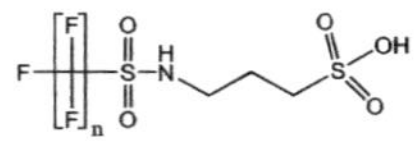	3M (branched and linear)
5	Perfluoroalkane sulphonamide (FASA)	C3–C6 (n=4)	none	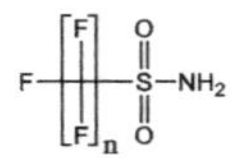	3M (branched and linear)

6	Perfluoroalkyl sulphinate (PFASi)	C4–C6, C8 (n=4)	none	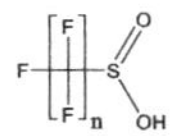	3M (branched and linear), maybe indicator of anaerobic conditions
7	n:2 tridecafluoroalkyl sulphonyl(SO_2) propanoamido-methylpropylsulphonate (n:2 FASO2PA-MePS)	6:2, 8:2 (n=2)	none	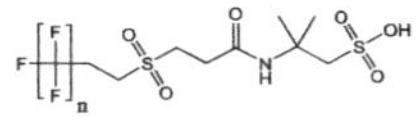	Telomer (linear only), Oxidized degradation product of Ansul active ingredient
8	n:2 Fluorotelomer sulphonamido betaine (N-CMAmP-n:2 FASA)	C4–C6 (n=3)	none	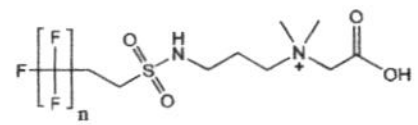	Telomer (linear only), active ingredient in National Foam AFFF
9	n:2 Fluorotelomer sulphonamido amine (N-AP-n:2 FASA)	C4–C6 (n=3)	none	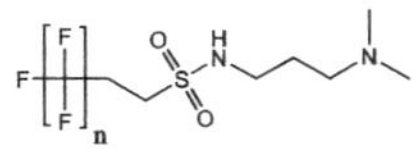	Telomer (linear only), active ingredient in National Foam

9.2.3 *Results*

9.2.3.1 *Upgradient groundwater sample characterisation*

An upgradient groundwater sample closer to the source zone than the influent sample in this study was analysed using LC-QTOF. A total of 52 individual PFAS were identified (Table 9.5). These 52 PFAS were therefore included in sample analyses using the LC-MS/MS method created for this study. Of these 52 PFAS detected, PFCA data suggested mixed use of ECF- and telomerisation-based AFFF at the site. The data also suggest the presence of various linear-only precursors likely from the telomerisation process.

9.2.3.2 *Influent characterisation*

When the GAC influent was screened using the 52-analyte LC-MS/MS method, only 19 PFAS were detected (Figure 9.6). The reduced number of PFAS detected in the groundwater influent (19 PFAS) compared to the groundwater samples collected from monitoring wells (52 PFAS) may contribute to PFAS composition and concentration variations in the groundwater extraction wells. For example, lower concentrations of PFAS precursors could be diluted and become non-detectable when

Table 9.5. PFAS groups detected in a groundwater sample collected upgradient from the FT-02 P&T system.

Groups	Chain Length	EC-Linear	ECF Branched	FT Linear
PFCAs	C6–C14	☑	☑	☑
PFSAs	C4, C6 and CR	☑	☑	
Fluorotelomer sulfonates	4:2, 6:2, 8:2			☑
N-SP-FASA	C4–C6	☑	☑	
FASA	C3–C6	☑	☑	
PFASi	C4–C6, C8	☑	☑	
n:2 FASO2PA-MePS	C6, C8			☑
N-CMAmP-n:2 FASA	C4–C6			☑
N-AP-n:2 FASA	C4–C6			☑

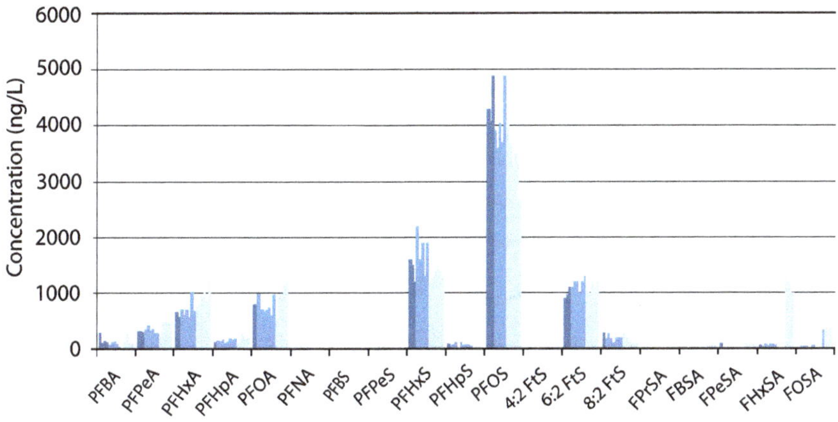

Figure 9.6. PFAS in the influent for GAC treatment.

groundwater was extracted from multiple extraction wells across the PFAS-contaminated area for treatment.

The most prevalent PFAS were PFOS, 6:2 FTS and PFHxS. The linear and branched isomers of PFCAs, PFSAs and FASAs in the influent indicate the use of ECF-based AFFF at this site (Table 9.6). The Br-PFAS/L-PFAS (Br/L) isomer ratios ranged from no detections of Br-PFAS to 0.63. A Br/L isomer ratio greater than 0.5 indicates that Br-PFAS concentrations were higher than those of L-PFAS isomers. A Br/L ratio greater than 0.43 suggests possible Br-PFAS enrichment based on an assumed ECF manufacturing process Br/L isomer ratio of 30/70 (Buck *et al.*, 2011).

The influent data show that shorter-chain Br-PFSAs (Br-PFBS and Br-PFPeS) were generally not detected due to their low abundance in groundwater samples. The Br/L PFOS isomer ratio of 0.63 (Br-PFOS 1200 ng/L and L-PFOS 1900 ng/L) indicates branched isomer enrichment, which can be attributed to the higher retardation of L-PFOS, resulting in enrichment of Br-PFOS in groundwater. The Br-PFAS enrichment was also observed for long-chain PFSA precursors FOSA and N-TAmP FHxSA, while shorter-chain FPeSA and FHxSA did not exhibit such Br-PFAS enrichment. FTS compounds were detected with 100% linear isomers.

We hypothesised that the enrichment of L-PFCAs can indicate the transformation of linear precursors into L-PFCAs. We also hypothesised that the Br/L ratios were influenced in this case by PFAS manufacturing processes (both ECF and telomerisation), PFAS chain length and the

Table 9.6. Branched/linear isomers in the influent.

	Average Influent Concentration (ng/L)		**Branched/Linear (30/70=0.43)**
PFAS	Branched	Linear	
PFBA	<LOQ	100 ± 22	<0.43
PFPeA	19 ± 2.5	350 ± 47	0.054
PFHxA	<LOQ	740 ± 70	<0.43
PFHpA	15 ± 2.0	150 ± 20	0.1
PFOA	81 ± 7.2	820 ± 73	0.1
PFNA	<LOD	18 ± 5.7	<0.43
PFBS	<LOD	17 ± 3.0	<0.43
PFPeS	<LOD	30 ± 9.0	<0.43
PFHxS	190 ± 29	1400 ± 210	0.14
PFHpS	29 ± 5.5	62 ± 11.9	0.47
PFOS	1200 ± 110	1900 ± 170	0.63
4:2 FtS	NA	13 ± 3.0	
6:2 FtS	NA	1000 ± 76	
8:2 FtS	NA	170 ± 35	
FPeSA	<LOQ	125 ± 5.8	<0.43
FHxSA	700 ± 48	2100 ± 140	0.33
FOSA	100 ± 9.4	200 ± 19	0.5
N-TAmP FHxSA	51 ± 21	97 ± 40	0.52

transformation of L-PFAS precursors, while the transformation rates of L-PFAS precursors into terminal L-PFAS are subject to natural and manipulated conditions at the site through remedial activities.

9.2.3.3 *PFAAs breakthrough from GAC pilot system*

The breakthrough curves of PFAS from treatment train #1 are illustrated in Figure 9.7. PFOA and PFOS began to break through at 7,500 BV, with PFOS showing breakthrough later compared to PFOA. By the project's termination, PFOA and PFOS concentrations (0.36 micrograms per litre [μg/L] for PFOA and 1.2 μg/L for PFOS) in the GAC effluent were

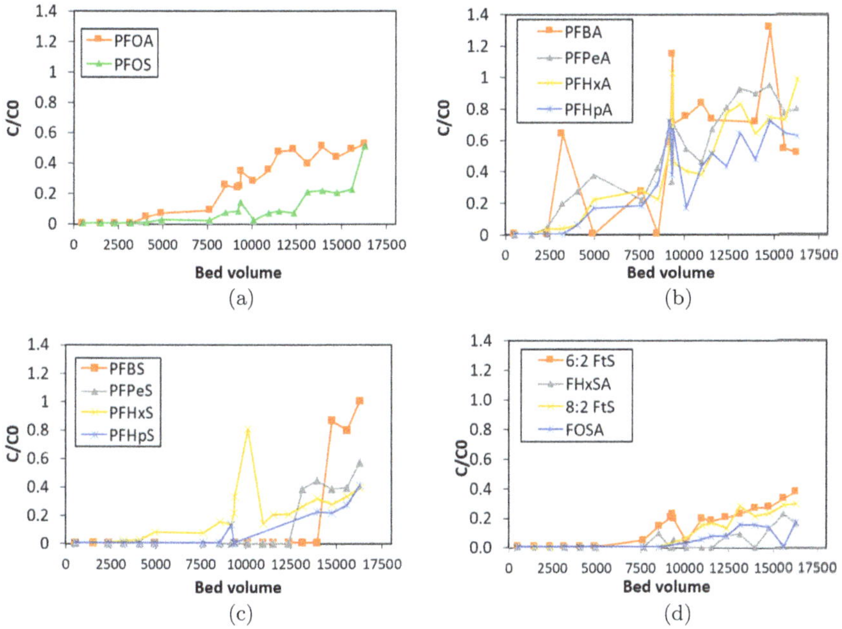

Figure 9.7. Breakthrough curves of PFOA and PFOS (a), short-chain PFCAs (b), short-chain PFSAs (c) and precursors (d).

approximately 50% of the influent concentrations. For the short-chain PFCAs, breakthrough was observed as early as 2,500 BV for PFBA, with breakthrough being delayed with increasing chain length. The breakthrough time of PFAS on GAC is primarily governed by the hydrophobic properties of each compound. Although expected to be similar, the breakthrough of PFSAs did not follow the same trend as PFCAs. The breakthrough of PFBS (C4) and PFPeS (C5) was actually slower than that of PFHxS (C6) and PFHpA (C7). Despite having a late breakthrough time, PFBA reached complete breakthrough much faster with a more pronounced concentration increase around 15,000 BV. In general, PFSAs tend to exhibit slower breakthrough than PFCAs with the same chain length. All precursors broke through after 7,500 BV, with the perfluorohexane sulphonamide (FHxSA) concentration remaining below the detection limit until 11,500 BV. The final concentrations of precursors in the effluent were still well below those in the incoming water ($C/C_0 < 0.4$). This result is important for the design of the GAC system for PFAS treatment, as most GAC systems presently operating were designed solely based on the influent concentrations of PFOA and PFOS, which obviously break through slower than other short-chain PFAS.

9.2.3.4 *PFAA precursors breakthrough from GAC pilot system*

The total oxidisable precursor analysis (TOPA) was performed on the influent and effluent samples collected at five sampling dates during the pilot study. For the influent samples, post-TOPA data showed significantly higher concentrations of PFBA (845%) and PFHxA (655%) than pre-TOPA data, which suggested the presence of large quantities of C4 and C6 precursors in the groundwater (Figure 9.8). Comparing the influent TOPA data with the breakthrough TOPA data (Figure 9.9), the concentrations of PFBA and PFHxA decreased after passing through the GAC vessel. This result indicated some removal of C4 and C6 precursors by GAC at the early stage of treatment. However, gradual breakthrough was observed for PFBA and PFHxA as the treatment continued, which was partially contributed to by the breakthrough of terminal PFBA and PFHxA, but also showed the breakthrough of C4 and C6 precursors. This is evidenced by the lower concentrations of PFBA and PFHxA in the pre-TOPA effluent samples. Therefore, TOPA data are critical for a comprehensive characterisation of the influent groundwater and the full evaluation of the technology in terms of its capability of removing PFAAs as well as their precursors.

Additionally, oxidation of TOPA did not significantly alter the Br/L isomer ratios, indicating that precursors had similar isomer ratios to those of the PFCAs, which were predominantly linear in nature. Changes in

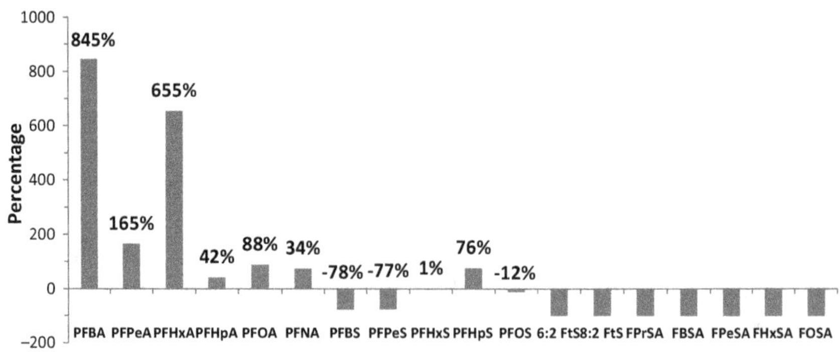

Figure 9.8. The percentage change in PFAS in post-TOP compared to pre-TOP data.

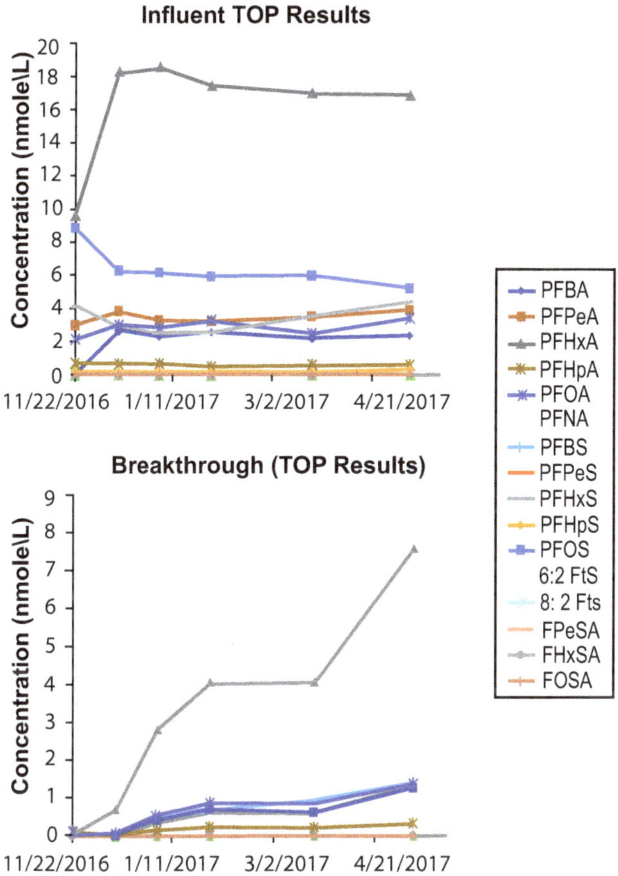

Figure 9.9. The influent and breakthrough concentrations from the treatment train #1 lead vessel (TOP data).

isomer ratios upon oxidation were rarely reported but can offer information on the Br/L characteristics of oxidisable precursors.

9.2.3.5 *Linear- and branched-PFOS sorption on GAC*

As described above, GAC is known to exhibit different adsorption capacities for branched and linear-structure PFAS compounds. In this pilot study, we conducted a preliminary evaluation of GAC performance in capturing branched and linear PFOS (Figure 9.10). Specifically, the

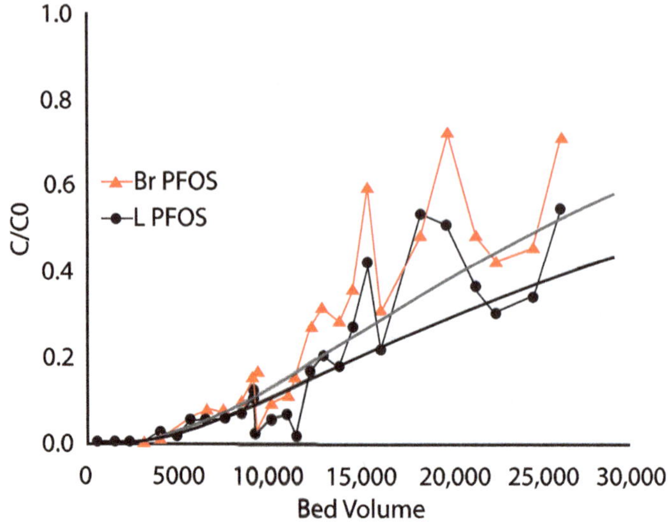

Figure 9.10. Breakthroughs of linear and branched PFOS.

Table 9.7. The branched-to-linear PFOS ratios in influent and lead vessel effluent samples.

Branched-PFOS to Linear-PFOS Ratios		
Sampling Date	**Influent**	**Effluent**
12/14/16	0.27	0.30
03/15/17	0.25	0.34
03/22/17	0.27	0.33
04/09/17	0.34	0.38
04/26/17	0.31	0.42
05/03/17	0.28	0.39

branched-to-linear PFOS ratios in six pairs of influent and effluent samples collected from the GAC lead vessel are shown in Table 9.7. In all six pairs of samples, the ratios increased after passing through the GAC vessel, suggesting enrichment of branched PFOS in the effluent. This provides evidence supporting the literature observation that GAC has a higher affinity for linear PFAS compounds than for branched.

9.2.4 *Significance of L/Br PFAS isomer characterisation to CSM development*

The case study shows an enrichment of linear PFCAs in the samples collected from the groundwater well and GAC influent. This linear PFCA enrichment can be attributed to the historical use of both ECF- and FT-based AFFF at the site. Given that the FT-based AFFF contains only linear isomers, the mixed uses of ECF and FT AFFF led to an enrichment of linear isomers. These linear isomers may consist of fluorotelomer surfactants or fluorotelomer derivatives that have undergone biotransformation to form linear PFCAs. While FTS and other fluorotelomer precursors can undergo biotransformation under different redox conditions, aerobic biotransformation is considered more efficient than other redox conditions. The presence of an air-sparging curtain immediately upgradient from groundwater extraction wells suggests that the linear enrichment may also be due to the biotransformation of linear FT precursors.

Branched isomers typically exhibit lower partition coefficients between soil and water (*K*d) or between soil organic matter and water (*K*oc). Br/L isomer ratios can exceed 30/70 downstream from the source when more linear isomers are retained in the source area. Groundwater influent data indicated enrichment of branched PFSA and FASA isomers in the extracted groundwater located 150 m downgradient from the source area. This observation confirms that linear PFSA and FASA have a higher sorption affinity in the aquifer materials, leading to enrichment of branched isomers. Additionally, the greater sorption of linear PFOS is demonstrated in Figure 9.10, where branched PFOS GAC breakthrough occurred faster than linear PFOS, resulting in an enrichment of branched PFOS in the effluent.

The study also confirmed the presence and abundance of PFAS precursors in the groundwater samples collected in an upgradient monitoring well and from the groundwater influent. Precursors were verified using target PFAS analysis and TOPA. The detections of precursors in the extracted groundwater indicate that precursors have migrated from the source area towards groundwater extraction wells. Through the TOPA oxidation process, precursors in the influent samples were oxidised into a range of PFCAs, including PFBA (10%), PFPeA (11%), PFHxA (73%), PFHpA (0.8%) and PFOA (5%). The predominant production of PFHxA (73%) generated through TOPA is consistent with the oxidation of the

C6-based precursors present in the influent, including 6:2 FTS, FHxSA and N-TamP-FHxSA. Another important finding is that the TOPA oxidation process did not alter the Br/L isomer ratios, indicating that precursors had similar isomer ratios to those of the PFCAs and were predominantly linear in nature. Changes in isomer ratios upon TOPA oxidation are rarely reported but can provide information on the Br/L compositions after the oxidation process. We suspect that *in situ* remediation techniques, such as air sparging, aerobic bioremediation or chemical oxidation, have the potential to accelerate the conversion of PFCA precursors into terminal PFCAs.

The case study focused on better understanding of "all" PFAS loading into a GAC treatment system and how L- and Br-PFAS will break through GAC. When considering "all" PFAS including PFAS isomers, we suggest that more studies may be needed to verify whether the enrichment of Br-PFAS over time would potentially diminish the effectiveness of PFAS removal in GAC treatment systems.

9.3 Summary

This study assessed both Br- and L-isomers of 15 individual PFAS: C4–C8 PFCAs, C4–C8 PFSAs and C6 and C8 FASAs, along with 4:2, 6:2 and 8:2 FtS. In the influent of the GAC system, most PFCAs, PFSAs and FASAs comprised both Br and L compositions, indicating the presence of ECF-based AFFF. However, it is important to note that PFCAs can be derived from either ECF or FT AFFF, so the Br/L ratio for PFCAs may vary when L-FtS are present and converted into L-PFCAs.

The detections of both Br- and L-PFAS and the predominance of L-PFCA isomers over Br-PFCA isomers indicate mixed use of both ECF- and FT-based AFFF. Additionally, the higher concentrations of L-PFCA isomers suggest that the AFFF products used underwent *in situ* transformation by biological processes and/or remedial activities at this demonstration site. TOPA was utilised to estimate the PFAS mass loading as opposed to using only PFOS and PFOA to assess the frequency of GAC changeout. By combining fingerprinting of the influent data with site investigation data, insights into precursor transformation can be obtained. We recommend that a CSM study include expanded PFAS analyte characterisation, along with an evaluation of fingerprinting data, to better design treatment systems and optimise operational costs.

This research study highlights the need to consider the following when studying PFAS that exhibit isomers:

(i) Standardised analytical approaches are needed to differentiate and quantify the PFAS isomers in the source areas and in the influent and effluent of PFAS treatment systems. The monitoring of isomers allows the system design and operation to cope with Br-PFAS, which are more difficult to remove using GAC.
(ii) When stringent federal and state regulatory limits are at low (single-digit) parts per trillion, higher costs of PFAS treatment system operation and maintenance than originally anticipated are possible if laboratory data are not processed accurately to troubleshoot the cause of more frequent changeouts. Exclusion of analyses of PFAS precursors and branched isomers limits the opportunity for a thorough assessment.
(iii) When PFAS treatment system optimisation is needed, the selection of optimisation approaches should consider and evaluate the isomeric distribution of PFAS in the water being treated.
(iv) More research is needed to assess degradation rates, reaction mechanisms and competitive sorption of specific isomers in environmentally realistic mixtures, which would be valuable in evaluating PFAS treatment options.
(v) For contaminated sites being remediated for non-PFAS contaminants, such as hydrocarbons and chlorinated solvents, current remedial actions may include mechanisms that alter the PFAS composition (e.g., precursor transformation and PFAS isomer distribution). Higher-resolution PFAS investigations and the establishment of a robust monitoring programme to understand the changes in PFAS composition over time will help establish a better CSM and ultimately optimise the selection and operation of PFAS treatment technologies for site cleanup.

Acknowledgements

The case study for pilot testing of PFAS removal using GAC was funded by the U.S. Air Force Broad Agency Announcement programme. The University of Georgia was the prime contractor for this work, and Dr. Jack Huang at the University of Georgia was the principal investigator of the project. AECOM was the subcontractor responsible for managing field activities and collaboration with university partners (including Dr. Huang at the University of Georgia and Dr. Jennifer Field at Oregon State University) on PFAS analysis. The author would also like to acknowledge the project team members from AECOM and Bay West for conducting the on-site pilot study. Additionally, Project Engineer Justin Gail at WSP provided additional site information for this chapter.

References

Alix E. Robel, Kristin Marshall, Margaret Dickinson, David Lunderberg, Craig Butt, Graham Peaslee, Heather M. Stapleton, and Jennifer A. Field, 2017, Closing the Mass Balance on Fluorine on Papers and Textiles, *Environ. Sci. Technol.* 51, 16, 9022–9032.

Beesoon, S. and Martin, J. W. (2015). Isomer-specific binding affinity of perfluorooctanesulfonate (PFOS) and perfluorooctanoate (PFOA) to serum proteins. *Environmental Science and Technology*, 49(9). Available at: https://doi.org/10.1021/es505399w.

Belkouteb, N., Franke, V., McCleaf, P., Köhler, S., and Ahrens, L. (2020). Removal of per- and polyfluoroalkyl substances (PFASs) in a full-scale drinking water treatment plant: Long-term performance of granular activated carbon (GAC) and influence of flow-rate. *Water Research*, 182. Available at: https://doi.org/10.1016/j.watres.2020.115913.

Buck, R. C., Franklin, J., Berger, U., Conder, J. M., Cousins, I. T., Voogt, P. de, Jensen, A. A., Kannan, K., Mabury, S. A., and van Leeuwen, S. P. J. (2011). Perfluoroalkyl and polyfluoroalkyl substances in the environment: Terminology, classification, and origins. *Integrated Environmental Assessment and Management*, 7(4). Available at: https://doi.org/10.1002/ieam.258.

Chaplin, B. P. (2014). Critical review of electrochemical advanced oxidation processes for water treatment applications. *Environmental Sciences: Processes and Impacts*, 16(6). Available at: https://doi.org/10.1039/c3em00679d.

Chen, X., Zhu, L., Pan, X., Fang, S., Zhang, Y., and Yang, L. (2015). Isomeric specific partitioning behaviors of perfluoroalkyl substances in water dissolved phase, suspended particulate matters and sediments in Liao River Basin and Taihu Lake, China. *Water Research*, 80. Available at: https://doi.org/10.1016/j.watres.2015.04.032.

Chu, S. and Letcher, R. J. (2009). Linear and branched perfluorooctane sulfonate isomers in technical product and environmental samples by in-port derivatization-gas chromatography-mass spectrometry. *Analytical Chemistry*, 81(11). Available at: https://doi.org/10.1021/ac8027273.

Fang, C., Chen, Z., Megharaj, M., and Naidu, R. (2016). Potentiometric detection of AFFFs based on MIP. *Environmental Technology and Innovation*, 5. Available at: https://doi.org/10.1016/j.eti.2015.12.003.

Gosetti, F., Bolfi, B., Robotti, E., Manfredi, M., Binotti, M., Ferrero, F., Bona, G., and Marengo, E. (2018). Study of endocrine disrupting compound release from different medical devices through an on-line SPE UHPLC-MS/MS method. *Analytica Chimica Acta*, 1042. Available at: https://doi.org/10.1016/j.aca.2018.07.028.

Greaves, A. K. and Letcher, R. J. (2013). Linear and branched perfluorooctane sulfonate (PFOS) isomer patterns differ among several tissues and blood of polar bears. *Chemosphere*, 93(3). Available at: https://doi.org/10.1016/j.chemosphere.2013.07.013.

Gu, Y., Dong, W., Luo, C., and Liu, T. (2016). Efficient reductive decomposition of perfluorooctanesulfonate in a high photon flux UV/sulfite system. *Environmental Science and Technology*, 50(19). Available at: https://doi.org/10.1021/acs.est.6b03261.

Houde, M., Czub, G., Small, J. M., Backus, S., Wang, X., Alaee, M., and Muir, D. C. G. (2008). Fractionation and bioaccumulation of perfluorooctane sulfonate (PFOS) isomers in a lake Ontario food web. *Environmental Science and Technology*, 42(24). Available at: https://doi.org/10.1021/es800906r.

Houtz, E. F. and Sedlak, D. L. (2012). Oxidative conversion as a means of detecting precursors to perfluoroalkyl acids in urban runoff. *Environmental Science and Technology*, 46(17). Available at: https://doi.org/10.1021/es302274g.

Jin, H., Zhang, Y., Zhu, L., and Martin, J. W. (2015). Isomer profiles of perfluoroalkyl substances in water and soil surrounding a Chinese fluorochemical manufacturing park. *Environmental Science and Technology*, 49(8). Available at: https://doi.org/10.1021/acs.est.5b00212.

Lehmler, H. J. (2005). Synthesis of environmentally relevant fluorinated surfactants — A review. *Chemosphere*, 58(11). Available at: https://doi.org/10.1016/j.chemosphere.2004.11.078.

Liu, J., Zhong, G., Li, W., and Mejia Avendaño, S. (2019). Isomer-specific biotransformation of perfluoroalkyl sulfonamide compounds in aerobic soil. *Science of the Total Environment*, 651. Available at: https://doi.org/10.1016/j.scitotenv.2018.09.214.

Londhe, K., Lee, C. S., McDonough, C. A., and Venkatesan, A. K. (2022). The need for testing isomer profiles of perfluoroalkyl substances to evaluate treatment processes. *Environmental Science and Technology*, 56(22). Available at: https://doi.org/10.1021/acs.est.2c05518.

McCleaf, P., Englund, S., Östlund, A., Lindegren, K., Wiberg, K., and Ahrens, L. (2017). Removal efficiency of multiple poly- and perfluoroalkyl substances (PFASs) in drinking water using granular activated carbon (GAC) and anion exchange (AE) column tests. *Water Research*, 120. Available at: https://doi.org/10.1016/j.watres.2017.04.057.

Mueller, R. and Yingling, V. (2017). History and use of per- and polyfluoroalkyl substances (PFAS). Interstate Technology Regulatory Council Sheets, KEMI 2015.

Ochoa-Herrera, V., Sierra-Alvarez, R., Somogyi, A., Jacobsen, N. E., Wysocki, V. H., and Field, J. A. (2008). Reductive defluorination of perfluorooctane sulfonate. *Environmental Science and Technology*, 42(9). Available at: https://doi.org/10.1021/es702842q.

Park, M., Daniels, K. D., Wu, S., Ziska, A. D., and Snyder, S. A. (2020). Magnetic ion-exchange (MIEX) resin for perfluorinated alkylsubstance (PFAS) removal in groundwater: Roles of atomic charges for adsorption. *Water Research*, 181. Available at: https://doi.org/10.1016/j.watres.2020.115897.

Radjenovic, J. and Sedlak, D. L. (2015). Challenges and opportunities for electrochemical processes as next-generation technologies for the treatment of contaminated water. *Environmental Science and Technology*, 49(19). Available at: https://doi.org/10.1021/acs.est.5b02414.

Rodowa, A. E., Knappe, D. R. U., Chiang, S. Y. D., Pohlmann, D., Varley, C., Bodour, A., and Field, J. A. (2020). Pilot scale removal of per- and polyfluoroalkyl substances and precursors from AFFF-impacted groundwater by granular activated carbon. *Environmental Science: Water Research and Technology*, 6(4). Available at: https://doi.org/10.1039/c9ew00936a.

Schulz, K., Silva, M. R., and Klaper, R. (2020). Distribution and effects of branched versus linear isomers of PFOA, PFOS, and PFHxS: A review of recent literature. *Science of the Total Environment*, 733. Available at: https://doi.org/10.1016/j.scitotenv.2020.139186.

Stefanac, T., McCrindle, R., McAlees, A. J., Riddell, N., Brazeau, A. L., and Chittim, B. C. (2018). Characterization of nine isomers in commercial samples of perfluoroethylcyclohexanesulfonate and of some minor components including PFOS isomers. *Environmental Science and Technology*, 52(17). Available at: https://doi.org/10.1021/acs.est.8b02369.

Stock, N. L., Furdui, V. I., Muir, D. C. G., and Mabury, S. A. (2007). Perfluoroalkyl contaminants in the Canadian arctic: Evidence of atmospheric transport and local contamination. *Environmental Science and Technology*, 41(10). Available at: https://doi.org/10.1021/es062709x.

Tenorio, R., Liu, J., Xiao, X., Maizel, A., Higgins, C. P., Schaefer, C. E., and Strathmann, T. J. (2020). Destruction of per-and polyfluoroalkyl substances (PFASs) in aqueous film-forming foam (AFFF) with UV-sulfite photoreductive treatment. *Environmental Science and Technology*, 54(11). Available at: https://doi.org/10.1021/acs.est.0c00961.

Trojanowicz, M., Bartosiewicz, I., Bojanowska-Czajka, A., Szreder, T., Bobrowski, K., Nałęcz-Jawecki, G., Męczyńska-Wielgosz, S., and Nichipor, H. (2020). Application of ionizing radiation in decomposition of perfluorooctane sulfonate (PFOS) in aqueous solutions. *Chemical Engineering Journal*, 379. Available at: https://doi.org/10.1016/j.cej.2019.122303.

Uwayezu, J. N., Carabante, I., Lejon, T., van Hees, P., Karlsson, P., Hollman, P., and Kumpiene, J. (2021). Electrochemical degradation of per- and polyfluoroalkyl substances using boron-doped diamond electrodes. *Journal of Environmental Management*, 290. Available at: https://doi.org/10.1016/j.jenvman.2021.112573.

Uwayezu, J. N., Yeung, L. W. Y., and Bäckström, M. (2019). Sorption of PFOS isomers on goethite as a function of pH, dissolved organic matter (humic and fulvic acid) and sulfate. *Chemosphere*, 233. Available at: https://doi.org/10.1016/j.chemosphere.2019.05.252.

Wang, Z., Dewitt, J. C., Higgins, C. P., and Cousins, I. T. (2017). A never-ending story of per- and polyfluoroalkyl substances (PFASs)? *Environmental Science and Technology*, 51(5). Available at: https://doi.org/10.1021/acs.est.6b04806.

Wang, Y., Pierce, R. "david," Shi, H., Li, C., and Huang, Q. (2020). Electrochemical degradation of perfluoroalkyl acids by titanium suboxide anodes. *Environmental Science: Water Research and Technology*, 6(1). Available at: https://doi.org/10.1039/c9ew00759h.

Xu, C., Xu, C., Zhou, Q., Shen, C., Peng, L., Liu, S., Yin, S., and Li, F. (2024). Spatial distribution, isomer signature and air-soil exchange of legacy and emerging poly- and perfluoroalkyl substances. *Environmental Pollution*, 343. Available at: https://doi.org/10.1016/j.envpol.2023.123222.

Yamamoto, T., Noma, Y., Sakai, S. I., and Shibata, Y. (2007). Photodegradation of perfluorooctane sulfonate by UV irradiation in water and alkaline 2-propanol. *Environmental Science and Technology*, 41(16). Available at: https://doi.org/10.1021/es0706504.

https://doi.org/10.1142/9781800617841_0010

Chapter 10

Preliminary Health Risk Assessment and Contaminated Sites Prioritisation

Marco Falconi[*,§], **Lana Kukobat**[†], **and Dragana Vidojević**[‡]

[*]*Institute for Environmental Protection and Research (ISPRA), Via Vitaliano Brancati, 48 - 00144 Rome, Italy*

[†]*University of Belgrade – Institute for Multidisciplinary Research, Kneza Višeslava 1, 11030, Belgrade, Serbia*

[‡]*Environmental Protection Agency, Ministry of Environmental Protection, Republic of Serbia, Ruže Jovanovića 27, Belgrade, Serbia*

[§]*marco.falconi@isprambiente.it*

Abstract

Contaminated sites present risks to public health, the environment and the economy, resulting in a complex set of challenges that require careful attention. The perception of risks from contaminated sites among decision- and policymakers varies across different contexts. Limitations regarding technical and financial capabilities at the state or regional level significantly impact the effectiveness of addressing this global issue, which is evident in the increasing number of these sites. Assessing risks to human health in areas with multiple industrial activities that impact soil and groundwater presents significant challenges from both economic and regulatory perspectives. The lack of financial resources allocated to these issues limits the ability to conduct thorough investigations, crucial for understanding the specific environmental and health impacts of each site. Consequently, this limitation makes the development of effective remedial strategies more difficult, leading to ongoing public health risks and environmental degradation. To address this, preliminary health risk assessment (PHRA) methodologies based on site characterisation data emerge as a practical approach. Adequate site characterisation is attainable, as many countries have adopted contaminated site management strategies and collect data, which can potentially support the application of PHRA methodologies, thus facilitating the incorporation of risk-based criteria. PHRAs can provide a robust framework for contaminated site risk management that further integrates social, political, technical and economic considerations. This serves as a foundation for decision-making on both detailed risk assessments and comparative risk analyses, guiding effective site management actions at the country or regional level. This chapter describes several examples of PHRAs, including a case study illustrating how this type of risk assessment was implemented in a developing country that lacks a developed legal framework and its own risk assessment methodology.

10.1 Introduction

Industrialisation poses significant risks to environmental and human health, particularly in developing nations where economic growth often prioritises progress over environmental protections. Despite the clear consequences observed in developed countries from past neglect of environmental practices, developing nations still struggle to effectively apply the "lessons learned." The consequence is a rise in the number of contaminated locations around the world, a widely known trend that highlights past inadequacies in managing industrial pollution (Pérez and Rodríguez, 2018). Addressing the issues of contaminated sites demands considerable investment in environmental forensics to determine the full scope of pollution and to decide on cleanup priorities (Troldborg *et al.*, 2008).

According to the World Health Organization (WHO), contaminated sites are those that either currently or historically have been host to human activities resulting in environmental pollution. These activities have resulted in specific environmental and health hazards, including soil contamination with hazardous chemicals, leaching of pollutants into surface and groundwater, degradation of air quality and the introduction of toxins into the food chain, all of which carry elevated public health risks, such as water-borne diseases and food contamination (WHO, 2014). Risks to human health, such as respiratory conditions, cancers and neurological disorders, necessitate thorough risk assessments and management to ensure public safety (USEPA, 2012). This complexity highlights an urgent need for holistic approaches to mitigate pollution, embracing sustainable practices and technological innovations to ensure a harmonious coexistence between a healthy environment and the well-being of bystanders, individuals and communities (FAO and UNEP, 2021).

Risk assessment, defined in 1983, is a systematic process of analysing the health consequences of exposure to hazardous substances, relying on scientific evidence (National Research Council, 1983). Over the past four decades, the health risk assessment process has been recognised as critical for evaluating exposure implications. It includes four components: hazard identification (identifying harmful substances), dose-response assessment (linking exposure levels to health outcomes), exposure assessment (quantifying exposure) and risk characterisation (synthesising data to estimate risks) (National Research Council, 1983; USEPA, 1992, 1998; Calabrese, 2004). Risk assessments require significant time, often years, and are primarily conducted through practical investigations utilising scientific knowledge, with most assessments being performed by consultancies. They must also take into account the legal, economic and policy factors specific to each country. Given the sheer number of contaminated sites worldwide, conducting in-depth risk assessments for each is usually impractical in terms of cost and time (von Stackelberg and Williams, 2021).

In response to the challenges mentioned above, the concept of preliminary health risk assessments (PHRAs) focuses on initial evaluations that help identify potential hazards and guide subsequent decision-making processes, utilising data gathered during the early stages of site examination. PHRAs provide initial risk identification by assessing the likelihood and severity of exposure and screening for further action to determine if detailed investigations are needed. PHRAs identify vulnerable populations who may be at greater risk. They involve data collection and analysis to assess contaminant levels and exposure pathways while identifying data gaps that may necessitate further research. Additionally, PHRAs establish exposure scenarios to estimate health impacts and aid in prioritising risks for resource allocation. The results also serve to communicate findings to stakeholders, inform public health interventions and support regulatory compliance. PHRAs also provide a framework for more comprehensive risk assessments if required and help integrate risk-based criteria into contaminated site management.

In recent years, many countries have adopted strategies for managing contaminated sites to facilitate the collection of necessary data that could be used for preliminary risk-based assessments, thereby facilitating risk-based decision-making (Pérez and Rodríguez, 2018). Risk-based management merges economic, regulatory, social and technical considerations, aligning them with management actions suggested by risk assessment findings (National Academies of Sciences, Engineering, and Medicine, 2018). By merging these varied considerations, contaminated site management becomes more responsive to the complex interactions between environmental and public health challenges, thereby facilitating the development of evidence-based strategies that protect both community health and ecosystem functionality.

Prioritisation is brought into practice by identifying locations with the most pressing health and environmental risks. The prioritisation of contaminated sites using PHRA models enables targeting areas where the contamination of environmental media endangers humans, often extending beyond the immediate surroundings. It supports decision-making and enhances the effective allocation of financial resources for addressing human health and environmental issues.

The preliminary risk assessment models' strength lies in their adaptability, allowing for the possibility to iterate the assessment as new data become available and facilitating the evaluation and further ranking of varied contaminated sites. These types of risk assessment models are indispensable for policymakers to systematically evaluate and prioritise contaminated sites, addressing complex environmental and health issues effectively (Altieri *et al.*, 2004).

10.2 Contaminated Sites Management

Contaminated sites, as defined by the World Health Organization (WHO), encompass areas where human activities have led to environmental pollution, posing risks to human health, ecosystems and the environment. These sites result from diverse contamination sources, including industrial activities, waste disposal and accidental releases. The challenge in managing contaminated sites lies in the complexity of environmental data due to a diverse range of pollutants, exposure pathways and complex ecosystem interactions (WHO, 2012). The global prevalence of contaminated sites is attributed to historical industrial operations, inadequate waste management and accidental pollution incidents (Oliveira *et al.*, 2019). The complexity of contamination is heightened not only by the proximity of contaminated sites to urban centres, fluvial systems and ecologically sensitive areas but also by the intricate behaviour of pollutants in soil, their heterogeneous distribution and their complex interactions with groundwater and surrounding media, necessitating systematic surveillance.

From an economic and social perspective, contaminated sites can be costly to remediate and may result in unequal pollutant exposure among communities. Compliance with environmental regulations is crucial for successful assessment, management and cleanup.

The structured management of contaminated sites follows regulatory frameworks and involves stakeholders such as property owners, regulatory agencies and the public. The process includes a feasibility study for selecting practical remediation strategies, community involvement, ongoing monitoring, validation and integration of emerging technologies (DOE, 2002; USEPA, 2012). Clear risk communication is imperative for transparency, trust, strategic resource allocation and exploration of legal frameworks (Freeman and Lewis, 2008). In the European context, the EEA's "Progress in management of contaminated sites" indicator LSI003 serves as a screening tool, offering insights into countries' initiatives in identifying, confirming, remediating and mitigating contamination risks (Pérez and Eugenio, 2018). Coordinated definitions and management strategies, supported by the JRC, aim to streamline the identification and management of contaminated sites globally, focusing on reducing risks to acceptable levels (Panagos, 2013).

10.2.1 *Risk-based prioritisation of contaminated sites*

The initial idea of risk-based prioritisation is that efforts and investments should be directed towards reducing the most significant risks to human health and the environment, adhering to the principles of environmental justice and sustainability. As a pragmatic approach, it helps to effectively manage the challenges posed by contaminated sites. Site prioritisation founded on human health risk-based frameworks ensures that the sites with the highest health and environmental risk will receive the adequate attention and resources required for effective remediation and risk management (Jardine *et al.*, 2003). Following data collection and site characterisation, risk-based prioritisation of contaminated sites is further based on the assessment and ranking of sites according to the identified health hazards, dominant exposure pathways and sensitiveness of receptors. The risk-based approach in contaminated site prioritisation should serve to efficiently allocate limited resources to address the most pressing sites.

10.2.2 *Data collection and site characterisation complexity*

Data collection is a primary aspect of the risk assessment process, as it serves as the fundamental building block upon which the entire assessment relies, ensuring a comprehensive and empirical understanding of health risks associated with contaminated sites (von Stackelberg *et al.*, 2021). Although risk assessment procedures among countries often have a similar layout, data collection methods and sources are usually very different from country to country. Data can be derived from various primary and secondary sources, each serving a unique scientific purpose. Primary data sources encompass direct measurements, environmental sampling, surveys and field investigations. These sources provide first-hand information, often gathered through scientific observation and experimentation, ensuring the accuracy and reliability of data. For example, collecting soil and water samples from a brownfield site to measure contaminant concentrations is a primary data source (ATSDR, 2015). Surveys and questionnaires, particularly in epidemiological studies, are valuable for gathering data on human exposure, health outcomes and demographic information. Secondary sources include historical records, government reports, environmental monitoring databases, geographical information systems (GIS) and literature. These sources complement primary data by offering historical context and a broader perspective, enhancing the scientific rigour of the assessment (Radojevic and Bashkin, 2006; USEPA, 2014). GIS enable the spatial analysis and visualisation of data, providing a scientific framework for identifying patterns and hotspots of contamination. Historical aerial photographs to assess land-use changes over time are an example of secondary data (USEPA, 2014). For instance, geophysical surveys can help identify subsurface contaminants or anomalies. Remote sensing technologies, such as satellite imagery and aerial photography, offer a scientifically robust means to assess land use, vegetation and surface conditions, providing valuable data for characterising a site.

Site characterisation relies on a methodical approach to collecting and analysing data about the site in question. This process entails the systematic gathering of a broad spectrum of information, including the identification and quantification of contaminants present, their distribution and concentrations, potential exposure pathways and the specific characteristics of the exposed population. Through the systematic integration of various data sources and scientific methods, site characterisation ensures that health risks are evaluated with accuracy and precision. By systematically assembling and analysing data from various sources using advanced methodologies, the process provides a strong scientific foundation for evidence-based decision-making and the formulation of effective risk management strategies (Mahammedi *et al.*, 2022).

10.2.3 *Financial aspects of site characterisation and risk assessment*

In the domain of contaminated site management, the challenge of the financial aspect lies in finding the optimal balance between environmental protection and economic efficiency. The judicious allocation of financial resources directly influences the ability to address contaminated sites promptly and comprehensively. Prioritising which sites to remediate is strongly influenced by available resources. Key financial aspects include cost–benefit analysis (CBA), resource allocation, funding sources, economic metrics and the nuanced trade-offs between environmental benefits and economic efficiency. CBA is a fundamental tool in the prioritisation of contaminated site cleanup. It involves assessing the costs and benefits of different remediation options to determine their economic feasibility (Mishan and Quah, 2020). Critical decisions on resource allocation hinge on a careful evaluation of potential costs weighed against expected benefits. Strategic prioritisation directs resources to sites with higher health and environmental risks, ensuring that the most critical issues are addressed first (USEPA, 2011). While site prioritisation is essential, obtaining funding for remediation is a separate challenge that significantly affects the ability to carry out these priorities. Government grants, private investments and public–private partnerships play a significant role in shaping the extent to which remediation efforts can be executed (USEPA, 2004). Economic metrics, such as the net present value (NPV) and return on investment (ROI), play an important role in assessing the financial viability of prioritisation. Decision-makers use these metrics to compare the long-term financial benefits against the immediate costs of remediation. However, a central challenge lies in reconciling environmental goals with economic efficiency. Instances may arise where the most cost-effective solution does not align with the most environmentally protective option, necessitating careful consideration of trade-offs (Gowdy, 2008).

10.3 Preliminary Health Risk Assessment Approach

A PHRA evolves from its ability to offer a data-driven, adaptable and scientifically grounded framework for the identification, prioritisation and mitigation of risks associated with contaminated sites. The dynamic nature of PHRAs supports continuous improvement and adaptive management, ensuring relevance and up-to-date assessments over time (USEPA, 2017). Serving as a crucial element in decision support, public health protection, regulatory alignment and the potential for continuous improvement, PHRAs establish the foundation for robust environmental and human health protection management. This approach fosters sustainable and health-conscious decision-making, ensuring a comprehensive approach to addressing the challenges posed by contaminated sites (USEPA, 2014). This holistic approach considers various exposure pathways, covering a wide spectrum of environmental media. It identifies and prioritises risks, guiding resource allocation and risk reduction efforts. Relying on scientific information, PHRAs facilitate the collection, analysis and interpretation of data related to contaminant levels, exposure pathways and toxicological profiles (USEPA, 2010). The PHRA approach mostly serves to protect public health by quantifying and characterising potential health risks, prioritising the health of people living near contaminated sites and preventing adverse health outcomes. Additionally, this approach facilitates the selection of adequate remediation and management strategies, aligning regulatory and health considerations with practical site-specific solutions. In this scenario, PHRA methods take precedence, leveraging data from initial site characterisations. Data acquisition is increasingly feasible as more countries adopt systematic strategies for managing contaminated locations, underpinning the application of PHRAs for site prioritisation (UNEP, 2016). Using PHRAs for sorting polluted sites provides a foundational structure for their administration based on associated hazards.

While different PHRAs usually share common goals, each methodology has unique features in its approach, such as engagement with stakeholders and adaptability to emerging challenges. The Canadian methodology emphasises a multi-phase process and Indigenous engagement, New Zealand's focuses on soil contamination with a collaborative and adaptive approach, and the U.S. HRS prioritises hazardous waste site evaluations with a structured scoring system. Each framework reflects the respective country's commitment to addressing environmental risks effectively and transparently.

10.4 Preliminary Health Risk Assessment Methodologies

10.4.1 *The Hazard Ranking System*

The Hazard Ranking System (HRS) is a vital tool utilised by the United States Environmental Protection Agency (USEPA) to assess and prioritise sites for potential inclusion in the National Priorities List (NPL) under the Comprehensive Environmental Response, Compensation, and Liability Act (CERCLA), commonly known as the Superfund programme (USEPA, 2023). This systematic methodology is designed to evaluate the relative risk posed by hazardous waste sites, aiding in the identification and remediation of severe environmental contamination issues nationwide. The HRS starts with the identification of potentially contaminated sites, prompted by factors such as citizen complaints, emergency responses or routine inspections. A preliminary assessment (PA) is then conducted to determine if further evaluation is necessary, considering site history, potential contaminant pathways and the presence of hazardous substances. If the PA indicates a potential threat, a more detailed site inspection (SI) is undertaken, involving on-site investigations, sample collection and assessment of exposure pathways.

The HRS assigns numerical scores based on three main components: likelihood of release, waste characteristics and potential for human exposure. The likelihood of release assesses the potential for hazardous substances to be released into the environment. Waste characteristics examine the toxicity, persistence and mobility of the hazardous substances. The potential for human exposure evaluates pathways through which humans may come into contact with hazardous substances. These factors contribute to an overall HRS score, ranking the site on a scale of 0–100. Sites with scores above a certain threshold are eligible for inclusion in the NPL, indicating a higher priority for cleanup. It evaluates pathways through which hazardous substances may be released, considering the toxicity and mobility of substances. Proximity to populated areas also contributes to the overall score.

Community involvement is an integral part of the HRS. The USEPA encourages public participation throughout the site assessment and ranking process, providing opportunities for community input. This ensures that local knowledge and concerns are considered in decision-making. The USEPA periodically reviews and updates the HRS to incorporate new scientific information and enhance its efficiency and effectiveness. This commitment to continuous improvement ensures that the HRS remains a robust and adaptive framework, reflecting the latest advancements in environmental science and risk assessment. The HRS plays a pivotal role in the USEPA's efforts to address hazardous waste sites and safeguard human health and the environment. Its transparent and science-based approach ensures a comprehensive assessment of environmental hazards, while community involvement incorporates local perspectives and concerns. As environmental challenges evolve, the USEPA adapts the HRS to meet changing demands and advance the goals of environmental protection and remediation.

10.4.2 *Canada's National Classification System for Contaminated Sites*

Canada's National Classification System for Contaminated Sites is a comprehensive framework designed to assess and prioritise contaminated sites based on the potential risks they pose to human health and the environment. The system operates within the broader framework of federal, provincial and territorial environmental regulations, aiming to ensure a consistent and science-based approach to managing environmental risks associated with contamination (Canadian Council of Ministers of the Environment, 2008). Central to the National Classification System is a risk-based approach, which involves evaluating the potential risks posed by contaminated sites to both human health and the environment. This approach allows for a more targeted allocation of resources for remediation, acknowledging that not all contaminated sites pose the same level of risk. The system typically involves multiple phases, starting with a preliminary site investigation (PSI) and progressing to a detailed site investigation (DSI) and, if necessary, remediation. Each phase plays a crucial role in the overall assessment and management of contaminated sites.

The PSI is the initial step in the classification process, aiming to identify potentially contaminated sites and assess whether further investigation is required. It involves the collection of historical data, site inspections and the evaluation of potential contamination sources. The DSI builds on the information gathered during the PSI, conducting a more in-depth assessment of the contamination and associated risks. This phase involves collecting samples from soil, water and air and analysing them for contaminants, helping to quantify the extent of contamination and support the risk assessment process. The risk assessment phase evaluates the potential risks posed by the contamination to human health and the environment. It considers exposure pathways, toxicity of contaminants and the sensitivity of the surrounding environment. The National Classification System often employs standardised risk assessment models to ensure consistency in the evaluation process. If the risk assessment identifies a site as posing significant risks, remediation activities are initiated to mitigate these risks and restore the site to an acceptable condition. Remediation strategies vary depending on the nature and extent of contamination but often include activities such as soil excavation, groundwater treatment and the installation of containment measures.

The National Classification System categorises sites into risk classes based on the findings of the risk assessment. The risk classes typically range from low to high, indicating the level of risk associated with a particular site. This classification serves as the foundation for prioritising sites for remediation efforts. Stakeholder engagement is a key aspect of the National Classification System, involving local communities, Indigenous groups and regulatory authorities. This ensures that local knowledge is considered in the decision-making process, promoting transparency and community involvement in contaminated site management. The system is subject to ongoing evaluation and updates to incorporate advancements in scientific knowledge and changes in environmental regulations. This commitment to continuous improvement ensures that the classification system remains robust, responsive and aligned with evolving environmental standards. Canada's National Classification System for Contaminated Sites is a vital tool for systematically assessing and prioritising sites based on the risks they pose to human health and the environment. The risk-based approach, multiphase investigation process and stakeholder engagement contribute to the effectiveness of the system, supporting the overarching goal of managing and mitigating the environmental impacts of contamination across the country.

10.4.3 *Risk Screening System for contaminated sites*

New Zealand has implemented a comprehensive Risk Screening System for contaminated sites, strategically designed to assess and manage environmental risks stemming from soil contamination. Within the broader framework of environmental management and regulatory compliance, this system aims to efficiently identify and prioritise contaminated sites, enabling targeted interventions and remediation efforts (Ministry for the Environment of New Zealand, 2004).

The Risk Screening System initiates with a preliminary site assessment (PSA), systematically evaluating available information such as historical records and potential contamination sources. This phase identifies sites meriting further investigation based on observed conditions and potential risks. Subsequently, a detailed risk assessment is conducted to quantify potential risks associated with the soil and groundwater contamination. Standardised risk assessment models may be employed to ensure consistency in the evaluation process. Contaminated sites are categorised into risk classes based on the assessment, ranging from low to high risk. This classification helps prioritise sites for remediation efforts, with low-risk sites posing minimal threats and high-risk sites indicating significant potential hazards. The Risk Screening System considers site-specific characteristics, recognising the uniqueness of each contaminated site. Legal and regulatory compliance is paramount, ensuring alignment with established standards and procedures. Stakeholder engagement is integrated into the Risk Screening System, involving collaboration with local communities, regulatory authorities and potentially affected parties to address local knowledge and community concerns.

An adaptive management approach is embraced, allowing continuous reassessment and refinement based on emerging scientific knowledge, changes in environmental regulations and stakeholder feedback. Remediation strategies are implemented for sites posing significant risks, encompassing soil excavation, groundwater treatment, containment measures or other site-specific interventions. Following remediation, ongoing monitoring verifies the effectiveness of interventions, ensuring compliance with standards and appropriate management of any residual risks. The Risk Screening System represents New Zealand's commitment to a systematic, risk-based approach for managing environmental risks associated with soil and groundwater contamination. This dedication to legal compliance and ongoing monitoring underscores New Zealand's commitment to environmental protection and the sustainable management of contaminated sites.

10.5 Case Study: Preliminary Health Risk Assessment Methodology Application in Serbian Contaminated Sites

In the Republic of Serbia, there are no well-established mandatory methodologies for risk assessments within the legal framework (Matić *et al.*, 2022). Contaminated sites are defined in legal documents as *sites with confirmed presence of hazardous and dangerous substances caused by human activity in concentrations that may cause a significant risk to human health and the environment.* The cadastre aims to provide organised information on pollution sources, such as the type, amount, method and location of pollutant discharges into the soil, facilitating the implementation of prevention, rehabilitation and remediation measures. In the Cadastre of Contaminated Sites in the Republic of Serbia, developed by the Serbian Environmental Protection Agency, only 213 sites are registered as either potentially contaminated or contaminated (Vidojević *et al.*, 2022). The data acquisition during the contaminated site investigation process is a task facilitated by national strategies for soil protection that provide guidelines for soil sampling and information gathering. The integration of such data could support the application of preliminary or comparative risk assessment methodologies and priority setting.

10.5.1 *Preliminary risk assessment model for the identification and assessment of problem areas for soil contamination in Europe—PRA.MS*

To streamline decision-making on management actions, a preliminary risk assessment methodology, specifically the EEA's "PRA.MS — Preliminary Risk Assessment Model for the identification and assessment of problem areas for Soil contamination in Europe," was applied to prioritise the industrial locations as a pilot study. The PRA.MS methodology, *a system for preliminary risk assessment for the individual locations in order to set a priority list of contaminated sites*, has been developed in the framework of the project "[t]owards an EEA Europe-wide assessment of areas under risk for soil contamination" since 2005 (Altieri *et al.*, 2004). This preliminary risk assessment model was adopted by the European Information and Observation Network (EIONET) and represents a review of 27 existing and documented international methodologies for preliminary and simplified risk assessment of (potentially) contaminated sites. It was applied in several EU member countries and overseas at the national or regional level for the prioritisation and planning of soil remediation and protection programmes. The PRA.MS methodology was developed to support national, regional or local programmes for the rehabilitation and remediation of contaminated sites (Gentile and Quercia, 2005). The basis of the PRA.MS methodology was built on the paradigm source–pathway–receptor, which has been adopted in the design of the conceptual site model and relative risk assessment. In this case, the "source" represents contaminated soil or disposed waste in the selected area/location. The human health risk exposure scenario and the generic conceptual model include four exposure pathways to potential human receptors: groundwater (GW), surface water (SW), air particulates (AIR) and direct contact (DC) (Fons, 2006). This methodology is a tiered approach that includes three ranks ("tier 0," "tier 1" and "tier 2"), and each consists of sets of data that, by the higher rank, are characterised by greater extent, detail and precision. It starts from a preliminary selection of sites that are potentially hazardous based on source and size criteria, followed by its implementation with fundamental risk elements, such as source hazard, pathways and receptor information (Fons, 2006). In this study, results obtained in the "tier 1" rank of the PRA.MS methodology are presented.

The PRA.MS model *adopts a mixed additive and multiplicative algorithm for calculating the total risk value of the site*. Parameters are aggregated and presented by factors that are composed of quantitative or semi-qualitative values of parameters (Fons, 2006). The factors are further grouped into indicators sorted according to the *source–pathway–receptor* paradigm, weighted depending on the relative importance in the description of the indicators and summarised in order to obtain the result for each indicator from this paradigm. The results of the source–pathway–receptor indicator are then multiplied for each of the four exposure pathways to human receptors (GW, SW, AIR and DC) (Altieri *et al.*, 2004). Human health (*V*tot) risk values are within the range of 0–100. When the required data are not available, the model expresses this deficiency through the uncertainty factor (Fons, 2006). During the development of the PRA.MS methodology, the French Simplified Risk Assessment (SRA) methodology was used to develop calculations for the accompanying uncertainty factors. The uncertainty factor assessment procedure proposed in the SRA methodology is adapted to the needs of the PRA.MS methodology and is based on the input of symbol "?" when one or more parameters that contribute to the factor score are not available and/or one or more parameters are not accurate because of the low resolution of required data or if the data do not meet the PRA.MS parameter requirements (proxy data) (Altieri *et al.*, 2004).

The risk classification of locations proposed by the PRA.MS methodology is based on the level of the overall risk to human health. After obtaining the results of the risk values to human health, the sites can be classified into four groups according to the level of risk reflected in the scores (Table 10.1). Regarding the uncertainty factor, there is also a proposed classification on the basis of which it can be determined whether it is in a very high, high, medium or low class of uncertainty factor (Table 10.2) (Falconi *et al.*, 2005; Fons, 2006).

This approach was applied within the UN Environment/GEF project "Enhanced cross-sectoral land management through land-use pressure reduction and planning" to 32 industrially contaminated sites that were explored and surveyed during the period 2015–2018. The goal was to

Table 10.1. Total risk classes.

Risk Class	Range
Low risk	0–20
Medium risk	20–40
High risk	40–60
Very high risk	>60

Table 10.2. Uncertainty factor classes.

Uncertainty Factor Class	Range
Low uncertainty factor	0–20
Medium uncertainty factor	20–40
High uncertainty factor	40–60
Very high uncertainty factor	>60

establish a prioritisation list based on the relative risk to human health and the environment, providing methodological support for data processing, systematisation and informed allocation of financial resources for location management (Kukobat *et al.*, 2018; UNEP, Evaluation Office, 2023).

Beyond presenting the results and relative-risk-based priority list, the application of PRA.MS methodology to the 32 contaminated sites in the Republic of Serbia aimed to examine the logistic and economic feasibility of applying the comparative risk-based methodology to other contaminated sites in the Cadastre of Contaminated Sites of the Republic of Serbia. Secondly, it was investigated whether the results obtained, including exposure pathways and relative risk values, can serve as relevant risk communication tools (Vidojević *et al.*, 2021).

10.5.2 *Data collection process*

In addition to site visits, data were gathered from local authorities, relevant institutions, operators and technical experts. For cases involving industrial complexes declared bankrupt, information and data were obtained from bankruptcy administrators responsible for the management of these sites. Some locations were extensively investigated, while others were only surveyed at a basic level. General site characteristics, such as climate, geology, hydrology, soil type and properties, seismic features and the presence of protected cultural and natural assets, were obtained from national monitoring systems, collaborating institutions and existing studies for selected sites. Climate and hydrological data for a broader location were obtained from annual reports of the Republic Hydrometeorological Institute of the Republic of Serbia. Additionally, an interactive map of protected areas in Serbia, established by the Institute for Nature Protection of Serbia (GDi Visios—https://cloud.gdi.net/visios/zzps), was also used to identify protected areas near the studied locations.

The PRA.MS model required information on soil contaminants and methods of waste storage at the locations as the source part of the source–pathway–receptor paradigm, essential for subsequent calculations for each exposure pathway and receptor. The results from laboratory analyses of 264 soil samples, collected during field visits from 2016 to 2018, were utilised during the application of the PRA.MS model, representing the most recent data on detected concentrations of contaminants. The identified contaminants entered into the model were heavy metals that exceeded the limit or intervention values specified in the regulation on the limit values of polluting, harmful and dangerous substances in soil ("Official Gazette of the RS" no. 30/18 and 64/19). The presence of heavy metals in soil samples was determined using the EPA 7000b: 2007 method for copper, nickel, lead and zinc and EPA 7010: 2007 for arsenic, cadmium and chromium. Sampling depths varied (10, 30 and 50 cm) based on soil type. For most of the 32 locations, detailed hydrogeological studies were either not conducted or not available. The information regarding the volume of the stored waste and volume of contaminated soil also remained unknown; however, the area extent of surface soil contamination was collected for each location and used as input data for the PRA.MS model. Within the PRA.MS methodology, these data suffice for further calculations in the source–pathway–receptor paradigm. Information on waste/contaminant disposal methods, the condition of the waste deposition area and monitoring systems for waste areas, critical for exposure calculations, were gathered during site visits conducted by the project's multidisciplinary team.

Concerning the dataset required to compute exposure pathway parameters and factors in the PRA.MS model, it is noteworthy that information on the release of contaminants into groundwater is unknown. Consequently, a value of −1 (indicating no known release into groundwater) was assigned for each location. For many locations, data from periods when significant changes occurred, affecting site infrastructure and potentially impacting climate and the subsurface dispersion of contaminants, were available. Recent updates on site conditions, including waste storage status and disposal methods, were acquired during comprehensive site visits. Regarding data necessary for estimating potential contaminant migration from the site, the PRA.MS model incorporates terrain slope data. Given that these are industrial sites prepared for production processes and factory infrastructure, the terrain slope generally did not exceed 5°. Additionally, soil texture data were obtained from chemical and physical analyses conducted on samples collected during field visits. For most locations, information on the thickness of the impermeable layer and the depth of the aquifer from the surface remains unknown due to the absence of detailed geological or hydrogeological studies. Data on groundwater depth were obtained from a few existing hydrogeological reports covering a broader territory (available for 26 out of 32 locations). For characterisation of the transportability of soil particles from contaminated sites, the particulate mobility indicator is used in the PRA.MS methodology. The information on mean annual precipitation (mm), temperature (°C) and wind speed (m/s) were taken from annual reports developed through national climate data monitoring.

The factors within the "receptor" aspect of the PRA.MS paradigm for risk calculation, encompassing details such as groundwater and surface water use, minimum distance from surface water bodies, on-site and off-site land use, distance to the nearest residential area and site accessibility, were available for all 32 selected industrially contaminated locations.

10.5.3 *Results of the PRA.MS model*

Table 10.3 shows the list of locations with exposure pathways and human health risk values for the 32 potentially contaminated sites, suspected to pose a risk to human health, ranked according to the overall risk values (*V*tot). Besides the overall risk to human health (*V*tot), it also lists the values for each individual exposure pathway, i.e., for air particulates (*V*air), direct contact (*V*dc), groundwater (*V*gw) and surface water (*V*sw). Risk is classified into four categories: low risk (0–20), medium risk (20–40), high risk (40–60) and very high risk (above 60) (Table 10.1) (Fons, 2006). Alongside the risk values, there are uncertainty factors for each exposure pathway (*V*air_*U*, *V*dc_*U*, *V*gw_*U*, *V*sw_*U*) as well as the value of the uncertainty factor for overall risk (*V*tot_*U*) (Table 10.3). Uncertainty factors are classified as follows: low uncertainty (0–20), medium uncertainty (20–40), high uncertainty (40–60) and very high uncertainty (greater than 60) (Table 10.2) (Fons, 2006).

10.5.4 *Case study key findings*

Based on the risk values obtained for human health and proposed classification according to the PRA.MS methodology, out of a total of 32 locations, four are in the high-risk class, one in the low-risk class, while the other sites are in the medium-risk class (Table 10.4). The table additionally displays the industry type associated with each surveyed site. The industrial site with the highest ranking (Site21), according to the PRA.MS methodology, is a location where the extraction and smelting of copper ore have been carried out for more than 100 years. The next location (Site26) is a thermal power plant and an ash dump along the banks of the Sava River. According to the values obtained for the exposure routes, the exposure route with the highest ranking is via surface water due to its

Table 10.3. The list of locations with exposure pathways and human health risk values.

Site ID	*Vair*	*Vair_U*	*Vdc*	*Vdc_U*	*Vgw*	*Vgw_U*	*Vsw*	*Vsw_U*	*Vtot*	*Vtot_U*
Site21	48.60	0.00	48.00	0.00	22.55	9.41	47.00	0.00	42.96	4.71
Site26	33.30	0.00	48.00	0.00	31.24	9.06	50.92	18.61	41.78	10.35
Site6	16.21	0.00	40.80	0.00	29.51	10.59	64.60	0.00	41.75	5.29
Site30	21.61	0.00	45.60	0.00	40.63	10.59	50.35	0.00	41.02	5.29
Site11	14.80	0.00	38.40	0.00	31.63	10.33	60.80	0.00	39.97	5.17
Site9	21.91	0.00	45.60	0.00	38.31	10.33	46.21	8.89	39.25	6.81
Site24	42.30	0.00	35.00	0.00	24.68	11.71	40.74	18.61	36.34	10.99
Site25	30.80	0.00	37.80	0.00	16.61	9.41	45.83	18.61	34.47	10.43
Site29	12.48	0.00	24.85	0.00	29.47	10.59	53.96	0.00	33.74	5.29
Site23	29.60	0.00	38.95	0.00	30.25	10.33	32.16	18.61	32.95	10.64
Site32	18.43	0.00	40.80	0.00	15.05	8.42	45.05	0.00	32.63	4.21
Site5	16.92	0.00	36.00	0.00	25.05	8.42	44.46	10.83	32.35	6.86
Site1	21.07	0.00	40.80	0.00	35.95	11.61	27.46	8.89	32.23	7.31
Site8	18.80	0.00	38.40	0.00	31.04	10.59	35.28	8.75	31.77	6.87
Site14	17.76	0.00	34.85	0.00	30.51	9.31	36.18	7.78	30.70	6.07
Site2	18.80	0.00	38.40	0.00	25.21	9.06	32.34	16.67	29.62	9.48
Site19	17.21	0.00	38.40	0.00	27.73	9.06	31.09	18.61	29.60	10.35
Site10	23.41	0.00	45.60	0.00	27.41	7.14	7.98	16.67	29.34	9.07
Site31	21.07	0.00	34.85	0.00	37.87	10.33	18.30	8.89	29.27	6.81
Site3	27.82	0.00	34.85	0.00	22.16	11.71	27.87	9.31	28.53	7.48
Site22	12.48	0.00	29.11	0.00	31.63	10.33	35.29	9.86	28.50	7.14
Site27	33.30	0.00	32.80	0.00	19.04	7.14	24.66	18.61	28.09	9.97
Site28	18.50	0.00	28.70	0.00	17.32	9.06	40.33	18.61	27.81	10.35
Site12	21.50	0.00	36.90	0.00	20.41	7.14	28.34	18.61	27.58	9.97
Site20	30.60	0.00	25.20	0.00	23.56	11.71	30.02	18.61	27.51	10.99
Site4	12.28	0.00	30.75	0.00	30.74	10.33	26.13	18.61	26.10	10.64
Site16	12.03	0.00	34.08	0.00	27.73	9.06	19.99	7.78	24.87	5.97
Site18	12.48	0.00	34.08	0.00	18.74	6.12	28.28	8.47	24.84	5.23
Site15	15.04	0.00	36.00	0.00	22.14	14.01	21.15	8.33	24.80	8.15
Site13	9.83	0.00	29.11	0.00	28.21	11.61	25.21	9.31	24.37	7.44
Site17	3.78	0.00	29.11	0.00	28.96	10.33	19.63	8.75	22.83	6.77
Site7	/	/	18.70	0.00	9.76	9.06	9.60	16.67	11.59	9.48

Table 10.4. The relative-risk-based priority list compiled using the PRA.MS model.

Site ID	Type of Industry	Human Health Risk Class
Site21	Mining industry	High risk
Site26	Thermal energy production	High risk
Site6	Paper industry	High risk
Site30	Chemical industry	High risk
Site11	Chemical industry	Medium risk
Site9	Metal industry	Medium risk
Site24	Thermal energy production	Medium risk
Site25	Thermal energy production	Medium risk
Site29	Metal industry	Medium risk
Site23	Thermal energy production	Medium risk
Site32	Chemical industry	Medium risk
Site5	Chemical industry	Medium risk
Site1	Metal industry	Medium risk
Site8	Metal industry	Medium risk
Site14	Metal industry	Medium risk
Site2	Waste disposal site	Medium risk
Site19	Metal industry	Medium risk
Site10	Chemical industry	Medium risk
Site31	Metal industry	Medium risk
Site3	Metal industry	Medium risk
Site22	Metal industry	Medium risk
Site27	Thermal energy production	Medium risk
Site28	Thermal energy production	Medium risk
Site12	Chemical industry	Medium risk
Site20	Mining industry	Medium risk
Site4	Metal industry	Medium risk
Site16	Paper industry	Medium risk
Site18	Metal industry	Medium risk
Site15	Metal industry	Medium risk
Site13	Chemical industry	Medium risk
Site17	Chemical industry	Medium risk
Site7	Rubber industry	Low risk

immediate proximity to the river. The third-ranked location (Site6) is a factory for paper and packaging, specifically the lagoons of wastewater located in the immediate vicinity of the South Morava River. Anticipating a high-risk value is reasonable, given the identified release of contaminants into the river, which, based on gathered information, is utilised for irrigation and is situated in close proximity to the site. In the fourth place is the site of a former prominent entity in the chemical industry, which has been inactive for several decades. The location is now in a state of devastation and is frequently targeted for unauthorised entry, often for theft. The only location classified as low risk (Site7) is a synthetic rubber factory in the northern part of the country. The factory has undergone renovation, and the area of the former waste disposal site is now covered with vegetation. This vegetative cover serves to prevent the dispersion of contaminated soil via air particulates.

The highest uncertainty factor values are observed for the groundwater exposure pathway. Due to a general lack of information about groundwater in most surveyed locations during the site investigations, this result was anticipated. Missing data for exposure calculations include thickness and presence of an impermeable layer (for all locations), depth of aquifer (for 6 locations), flooding return period and surface water flow rate (for 15 locations). In contrast, a full set of information has been gathered for evaluating parameters for calculating the exposure to contaminated soil via air particulates. Although the results for risk classes were accompanied by low uncertainty factor values, which implies data sufficiency, further detailed investigations are necessary at specific locations. This is particularly important if more accurate comparisons of the assessed sites is needed. While the priority list generated through the use of the PRA. MS model aligns with expectations, it is imperative to implement high-resolution sampling in specific locations. Moreover, a subsequent risk assessment for areas in close proximity to human settlements is highly advisable. In light of the foregoing, a more comprehensive investigation of selected locations is recommended to acquire updated data for the PRA.MS model. Given that risk assessment is inherently an iterative process, this methodology also accommodates the inclusion of newly collected or further investigated data, enhancing the refinement of results. The application of the PRA.MS methodology, with the purpose of generating a relative-risk-based list of locations, has also provided insights into its potential applicability to other registered sites in the Cadastre of

Contaminated Locations of the Republic of Serbia. The majority of data required for calculating the relative health risk using the PRA.MS model are integral components of the forms and questionnaires that operators, as mandated by law, need to submit to the relevant authorities. The PRA.MS methodology, chosen for its capacity to identify dominant exposure pathways and receptors for each location, also addresses the population's frequent lack of awareness of the adverse effects of contaminated sites in close proximity. Abandoned industrial sites, subject to unauthorised access, pose potential health risks, emphasising the need for effective risk communication and prevention measures. During the project implementation, the engagement of relevant institutions and regulatory bodies facilitated the completion of the tier 1 level of the PRA.MS methodology in terms of supporting the data collection process. This demonstrated the ability of the country's institutions—such as governmental agencies, regulatory bodies and local authorities—to carry out, support and enforce implementation of these types of methodologies.

10.6 Discussion

Contaminated site management is a complex interplay of scientific procedures and political decision-making, where the need for rigorous analysis sometimes clashes with the urgency perceived by decision-makers. In this dynamic, PHRAs emerge as a critical tool, navigating the delicate balance between scientific integrity and expeditious decision-making.

Scientific procedures, often grounded in comprehensive data collection and analysis, face the challenge of aligning with the occasional limited timelines of decision-makers. The inherent tension lies in the fact that comprehensive risk assessments, vital for informed decision-making, can be time-consuming. This conflict underscores the importance of fostering a mutual understanding between the scientific community and policymakers. While political expediency is crucial, it should not compromise the application and quality of risk assessments.

The foundation of effective PHRA lies in the use of a robust methodology capable of comparing sites with varying quantities and levels of data, each characterised by different uncertainties. A one-size-fits-all approach is not feasible in the realm of contaminated sites. A well-defined methodology not only facilitates objective comparisons but also ensures that sites are evaluated consistently, irrespective of the available data. This methodology should be adaptable to the unique characteristics of each site, allowing for a nuanced understanding of potential risks.

Additionally, implementing a tiered approach to risk assessments allows the complexity of evaluations to scale according to site characteristics; low-risk sites can undergo basic analyses, while higher-risk sites require more detailed assessments (Swartjes, 2008). In cases requiring a higher-tier approach, it is advisable to combine it with a site-specific risk assessment and a comprehensive conceptual site model. These components provide the necessary depth to risk evaluations, considering the specific characteristics of each site.

A smart combination of model calculations and measurements, e.g., in contact media such as indoor air, crops and aquatic organisms, would greatly enhance the quality of assessments (Swartjes, 2015). Simultaneously, a comprehensive conceptual site model acts as a visual representation, aiding in the understanding of the complex interactions at play.

Uncertainty is an inherent aspect of contaminated site management due to incomplete or evolving data (USEPA, 2014). The PHRA framework application would greatly benefit from a thorough uncertainty analysis. Understanding the limitations and potential variations in the data is crucial for providing decision-makers with the necessary insights to make balanced choices in the face of incomplete or uncertain information. It is here that the intersection of scientific precision and decision-making pragmatism becomes evident. Incorporating thorough uncertainty analyses into PHRA processes is vital, as this helps decision-makers understand the limitations of the available data and make informed choices.

For a comprehensive analysis of risks associated with contaminated sites, PHRAs must be complemented by a further site-specific assessment and a comprehensive conceptual site model. These components provide the necessary depth to risk evaluations, considering the specific characteristics of each site. A site-specific risk assessment delves into the intricacies of contaminant behaviour, exposure pathways and site-specific conditions. Simultaneously, a comprehensive conceptual site model acts as a visual representation, aiding in the understanding of the complex interactions at play.

Identifying key stakeholders and developing an effective stakeholder engagement plan is essential for aligning risk assessment processes with practical decision-making while preserving scientific integrity. Fostering strong communication between scientists and policymakers is a critical component in this process, ensuring that scientific rigour is upheld throughout. Moreover, cross-border collaboration promotes a global community dedicated to refining risk assessment practices and outcomes.

To improve the applicability of PHRAs at the national level, standardising the approach is essential for achieving consistent, accurate and effective management of contaminated sites and their associated risks while ensuring greater precision and comparability (USEPA, 2017). Capacity-building and knowledge transfer can address technical resource constraints. Recently, integration of advanced technologies, such as remote sensing and AI, to expedite data analysis and decision-making for efficient site prioritisation was also proposed (Mahammedi *et al.*, 2022). Empowering local communities in risk assessments fosters sustainable solutions and inclusivity (Till and Sharp, 2000). The application of existing methodologies should be encouraged both at the national and regional levels. Even the preparation for their implementation enables the systematisation of available data, as well as an insight into any data gaps.

PHRAs can not only serve as a technical tool but also act as a bridge that connects the demands of robust scientific analysis with the need for timely decision-making.

10.7 Conclusion

In the complex landscape of contaminated site management, the role of PHRAs stands out as a beacon guiding decision-makers through the intricate process of assessment of risks, prioritisation of risks, characterisation of risks and contaminated site management. Their importance is particularly pronounced when resources are limited and political decisions necessitate swift and concrete action. PHRAs play a pivotal role in prioritising characterisation and remediation activities. In situations where resources are constrained, this tool becomes the guiding compass, directing efforts towards sites with the most significant potential risks. The strategic allocation of resources based on the insights provided by PHRAs ensures that environmental concerns are addressed efficiently.

In scenarios where the application of the polluter pays principle encounters obstacles and taxpayers' money is utilised for remediation, PRA acts as a safeguard. Its systematic approach aids decision-makers in judiciously allocating public funds, ensuring that taxpayer money is directed towards addressing the most pressing environmental threats. This strategic utilisation of resources aligns with the broader goal of safeguarding public health and environmental well-being. Access to funding is a critical consideration in contaminated site management. PHRAs also support the prioritised funding allocation for remediation efforts. The recognition of high-risk sites as priorities in the funding allocation process further strengthens the impact of PHRAs on effective decision-making.

In conclusion, PHRAs emerge as a linchpin in the comprehensive strategy for contaminated site management. Its important role in prioritisation, resource allocation and decision-making is especially evident when faced with constraints such as limited resources and the complexities of funding sources characteristic of developing countries. The synergistic relationship between scientific precision, pragmatic decision-making and resource optimisation underscores the enduring significance of PHRAs in the pursuit of environmental health and sustainability.

References

Agency for Toxic Substances and Disease Registry (ATSDR). (2015). Evaluating Contaminated Groundwater by Dermal Exposure. U.S. Department of Health and Human Services.

Altieri, A., Falconi, M., Quercia, F., and Vecchio, A. (2004). Towards an EEA Europe-Wide Assessment of Areas under Risk for Soil Contamination. Vol II. Review and Analysis of Existing Methodologies for Preliminary Risk Assessment. European Environmental Agency.

Calabrese, E. J. (2004). Hormesis: from marginalization to mainstream. A case for hormesis as the default dose-response model in risk assessment. *Toxicology and Applied Pharmacology*, 197(2), 125–136. DOI: 10.1016/j.taap.2004.02.007. PMID: 15163548.

Canadian Council of Ministers of the Environment. (2008). National Classification System for Contaminated Sites Guidance Document. Available at: https://www.ccme.ca/en/res/ncscs_guidance_e.pdf.

DOE. (2002). In Situ Remediation Integrated Program: Technology Overview. U.S. Department of Energy.

European Chemicals Agency (ECHA). (2020). Registration, Evaluation, Authorization, and Restriction of Chemicals (REACH).

Ewing, R., Bartholomew, K., Winkelman, S., Walters, J., and Chen, D. (2018). Growing Cooler: The Evidence on Urban Development and Climate Change.

Falconi, M., Quercia, F., Schamann, M., Tarvainen, T., Vecchio, A., and Wepner, M. (2005). PRA.MS: Scoring Model and Algorithm. Vol. III. Towards an EEA Europe-wide Assessment of Areas under Risk for Soil Contamination. European Environmental Agency.

FAO and UNEP. (2021). Global Assessment of Soil Pollution: Report. Rome. Available at: https://doi.org/10.4060/cb4894en.

Fons, J. PRA.MS Application (Tier 2). (2006). Towards an EEA Europe-wide Assessment of Areas under Risk for Soil Contamination. European Environmental Agency.

Freeman, A. M. and Lewis, N. (2008). Government incentives for environmental protection. *Ecological Economics*, 65(2), 245–259.

Gentile, A. R. and Quercia, F. (2005). Background and Outcomes of the Project. Vol. I. Towards an EEA Europe-wide Assessment of Areas under Risk for Soil Contamination. European Environmental Agency.

Gowdy, J. M. (2008). Behavioral economics and climate change policy. *Journal of Economic Behavior & Organization*, 68(3–4), 632–644.

International Atomic Energy Agency (IAEA). (2014). Radiation Protection and Safety of Radiation Sources: International Basic Safety Standards.

International Council for Harmonisation of Technical Requirements for Pharmaceuticals for Human Use (ICH). (1997). ICH Harmonised Tripartite Guideline: Safety Pharmacology Studies for Human Pharmaceuticals.

Jardine, C., Hrudey, S., Shortreed, J., Craig, L., Krewski, D., Furgal, C., and McColl, S. (2003). Risk management frameworks for human health and environmental risks. *Journal of Toxicology and Environmental Health, Part B: Critical Reviews*, 6(6), 569–720. DOI: 10.1080/10937400390208608. PMID: 14698953.

Kukobat, L., Vidojevic, D., and Siljic Tomic, A. (2018). Towards Soil Decontamination in the Republic of Serbia. Ministry of Environmental Protection, Environmental Protection Agency.

Mahammedi, C., Mahdjoubi, L., Booth, C. A., and Butt, T. E. (2022). Framework for preliminary risk assessment of brownfield sites. *Science of the Total Environment*, 807(Part 3), 151069. Available at: https://doi.org/10.1016/j.scitotenv.2021.151069.

Matić, B., Živković Perišić, S., Dragana Jovanović, D., Dejanović, S., Miljuš, D., and Kukobat, L. (2022). Mapping health risks of criteria chemical(s) to demonstrate benefits of risk assessment for decision-makers. In P*roceedings. 29th International Conference Ecological Truth and Environmental Research — EcoTER'22*, Sokobanja, Serbia, 21–24 June 2022.

Ministry for the Environment of New Zealand. (2004). Risk Screening System Contaminated Land Management Guidelines No. 3. Available at: https://environment.govt.nz/assets/Publications/Files/Contaminated-land-management-guidelines-No.-3-Risk-Screening-System.pdf.

Mishan, E. J. and Quah, E. (2020). *Cost-Benefit Analysis*. 6th ed. Routledge. https://doi.org/10.4324/9781351029780.

Morgan, M. G. and Henrion, M. (1990). *Uncertainty: A Guide to Dealing with Uncertainty in Quantitative Risk and Policy Analysis*. Cambridge: Cambridge University Press.

National Academies of Sciences, Engineering, and Medicine. (2018). *Advances in Causal Understanding for Human Health Risk-Based Decision-Making: Proceedings of a Workshop—in Brief.* Washington, DC: The National Academies Press. Available at: https://doi.org/10.17226/25004.

National Research Council. (1983). *Risk Assessment in the Federal Government: Managing the Process*. Washington, DC: The National Academies Press. DOI: 10.17226/366.

Panagos, P., Van Liedekerke, M., Yigini, Y., and Montanarella, L. (2013). Contaminated sites in Europe: Review of the current situation based on data collected through a European network. *Journal of Environmental and Public Health*, 2013, 11 p. Article ID 158764. Available at: https://doi.org/10.1155/2013/158764.

Pérez, A. P. and Rodríguez, N. E. (2018). Status of local soil contamination in Europe. Revision of the Indicator "Progress in the management Contaminated Sites in Europe", EUR 29124 EN. Luxembourg: Publications Office of the European Union. DOI: 10.2760/093804, JRC107508.

Radojevic, M. and Bashkin, V. N. (2006). *Practical Environmental Analysis*. 2nd ed. The Royal Society of Chemistry, Thomas Graham House, Cambridge.

Swartjes, F. A. (2015). Human health risk assessment related to contaminated land: State of the art. *Environmental Geochemistry and Health*, 37(4), 651–673. DOI: 10.1007/s10653-015-9693-0.

Swartjes, F. A. and Tromp, P. C. (2008). A tiered approach for the assessment of the human health risks of asbestos in soils. *Soil and Sediment Contamination: An International Journal*, 17(2), 137–149. Available at: https://doi.org/10.1080/15320380701870484.

Till, J. E. and Sharp, S. L. (2000). Public participation in decision-making for contaminated sites.

Troldborg, M., Lemming, G., Binning, P. J., Tuxen, N., and Bjerg, P. L. (2008). Risk assessment and prioritisation of contaminated sites on the catchment scale. *Journal of Contaminant Hydrology*, 101(1–4), 14–28. Available at: https://doi.org/10.1016/j.jconhyd.2008.07.006.

U.S. Consumer Product Safety Commission (US CPSC). (2021). Consumer Product Safety Act.

U.S. Department of Transportation (DOT). (2020). Hazardous Materials Regulations.

U.S. Environmental Protection Agency (USEPA). (1989). Risk Assessment Guidance for Superfund: Human Health Evaluation Manual (Part A). Washington, DC: U.S. Environmental Protection Agency.

U.S. Environmental Protection Agency (USEPA). (1992). Role of the Baseline Risk Assessment in Superfund Remedy Selection Decisions. Washington, DC: U.S. Environmental Protection Agency.

U.S. Environmental Protection Agency (USEPA). (1998). Guidelines for Ecological Risk Assessment. EPA/630/R-95/002F. Washington, DC: U.S. Environmental Protection Agency, Risk Assessment Forum.

U.S. Environmental Protection Agency (USEPA). (2004). Risk Assessment Principles and Practices: Staff Paper. EPA/100/B-04/001. Washington, DC: Office of the Science Advisor, U.S. Environmental Protection Agency.

U.S. Environmental Protection Agency (USEPA). (2010). Framework for Human Health Risk Assessment to Inform Decision Making. U.S. Environmental Protection Agency.

U.S. Environmental Protection Agency (USEPA). (2011). Superfund Remedial Program: Optimizing Superfund Cleanups. U.S. Environmental Protection Agency.

U.S. Environmental Protection Agency (USEPA). (2012). Superfund Remedial Program: Feasibility Study (FS) Field Study Guidance. U.S. Environmental Protection Agency.

U.S. Environmental Protection Agency (USEPA). (2014). Data Quality Assessment: A Reviewer's Guide. U.S. Environmental Protection Agency.

U.S. Environmental Protection Agency (USEPA). (2017). Preliminary Remediation Assessment Guidance. U.S. Environmental Protection Agency.

U.S. Environmental Protection Agency (USEPA). (2021). Clean Air Act; Clean Water Act.

U.S. Environmental Protection Agency UNEP, Evaluation Office, (2023). Terminal Evaluation of the UNEP/GEF Project "Enhanced cross-sectoral land management through land use pressure reduction planning" (GEF ID: 5822). Nairobi, Kenya: Evaluation Office of UNEP.

United Nations Environment Programme (UNEP). (2004). Guidance on the Global Monitoring Plan for Persistent Organic Pollutants. United Nations Environment Programme.

United Nations Environment Programme (UNEP). (2016). Global Waste Management Outlook.

USEPA. (n.d.). Hazard Ranking System. Available at: https://www.epa.gov/superfund/hazard-ranking-system-hrs (Consulted on 21 November 2023).

Vidojevic, D. and Jevtic, N. (2022). Progress in management of contaminated sites in the Republic of Serbia 2021. In *International Conference Contaminated Sites. 2022-10-12. Conference Paper*.

Vidojević, D., Kukobat, L., Damjanović, D., Jevtić, N., and Šiljić-Tomić, A. (2021). Enhancing management of contaminated sites using environmental monitoring data and preliminary risk assessment methodology in Serbia. In *3rd International and 15th National Congress of Serbian Society of Soil Science and Faculty of Agriculture*, University of Belgrade, Sokobanja, Serbia, 21–24 September 2021.

von Stackelberg, K. and Williams, P. R. D. (2021). Evolving science and practice of risk assessment. *Risk Analysis*, 41, 571–583. Available at: https://doi.org/10.1111/risa.13647.

von Stackelberg, K., Williams, P. R. D., and Sánchez-Triana, E. (2021). A systematic framework for collecting site-specific sampling and survey data to support analyses of health impacts from land-based pollution in low- and middle-income countries. *International Journal of Environmental Research and Public Health*, 18(9), 4676. DOI: 10.3390/ijerph18094676. PMID: 33924797; PMCID: PMC8125743.

World Health Organization (WHO). (2012). Contaminated Sites and Health: Report of Two WHO Workshops: Syracuse, Italy, 18 November 2011, Catania, Italy, 21–22 June 2012. Copenhagen: WHO Regional Office for Europe.

World Health Organization (WHO). (2014). Human Health in Areas with Industrial Contamination. Copenhagen, Denmark: WHO Regional Office for Europe.

World Health Organization (WHO). (2020). Food Safety. Copenhagen, Denmark: WHO Regional Office for Europe.

https://doi.org/10.1142/9781800617841_0011

Chapter 11

Assessing Health Risk of the Population Living in the Vicinity of the Bor Industrial Site

Branislava Matić[*,‡], Snežana Živković Perišić[†,§], and Dragan Miljuš[†,¶]

[*]*Institute of Public Health of Serbia, Environmental Health Department*

[†]*Institute of Public Health of Serbia, Department of Control of Non-communicable Diseases*

[‡]*brankicam@batut.org.rs*

[§]*snezana_zivkovic@batut.org.rs*

[¶]*dragan_miljus@batut.org.rs*

Abstract

Industrially contaminated sites are often located close to urban areas and/or socially deprived neighbourhoods; this makes exposure patterns more complex and results in interactions with other health determinants. There are several approaches and methods for assessing the health impact of National Priority Contaminated Sites (NPCS). The key reference and model for the Bor study in Serbia was the Italian SENTIERI project. The SENTIERI project is an example of a first-level descriptive approach, adopting an ecological study design, looking at the aggregate population level rather than at the individual level. In this Serbian pilot study, the SENTIERI epidemiological method was used to evaluate the industrial pollution exposure of the Bor population in order to describe the health profile of its residents through the analysis of cancer rates and deaths, as well as cardiovascular, respiratory, digestive and genitourinary diseases, together with mortality from all causes of death among residents of the Town of Bor (hereinafter, Bor).

The aetiological role of exposure to environmental factors is present, but available environmental quality data are insufficient for systematically monitoring the health status of the population in contaminated sites. The results presented are based on a modified study design and representative data, which reduces the study's potential for bias.

11.1 Introduction

The term "contaminated sites," as a general definition considering the public health perspective, is as follows: "areas hosting or having hosted human activities which have produced or might produce environmental contamination of soil, surface or groundwater, air, and food chain, resulting or being able to result in human health impacts" (WHO, 2012). Given this definition, an area affected by a single chemical contamination in a single environmental matrix (e.g., soil contamination caused by a given pesticide) and a large area with soil, water, air and food chain contamination by multiple chemicals (e.g., contamination caused by long-term emissions of a petrochemical complex) can be both considered contaminated sites (Pirastu *et al.*, 2013). Furthermore, industrially contaminated sites are often located close to urban areas and/or socially deprived neighbourhoods; this makes exposure patterns more complex and results in interactions with other health determinants (WHO, 2013).

The public health sector is more concerned with ICSs where defined vulnerable population groups and general population categories are exposed to hazardous pollutants present in all environmental fractions during their regular life course activities: home, school environment, public places and surroundings, and not only with environmental hot spots with contaminated layers of soil, as such happens on industrial landfills, presumably (Mudu *et al.*, 2014).

There are several approaches and methods for assessing the health impact of National Priority Contaminated Sites (NPCSs). On the one hand, one can apply risk assessment techniques, where available data on the degree of contamination (typically measures of concentration of specific hazardous chemical agents) are used to quantitatively estimate the risk of occurrence of health endpoints causally associated with the agents; on the other hand, epidemiological approaches applicable to NPCSs involve the inclusive collection of available health information of the resident population and various agents and exposures. In these assessments, a first descriptive level is based on epidemiological tools that do not require an *ad hoc* collection of data and aims at describing the health profile of populations by documenting ascertained or suspected associations with local environmental risks and the potential for providing efficient answers. More detailed analyses can be carried out, at a higher level of definition, by collecting specific data on health outcomes and/or exposure (WHO, 2013).

11.2 Methods

The key reference and model for the Bor study in Serbia was the Italian SENTIERI project. The SENTIERI project (or the Epidemiological Study of Residents in Italian Contaminated Sites) is an example of a first-level descriptive approach that adopts an ecological study design, looking at the aggregate population level rather than at the individual level (Pirastu *et al.*, 2010, 2011). The SENTIERI project is a national initiative developed to evaluate the health profiles of populations residing in the Italian sites of national interest for environmental remediation—NPCSs. These sites were labelled as sites of national interest due to their substantial contamination, documented in qualitative and/or quantitative terms, and the consequent potential impact on the health of residents. The methods proposed under the approach, exemplified by SENTIERI can be generalised and applied to other NPCSs (Pirastu *et al.*, 2013).

When studying how the environment can adversely affect human health, it is usually very difficult to identify clear cause–effect relationships because they are characterised by multicausality with different strengths of association. In addition, these relationships are influenced by individual factors (e.g., genetic makeup, diet, lifestyle, occupation and socioeconomic status) that can also play a role in determining both exposure levels and disease development.

SENTIERI project (Epidemiological Study of residents in National Priority Contaminated Sites—NPCSs) is an example of first level descriptive approach adopting an ecological study design, looking at the aggregate population level rather than at individual level (Pirastu *et al.*, 2011). In this study of the health profiles of the residents in NPCSs, we first consider what is done and why, as follows:

(1) site selection/study area,
(2) environmental data,
(3) study population,
(4) reference population,
(5) outcome selection,
(6) socioeconomic confounding.

11.2.1 *Site selection/study area/environmental data*

Bor was selected since it was an environmental and health hotspot. We first provide a historical preview of copper mining and metal-processing activities in the Bor District, as well as its institutional capacities.

11.2.2 *General description of the Bor District and Town of Bor*

The Bor District is located in the northeast region of the Republic of Serbia (Figure 11.1). It consists of four urban settlements: Bor, Majdanpek, Negotin and Kladovo. Its administrative centre is Bor, covering an area of 3,507 km^2. Its overall population is 124,992 (1.8% population of the Republic of Serbia), and its population density is 35.6 inhabitants/km^2.

Bor is one of the most important centres of the mining and metal-processing industry in Southeast Europe. It is located in eastern Serbia, 240 km from Belgrade and some 30 and 100 km away from the borders with Bulgaria and Romania, respectively. Also, it is located at the southern part of the Bor District. The district is rich in copper and gold deposits, especially in the Bor and Majdanpek areas. Bor is the administrative centre of the Bor District. It consists of 14 settlements (Table 11.1), covering an area of 856 km^2. According to the 2011 census, 50,159 inhabitants live in Bor, which is 0.7% of the total population of the Republic of Serbia and 38.1% of that of the Bor District. Close to 34,000 inhabitants live in the Town of Bor. The average annual population drop is 2.21% (official website of Bor, 2018).

Bor is located in the valley of the Bor River at an altitude of 360 m. Its main economic activity comprises mining and metal processing. Some 10,000–15,000 inhabitants are reportedly employed in this sector.

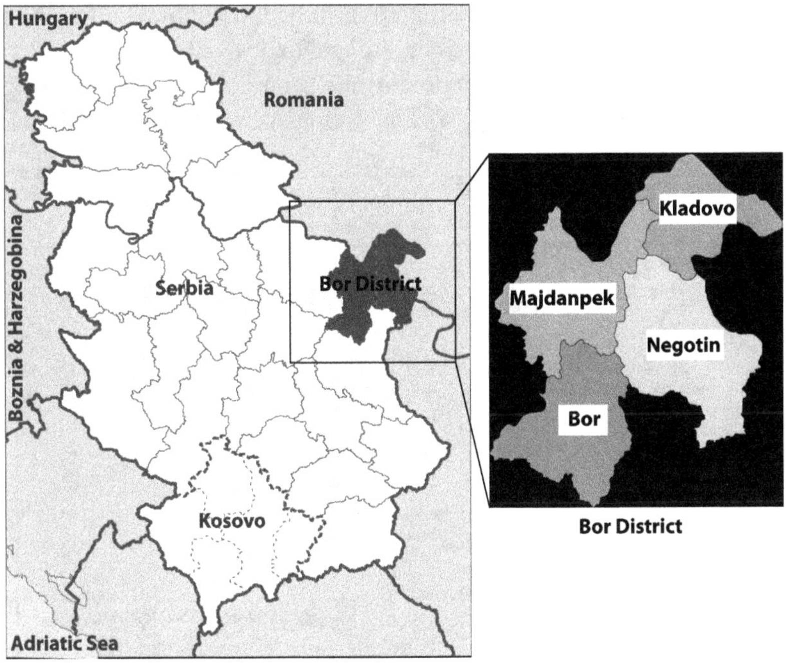

Figure 11.1. Bor District, easternmost region of the Republic of Serbia.

Table 11.1. Bor: settlements and demographic characteristics, 2002–2011.

Settlement	Population (2002)	Population (2011)	Population Migration (%)	Area (km^2)	Density (2002)
Bor	39387	34160	−13.27	47.621	827.1
Brestovac	295	269	−8.81	69.801	42.7
Bučje	666	579	−13.06	30.633	21.7
Gornjane	1.114	930	−16.52	90.349	12.3
D. Bela Reka	823	741	−9.96	40.674	20.2
Zlot	3757	3299	−12.19	229.867	16.3
Krivelj	1316	1052	20.06	99.207	13.3
Luka	612	537	−12.25	44.995	13.6
Metovnica	1331	1111	−16.53	44.131	30.2
Oštrelj	654	586	−10.40	19.6	33.4
Slatina	921	890	−3.37	38.575	23.9
Tanda	350	319	−8.86	40.764	8.6
Topla	100	97	−3.00	10.404	9.6
Šarbanovac	1836	1624	−11.55	50.415	36.4

The mining industry in eastern Serbia started in the first half of the 5th millennium BC, as verified by the archaeological site "Rudna Glava," which is noted in scientific literature as the centre of early European copper metallurgy (Filipović, 2015). The copper mine "Bor" (Figure 11.2)

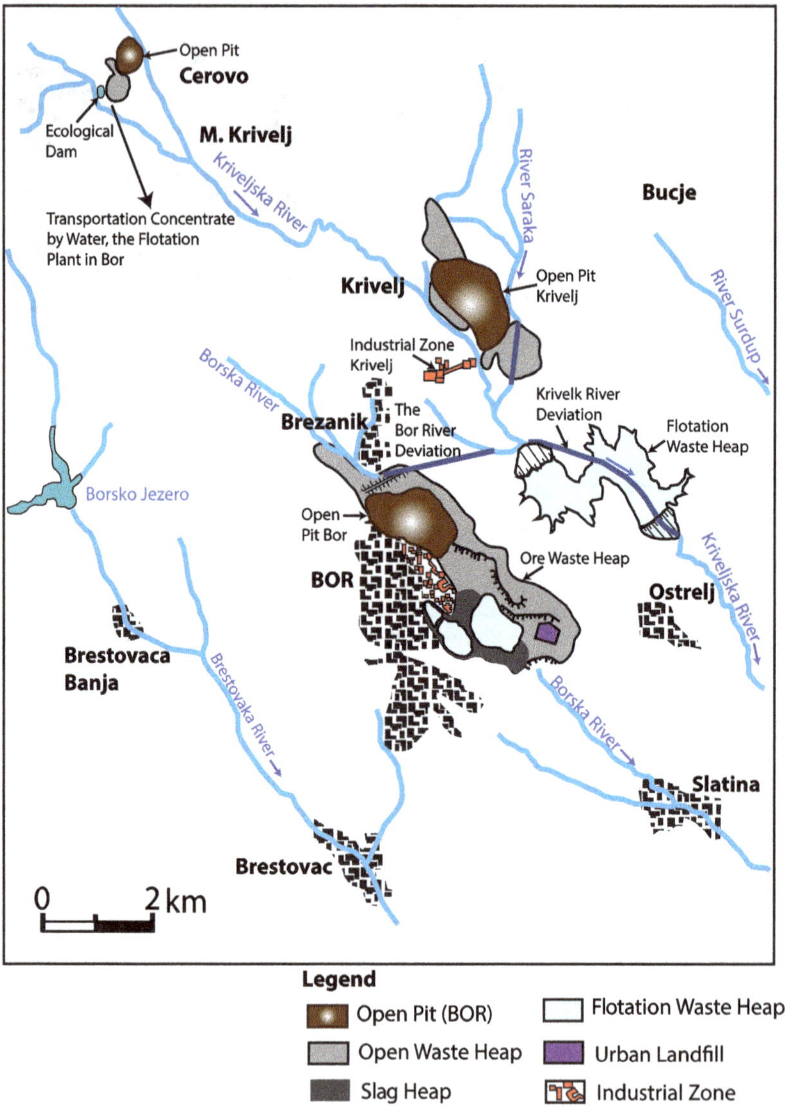

Figure 11.2. Bor copper mines.

Source: Tasić *et al.* (2017).

began in 1903, being funded by foreign capital. Metal production has increased over time, producing large quantities of waste products and therefore creating complex ecological problems (Tasić, 2017).

The Bor metallogenic zone is one of the most important metallogenic units in the Republic of Serbia. Volcanic processes in this unit are characterised by the domination of extrusive volcanic activity and the change in depositional environment during the numerous volcanic cycles, as well as facial transitions and huge deposition of syn- and post-eruptive resedimented volcanoclastic. The predominant metals in the Bor metallogenic zone are copper and gold, accompanied by iron, base metals, silver, molybdenum and minor platinum-group elements. The most prominent morphogenetic types of deposits comprise porphyry copper-gold, cupriferous pyrite, massive base-metal sulphides and hydrothermal veins, iron oxide skarns, carbonate replacement polymetallic deposits, volcanogenic epithermal gold mineralisation of the high sulphidation type and exceptionally rare clasts of copper sulphide ore mechanically accumulated in small sedimentary basins filled by pyroclastic materials. The total production of the Bor metallogenic zone since 1902 has been approximately 652 Mt of ore, with 4.93 Mt of copper and 280 tonnes of gold. Mineral resources of the Bor metallogenic zone are estimated at over 20 million of tonnes of copper and 1,000 tonnes of gold (Jelenković *et al.*, 2016).

11.2.3 *The history of metal mining in the municipality of Bor (1900–present)*

Underground mining of copper ores in Bor began in 1903. Almost a decade later, open-pit Bor operations started in 1912 and continued until 1986. A few decades ago, two more copper mines were opened in the vicinity of Bor, at Veliki Krivelj (1979) and Cerovo (1990). These mines are situated on the northeastern rim of the town, so that the open pit, the metallurgical smelting complex and the flotation tailings pond make a boundary between the urban and the industrial zones (Stanković *et al.*, 2009).

Continued immigration over the decades of workers for the mining and smelting complex resulted in the gradual transformation of the originally agricultural-based village of Bor into today's Town of Bor, with most of the urban residential area built close to and around its main employer, i.e., the mining and smelting complex (the city's centre is less than 1 km away from pits and smelter). The industrial activities in Bor, particularly those by the mining and smelter complex, have resulted in substantive negative impacts on the environment in the region (including air, water and soil) as well as having raised serious concerns about associated health effects on the population at large (Matić *et al*., 2017). The fact that the main polluter (located in the urban centre of Bor) is also the main employer in the area highlights the need to solve the environmental problems in a wider economic and social context.

11.3 Continuous Environmental Pollution in the Bor Municipality

The main source of continuous environmental pollution in Bor is the Mining & Smelter Complex Bor; in particular, the following activities are the main contributors:

- flotation (water pollution),
- smelting (air, water and soil pollution),
- open-cast pits and surrounding waste heaps (air and water pollution),
- underground mining (water pollution).

Other polluting sources include a thermal power/heating plant (air pollution) and the municipal sewage system (air, water and soil pollution).

Air emissions from the Copper Smelter Bor are a major environmental problem for the region. The main emission sources are the smelting process, open-cast pits and tailing ponds. The central heating plant in Bor is another source of air pollution. In 2002, it was estimated that the plant emitted approximately 70,000 tonnes of sulphur dioxide, several hundred tonnes of heavy metal-contaminated particulates, 360 tonnes of arsenic, 83 tonnes of lead, 830 tonnes of zinc and 0.1–0.2 tonnes of mercury. These high figures are, nevertheless, lower than the levels of emissions during the plant's peak production periods (1990), during which some 250,000 tonnes of sulphur dioxide, 1,000 tonnes of heavy-metal contaminated particulates, 1,000 tonnes of arsenic, 500 tonnes of lead, 2,500 tonnes of zinc and 1.6 tonnes of mercury were emitted annually.

The content of Arsenic (As) and heavy metals (Pb, Cd, Ni and Cu) in total suspended particles (TSPs) and PM_{10} at three locations (Park, Institute, and Jugopetrol) near the copper smelter in Bor (Serbia) has been analysed for the period 2004–2015 with the aim of investigating the seasonal and spatial changes of those pollutants in the suspended particles. The content of As in PM_{10} was over the annual EU limit value at all measuring points during the entire period of observation, while the contents of Cd and Pb were periodically above the annual EU limits (Tasić *et al.*, 2017).

High levels of SO_2 and metals/metalloids were measured from 2009 to 2015 in Bor. Enormously high daily SO_2 concentration measured at the suburban zone (3734 μg/m^3) was 187 times higher than the limit values (LV) given by the World Health Organization. Annual arsenic concentrations exceeded the LV at all the measuring sites during the study period. Extremely high annual As level in 2012 was 21 times higher than the LV proposed by the European Union. The annual lead and cadmium concentrations frequently exceeded the LV (Serbula *et al.*, 2014). It was observed that the soil in the tourist zone of Bor (Bor Lake and Brestovac Spa) was greatly polluted with Cu and As (Kalinović *et al.*, 2017). Mean annual As concentrations in PM_{10} in Bor for the period 2012–2017 are shown in Figure 11.3 (courtesy of the Mining and Metallurgy Institute Bor).

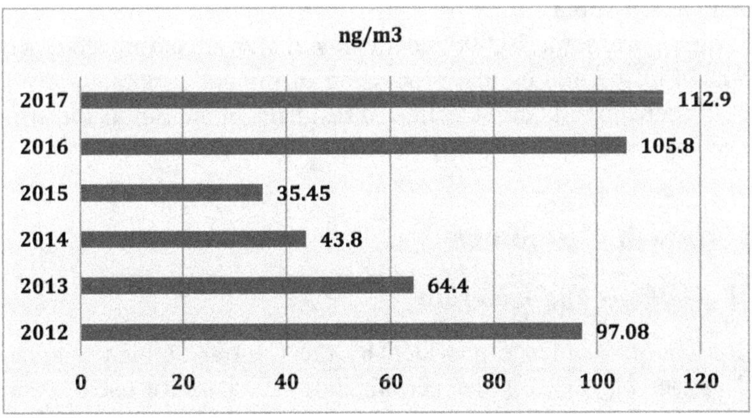

Figure 11.3. Mean annual concentrations of As (ng/m^3) in PM_{10} in Bor (2012–2017).

Based on the estimate obtained from the Air Quality Plan for Agglomeration Bor (2013), areas under the influence of polluted air comprise one-third of the surface of the agglomeration in Bor (298 km^2), with about 40,000 of inhabitants.

A number of environmental impact factors in this area may be recognised, such as air pollution, biodiversity loss (wildlife, agro-diversity), food insecurity (crop damage), loss of landscape (aesthetic degradation), noise pollution, soil contamination (approximately 22,000 ha) (Dimitrijević *et al.*, 2016), soil erosion (due to acid mine drainage), deforestation and loss of vegetation cover, surface water pollution, decreasing water quality, groundwater pollution or depletion and huge waste production.

11.4 Study Population/Reference Population

The study population at critical risk from exposure to environmental pollution (especially to air pollution) related to copper mining and smelting activities is the entire population of Bor. According to the 2011 census, it includes 50,159 inhabitants (0.7% of the total Serbian population and 38.1% of the Bor District). A large part of the population (about 34,000 inhabitants) lives in urban settlements in Bor, while the rest of the population lives in rural areas in 13 villages (local communities and settlements). Bor consists of the central settlement and the administrative seat—Bor, and 13 villages: Gornjane, Tanda, Luka, Krivelj, Bučje, Oštrelj, Donja Bela Reka, Brestovac, Slatina, Zlot, Šarbanovac, Metovnica and Topla.

Furthermore, some 10,000 inhabitants of Bor are employed in activities related to the mining and processing of mineral raw materials. This aspect further supports the selection of residents in the Bor as the primary exposed population.

11.5 Health Outcomes

11.5.1 *Methods and materials*

Data on cancer incidence provided by the Central Serbian population-based cancer registry for the period 2001–2015 were used. Data are available for all ages, all cancer types, by gender and at settlement, municipality, district and region levels (Central Serbia) for the period 1999–2015. The analysis concerned health indicators in the Bor District and Bor.

Incidence indicators. Incidence indicators are crude incidence, standardised incidence and standardised incidence ratio (SIR) with 90% confidence intervals (CIs) of all new cancer cases and selected cancer cases (ICD-10) by gender, all ages (16 classes) and geographical area (settlement or municipality) for the period 2001–2015.

Mortality indicators. Mortality indicators are crude mortality, standardised mortality and standardised mortality ratio (SMR) with 90% CIs of all cancers and selected cancer sites (ICD-10 and ICD-O3) by gender, all ages (16 classes) and geographical area (municipality) in Central Serbia for the period 2001–2015. The methodology for this study on major health outcomes in residents living in Bor followed the principles of the SENTIERI methodology used in studying the major health outcomes for residents in the Bor District and the Town of Bor.

The data source for this study is the National Cancer Registry, generated on a regular basis in the Institute of Public Health of Serbia "Dr. Milan Jovanović Batut," Department for prevention and control of non-communicable diseases. Besides this Department, Institute holds the Department for prevention and control of communicable diseases. Both departments mostly employ medical doctors, epidemiologists.

11.5.2 *Goals*

The following are the goals of the study:

- to describe the population health profiles in the Town of Bor;
- to analyse cancer incidence and mortality indicators: crude rate, standardised rate and standardised incidence and mortality ratios (SIR and SMR) with 90% CIs, for selected cancer sites and all cancer localisation in Bor;
- to analyse other mortality indicators: all causes of death, including deaths from cardiovascular, respiratory, digestive and genitourinary diseases;
- to compare the observed cases with the expected cases based on standard age classes, gender, geographical area and cancer sites' specific incidence and mortality rates.
- the analysis of health indicators of incidence and mortality in contaminated sites (CSs)—health risk estimations based on reliable SENTIERI methodology;
- to list the causes of mortality (all causes, cardiovascular diseases, respiratory, digestive, genitourinary, etc.).

11.5.3 *Results*

Following the principles of the SENTIERI methodology in studying the major health outcomes for residents of settlements close to CSs, the first-level approach in this study was to describe the population health profiles in contaminated areas for residents in selected CSs in the northeast area of the Republic of Serbia (totalling eight municipalities: four in the Bor District and four neighbouring municipalities of this district).

The analysis concerned the list of 35 cancer sites for all ages and both genders in the Bor District and Bor for the period 2006–2015. Results from this analysis clearly show that increase of risk has been confirmed, not only in Bor, but in the entire Bor District (four municipalities, 125,000 inhabitants) are, mostly confirmed and even raised in the analysis including only the Town of Bor (50,000 inhabitants). For example, in the overall district, the risk from all cancer sites is 1.02 (3,269 obs., 90% CI 1.00–1.05) in males, with a defect in females SIR = 0.88 (2,990 obs., 90% CI 0.85–0.90), while in Bor it was observed as an excess risk in both males (1,310 obs., SIR = 1.18, 90% CI 1.13–1.23) and females (1,246 obs., SIR = 1.05, 90% CI: 1.00–1.10). Also, for some targeted sites (liver, trachea, bronchus and lung), an excess risk in both males and females was observed. The results have shown that the cancer risk profile of the exposed population should be analysed by choosing the most suitable reference population, i.e., the population of Southeast Serbia, which is as much as possible similar to the study population of Bor and with the exception of the exposure of interest. According to the principles of the SENTIERI methodology for residents in the Bor District and Bor, the incidence indicators were analysed and computed for each of the 35 cancer sites, which is given in Tables 11.2–11.10).

The analysis showed that all malignant cancers except for skin exhibited a certain pattern in their risk, going from area to area in the Bor District, with the highest SIR detected in Bor in both males and females (SIR = 1.18 in men and 1.05 in women), while a weakening of the risk is observed when the target study population includes surrounding municipalities. This pattern is also observed for specific cancer sites, including cancer of the liver (SIR = 1.40 in men and 1.23 in women), colon rectum (SIR = 1.28 in men), pancreas (SIR = 2.67 in men and 1.03 in women), kidney (SIR = 1.28 in men and 1.39 in women), bladder (SIR = 1.44 in men), lymphopoietic tissue (SIR = 1.30 in men) and non-Hodgkin's lymphoma (SIR = 1.68 in men) and mesothelioma (SIR = 5.40 in men and

Table 11.2. New cancer cases and incidence in the Bor District, by primary site (ICD-10) and sex, 2006–2015.

ICD-10		Men			Women		
Code	Primary Site	Number	CR	ASR(W)	Number	CR	ASR(W)
C00–C96	All sites	3269	521.5	264.5	2990	456.6	235.4
C15	Oesophagus	50	8.0	4.1	17	2.6	1.4
C16	Stomach	129	20.6	9.4	69	10.5	4.3
C18–C21	Colon, rectum, anus	478	76.3	36.1	301	46.0	19.5
C22	Liver, intrahepatic bile ducts	85	13.6	6.5	72	11.0	4.2
C23	Gallbladder and biliary tract	20	3.2	1.3	16	2.4	1.2
C25	Pancreas	128	20.4	10.2	73	11.1	4.6
C32	Larynx	111	17.7	9.6	17	2.6	1.4
C33–C34	Trachea, bronchi, lungs	686	109.4	55.3	244	37.3	18.9
C40–C41	Bone and articular limb cartilage	27	4.3	3.2	15	2.3	1.3
C43	Melanoma cutis	38	6.1	3.4	40	6.1	3.0
C45	Mesothelioma	12	1.9	0.9	6	0.9	0.4
C47, C49	Connective and soft tissues	24	3.8	2.5	23	3.5	2.5
C50	Breast	22	3.5	1.6	727	111.0	58.1
C53–C55	Cervix uteri, uterus				527	80.5	47.9
C53	Cervix uteri				331	50.5	32.8
C54	Corpus uteri				162	24.7	12.5
C55	Uterus, unspecified part				34	5.2	2.6
C56	Ovary				176	26.9	14.3
C61	Prostate	455	72.6	29.6			
C62	Testis	42	6.7	6.9			

(*Continued*)

Table 11.2. (*Continued*)

ICD-10 Code	Primary Site	Men			Women		
		Number	CR	ASR(W)	Number	CR	ASR(W)
C64–C66	Kidney, renal pelvis, ureter	102	16.3	9.3	68	10.4	5.4
C67	Bladder	250	39.9	19.8	83		
C70–C72	Meninges, brain, central nervous system	103	16.4	10.3	78	11.9	7.3
C73	Thyroid gland	12	1.9	1.2	29	4.4	3.1
C81–C96	Lymphopoietic tissue	179	28.6	17.7	128	19.5	11.5
C81	Hodgkin's disease	19	3.0	2.1	16	2.4	2.4
C82–C85, C96	Non-Hodgkin's lymphoma	70	11.2	7.1	41	6.3	3.7
C88–C90	Multiple myeloma	26	4.1	2.3	27	4.1	2.0
C91–C95	Leukaemia	72	11.5	6.9	54	8.2	4.2
C91.0–C91.1	Lymphoid leukaemia	25	4.0	2.2	27	4.1	2.0
C91.0	Acute lymphoid leukaemia	6	1.0	0.7	8	1.2	1.0
C91.1	Chronic lymphoid leukaemia	19	3.0	1.5	24	3.7	1.2
C92.0–C92.1	Myeloid leukaemia	33	5.3	3.2	20	3.1	1.7
C92.0	Acute myeloid leukaemia	6	1.0	1.2	2	0.3	0.5
C92.1	Chronic myeloid leukaemia	27	4.3	2.0	18	2.7	1.2

Notes: CR: crude rate, ASR(W): standardised incidence rate (world population).

Table 11.3. Standardised incidence ratio, 90% CI, by primary site (ICD-10) and sex, 2006–2015.

ICD-10 Code	Primary Site	Men				Women			
			SIR	90% CI			SIR	90% CI	
C00–C96	All sites but skin	3269	1.02	1.00	1.05	2990	0.88	0.85	0.90
C15	Oesophagus	50	1.20	0.93	1.51	17	1.31	0.83	1.96
C16	Stomach	129	0.85	0.73	0.98	69	0.65	0.53	0.80
C18–C21	Colon, rectum, anus	478	1.02	0.95	1.10	301	0.79	0.72	0.87
C22	Liver and intrahepatic bile ducts	85	1.15	0.95	1.37	72	1.10	0.90	1.34
C23	Gallbladder and biliary tract	20	2.01	1.33	2.93	16	0.59	0.37	0.90
C25	Pancreas	128	1.29	1.11	1.49	73	0.72	0.59	0.88
C32	Larynx	111	0.84	0.72	0.99	17	1.29	0.82	1.93
C33–C34	Trachea, bronchus and lungs	686	0.88	0.82	0.93	244	0.78	0.70	0.86
C40–C41	Bone and articular limb cartilage	27	1.48	1.05	2.04	15	0.84	0.52	1.30
C43	Melanoma cutis	38	0.66	0.49	0.86	40	0.63	0.47	0.81
C45	Mesothelioma	12	3.82	2.20	6.18	6	3.74	1.63	7.38
C47, C49	Connective and soft tissue	24	1.27	0.88	1.79	23	1.83	1.25	2.60
C50	Breast	22	1.34	0.90	1.91	727	0.81	0.76	0.86
C53–C55	Cervix uteri and uterus					527	1.07	1.00	1.15
C53	Cervix uteri					331	1.25	1.14	1.37
C54	Corpus uteri					162	0.75	0.65	0.85
C55	Uterus, unspecified part					34	3.53	2.60	4.70
C56, C57	Ovary					176	1.21	1.07	1.37
C61	Prostate	455	1.13	1.04	1.22				
C62	Testis	42	1.06	0.81	1.37				

(*Continued*)

Table 11.3. (*Continued*)

ICD-10 Code	Primary Site	Men				Women			
			SIR	90% CI			SIR	90% CI	
C64–C66	Kidney, renal pelvis, ureter	102	1.08	0.91	1.28	68	0.97	0.78	1.19
C67	Bladder	250	1.12	1.01	1.25	83	0.88	0.73	1.06
C70–C72	Meninges, brain and central nervous system	103	1.11	0.93	1.30	78	0.88	0.72	1.06
C73	Thyroid gland	12	0.73	0.42	1.18	29			
C81 C96	Lymphopoietic system	179	1.02	0.90	1.15	128	0.77	0.55	0.90
C81	Hodgkin's disease	19	1.09	0.71	1.59	15	1.20	0.75	1.83
C82–C85, C96	Non-Hodgkin's myeloma	70	1.14	0.93	1.39	41	0.65	0.49	0.84
C88–C90	Multiple myeloma	26	0.68	0.48	0.95	27	0.89	0.63	1.23
C91–C95	Leukaemia	72	1.02	0.83	1.25	54	0.90	0.71	1.13
C91.0–C91.1	Lymphoid leukaemia	25	0.72	0.50	1.00	27	0.97	0.68	1.34
C91.0	Acute lymphoid leukaemia	6	1.22	0.53	2.40	8	4.67	2.32	8.43
C91.1	Chronic lymphoid leukaemia	19	0.64	0.42	0.93	24	0.97	0.67	1.37
C92.0–C92.1	Myeloid leukaemia	33	1.24	0.91	1.66	20	0.88	0.58	1.28
C92.0	Acute myeloid leukaemia	6	1.04	0.45	2.06	2	0.38	0.07	1.20
C92.1	Chronic myeloid leukaemia	27	1.30	0.92	1.79	18	1.03	0.66	1.52

Notes: SIR: standardised incidence ratio, 90% CI: 90% confidence interval. Reference: Central Serbia, number of cancer cases: 6.259, area: Bor District.

1.84 in women). Also, a similar pattern is observed for cervical cancer (SIR = 1.33) and ovarian cancer (SIR = 1.32) among females and for prostate cancer (SIR = 1.28) among males. The lowest risk is observed for almost all cancer sites in people living in areas outside the Bor District

Table 11.4. New cancer cases and incidence in Bor, by primary site (ICD-10) and sex, 2006–2015.

ICD-10 Code	Primary Site	Men			Women		
		Number	CR	ASR(W)	Number	CR	ASR(W)
C00–C96	All sites but skin	1310	536.7	303.7	1246	493.8	264.5
C15	Oesophagus	9	3.7	2.2	5	2.0	0.9
C16	Stomach	44	18.0	9.3	27	10.7	4.6
C18–C21	Colon, rectum, anus	205	84.0	45.2	127	50.3	23.5
C22	Liver, intrahepatic bile ducts	35	14.3	7.5	26	10.3	4.2
C23	Gallbladder and biliary tract	8	3.3	1.7	4	1.6	1.1
C25	Pancreas	48	19.7	11.1	34	13.5	6.8
C32	Larynx	48	19.7	11.0	9	3.6	2.1
C33–C34	Trachea, bronchi, lungs	284	116.4	63.2	110	43.6	22.0
C40–C41	Bone and articular limb cartilage	7	2.9	2.1	4	1.6	0.7
C43	Melanoma cutis	20	8.2	4.7	17	6.7	3.4
C45	Mesothelioma	6	2.5	1.3	1	0.4	0.1
C47, C49	Connective and soft tissues	10	4.1	3.3	11	4.4	3.0
C50	Breast	14	5.7	2.7	317	125.6	68.0
C53–C55	Cervix uteri, uterus				219	86.8	52.0
C53	Cervix uteri				131	51.9	33.6
C54	Corpus uteri				68	26.9	13.9
C55	Uterus, unspec. part				20	7.9	4.3
C56, C57	Ovary				69	27.3	14.8
C61	Prostate	165	67.6	34.3			
C62	Testis	18	7.4	6.9			

(*Continued*)

Table 11.4. (*Continued*)

ICD-10 Code	Primary Site	Men Number	Men CR	Men ASR(W)	Women Number	Women CR	Women ASR(W)
C64–C66	Kidney, renal pelvis, ureter	43	17.6	10.9	33	13.1	6.3
C67	Bladder	109	44.7	25.8	31	12.3	6.1
C70-C72	Meninges, brain, central nervous system	41	16.8	12.1	28	11.1	6.1
C73	Thyroid gland	5	2.0	1.3	14	5.5	3.7
C81–C96	Lymphopoietic tissue	81	33.2	20.9	51	20.2	12.6
C81	Hodgkin's disease	7	2.9	1.9	6	2.4	1.8
C82–C85, C96	Non-Hodgkin's lymphoma	37	15.2	9.8	20	7.9	4.5
C88–C90	Multiple myeloma	9	3.7	2.1	5	2.0	1.3
C91–C95	Leukaemia	29	11.9	7.3	19	7.5	5.1
C91.0–91.1	Lymphoid leukaemia	8	3.3	1.8	5	2.0	1.3
C91.0	Acute lymphoid leukaemia	1	0.8	0.4	2	0.8	0.8
C91.1	Chronic lymphoid leukaemia	4	2.5	1.4	2	1.2	0.5
C92.0–C92.1	Myeloid leukaemia	13	5.3	3.7	10	4.0	2.9
C92.0	Acute myeloid leukaemia	3	1.2	1.5	2	0.8	1.3
C92.1	Chronic myeloid leukaemia	6	4.1	2.2	5	3.2	1.6

Notes: CR: crude rate; ASR(W) – standardised incidence sate (world population).

Table 11.5. Standardised incidence ratio, 90% CI, by primary site (ICD-10) and sex, Bor, 2006–2015.

ICD-10 Code	Primary Site	Men				Women			
			SIR	90% CI			SIR	90% CI	
C00–C96	All sites but skin	1310	1.18	1.13	1.23	1246	1.05	1.00	1.10
C15	Oesophagus	9	0.60	0.31	1.04	5	1.13	0.45	2.38
C16	Stomach	44	0.85	0.65	1.09	2.7	0.77	0.55	1.07
C18–C21	Colon, rectum, anus	205	1.28	1.14	1.44	127	0.99	0.85	1.15
C22	Liver and intrahepatic bile ducts	35	1.40	1.03	1.85	26	1.23	0.86	1.70
C23	Gallbladder and biliary tract	8	2.43	1.21	4.39	4	0.46	0.16	1.05
C25	Pancreas	128	2.67	2.29	3.09	34	1.03	0.76	1.37
C32	Larynx	48	1.00	0.78	1.27	9	1.15	0.60	2.01
C33–C34	Trachea, bronchus and lungs	284	1.02	0.92	1.13	110	1.00	0.85	1.17
C40–C41	Bone and articular limb cartilage	7	1.03	0.48	1.94	4	0.62	0.21	1.43
C43	Melanoma cutis	20	0.97	0.64	1.41	17	0.75	0.48	1.13
C45	Mesothelioma	6	5.40	2.35	10.65	1	1.84	0.07	8.69
C47, C49	Connective and soft tissue	10	1.44	0.78	2.45	11	1.44	0.80	2.38
C50	Breast	5	0.87	0.34	1.82	317	0.98	0.89	1.08
C53–C55	Cervix uteri and uterus					219	1.22	1.09	1.37
C53	Cervix uteri					131	1.33	1.14	1.53
C54	Corpus uteri					68	0.88	0.71	1.08
C55	Uterus, unspecified part					20	6.20	4.11	9.00
C56, C57	Ovary					69	1.32	1.07	1.62
C61	Prostate	165	1.28	1.12	1.45				
C62	Testis	18	1.13	0.73	1.68				
C64–C66	Kidney, renal pelvis, ureter	43	1.28	0.98	1.66	33	1.39	1.01	1.85
C67	Bladder	109	1.44	1.22	1.69	31	0.98	0.71	1.32

(*Continued*)

Table 11.5. (*Continued*)

ICD-10 Code	Primary Site	Men				Women			
			SIR	90% CI			SIR	90% CI	
C70–C72	Meninges, brain and central nervous system	41	1.21	0.92	1.57	28	0.89	0.63	1.22
C73	Thyroid gland	5	0.82	0.32	1.73	14	0.74	0.45	1.15
C81 C96	Lymphopoietic system	81	1.30	1.07	1.56	51	0.89	0.69	1.12
C81	Hodgkin's disease	7	1.04	0.49	1.95	6	1.19	0.52	2.35
C82–C85, C96	Non-Hodgkin's myeloma	37	1.68	1.25	2.21	20	0.90	0.60	1.31
C88–C90	Multiple myeloma	9	0.68	0.35	1.19	5	0.49	0.19	1.02
C91–C95	Leukaemia	29	1.18	0.85	1.61	19	0.93	0.61	1.36
C91.0–C91.1	Lymphoid leukaemia	8	0.66	0.33	1.19	5	0.53	0.21	1.12
C91.0	Acute lymphoid leukaemia	2	0.99	0.17	3.11	2	1.60	0.28	5.03
C91.1	Chr. lymphoid leukaemia	6	0.59	0.26	1.17	3	0.37	0.10	0.96
C92.0–C92.1	Myeloid leukaemia	13	1.39	0.82	2.21	10	1.26	0.68	2.13
C92.0	Acute myeloid leukaemia	3	1.33	0.36	3.43	2	0.97	0.17	3.05
C92.1	Chr. myeloid leukaemia	10	1.41	0.77	2.40	8	1.36	0.68	2.45

Notes: SIR: standardised incidence ratio, 90% CI: 90% confidence interval. Reference: All Central Serbia, number of cancer cases: 2,556, area: Bor District and Bor.

and in the overall region consisting of eight municipalities (the Bor District plus four other municipalities outside Bor).

The results showed that the cancer risk profile of the exposed population should be analysed by choosing the most suitable reference population, i.e., the population of the southeast region of Serbia, which is as much as possible similar to the study population of Bor and with the exception of the exposure of interest.

The new analysis concerns a 15-year period (2001–2015) using the cancer rates from Southeast Serbia as the reference. As compared to the

Table 11.6. New cancer cases and incidence in Bor, by primary site (ICD-10) and sex, 2001–2015.

ICD-10		Men			Women		
Code	Primary Site	Number	CR	ASR(W)	Number	CR	ASR(W)
C00–C96	All sites	1818	480.0	284.4	1746	446.7	252.3
C15	Oesophagus	12	3.2	1.9	7	1.8	0.8
C16	Stomach	66	17.4	9.6	45	11.5	5.7
C18–C21	Colon, rectum, anus	275	72.6	40.7	168	43.0	20.9
C22	Liver, intrahepatic bile ducts	48	12.7	7.1	33	8.4	3.7
C23	Gallbladder and biliary tract	8	2.1	1.2	9	2.3	1.4
C25	Pancreas	58	15.3	9.1	50	12.8	6.5
C32	Larynx	64	16.9	10.4	12	3.1	1.8
C33–C34	Trachea, bronchi, lungs	421	111.1	63.3	141	36.1	19.4
C40–C41	Bone and articular limb cartilage	14	3.7	3.4	6	1.5	1.0
C43	Melanoma cutis	24	6.3	3.8	20	5.1	2.7
C45	Mesothelioma	6	1.6	0.9	1	0.3	0.1
C47, C49	Connective and soft tissues	17	4.5	3.3	17	4.3	3.4
C50	Breast	22	5.8	3.0	446	114.1	64.8
C53–C55	Cervix uteri, uterus				349	89.3	54.2
C53	Cervix uteri				227	58.1	37.6
C54	Corpus uteri				94	24.0	12.6
C55	Uterus, unspec. part				28	7.2	4.1
C56, C57	Ovary				88	22.5	12.7
C61	Prostate	225	59.4	31.6			
C62	Testis	25	6.6	6.1			

(*Continued*)

Table 11.6. (*Continued*)

ICD-10 Code	Primary Site	Men			Women		
		Number	CR	ASR(W)	Number	CR	ASR(W)
C64–C66	Kidney, renal pelvis, ureter	61	16.1	9.9	44	11.3	5.8
C67	Bladder	160	42.2	24.8	41	10.5	5.5
C70–C72	Meninges, brain, central nervous system	57	15.0	10.9	42	10.7	6.1
C73	Thyroid gland	9	2.4	1.5	19	4.9	3.4
C81–C96	Lymphopoietic tissue	107	28.2	19.9	65	16.6	11.6
C81	Hodgkin's disease	10	2.6	2.0	12	3.1	3.0
C82–C85, C96	Non-Hodgkin's lymphoma	47	12.4	8.5	25	6.4	3.9
C90	Mult. myeloma	15	4.0	3.3	6	1.5	1.1
C91–C95	Leukaemia	40	10.6	7.7	22	5.6	4.0
C91.0–C91.1	Lymphoid leukaemia	14	3.7	3.1	6	1.5	1.1
C91.0	Acute lymphoid leukaemia	5	1.3	1.8	2	0.5	0.5
C91.1	Chron. lymphoid leukaemia	9	2.4	1.3	4	1.0	0.6
C92.0–C92.1	Myeloid leukaemia	14	3.7	2.5	10	2.6	1.9
C92.0	Acute myeloid leukaemia	4	1.1	1.1	3	0.8	0.9
C92.1	Chronic myeloid leukaemia	10	2.6	1.5	7	1.8	1.0

Notes: CR: crude rate, ASR(W): standardised incidence rate (world population).

Table 11.7. Standardised incidence ratio, 90% CI, by primary site (ICD-10) and sex, Bor, 2001–2015.

ICD-10 Code	Primary Site	Men				Women			
			SIR	90% CI			SIR	90% CI	
C00–C96	All sites but skin	1818	1.21	1.16	1.26	1746	1.23	1.18	1.28
C15	Oesophagus	12	0.60	0.35	0.98	7	1.23	0.58	2.31
C16	Stomach	66	0.79	0.64	0.97	45	0.97	0.74	1.24
C18–C21	Colon, rectum, anus	275	1.37	1.23	1.61	168	1.19	1.04	1.38
C22	Liver and intrahepatic bile ducts	48	1.02	0.79	1.30	33	0.98	0.72	1.31
C23	Gallbladder and biliary tract	8	1.46	0.72	2.63	9	0.72	0.38	1.26
C25	Pancreas	58	1.21	0.96	1.50	50	1.38	1.07	1.74
C32	Larynx	64	0.89	0.71	1.09	12	1.35	0.78	2.18
C33–C34	Trachea, bronchus and lungs	421	1.15	1.06	1.25	141	1.27	1.10	1.46
C40–C41	Bone and articular limb cartilage	14	1.22	0.74	1.90	6	0.64	0.28	1.26
C43	Melanoma cutis	24	1.17	0.81	1.65	20	0.93	0.62	1.36
C45	Mesothelioma	6	3.66	1.59	7.23	1	1.98	0.08	9.37
C47, C49	Connective and soft tissue	17	1.72	1.10	2.58	17	2.24	1.43	3.36
C50	Breast	22	2.56	1.73	3.66	446	1.24	1.14	1.34
C53–C55	Cervix uteri and uterus					349	1.32	1.21	1.44
C53	Cervix uteri					227	1.52	1.36	1.70
C54	Corpus uteri					94	0.89	0.74	1.05
C55	Uterus, unspecified part					28	2.99	2.13	4.10
C56, C57	Ovary					88	1.26	1.05	1.50
C61	Prostate	225	1.44	1.29	1.61				
C62	Testis	25	1.08	0.75	1.51				
C64–C66	Kidney, renal pelvis, ureter	6.1	1.55	1.24	1.91	44	1.67	1.28	2.15
C67	Bladder	160	1.54	1.35	1.76	41	1.24	0.94	1.61

(Continued)

Table 11.7. (*Continued*)

ICD-10 Code	Primary Site	Men				Women			
			SIR	90% CI			SIR	90% CI	
C70–C72	Meninges, brain and central nervous system	57	0.99	0.78	1.23	42	0.94	0.72	1.22
C73	Thyroid gland	9	1.62	0.85	2.83	19	1.10	0.72	1.62
C81 C96	Lymphopoietic system	107	1.58	1.34	1.88	65	1.28	1.03	1.57
C81	Hodgkin's disease	10	1.10	0.60	1.87	12	1.87	1.08	3.03
C82–C85, C96	Non-Hodgkin's lymphoma	47	2.08	1.61	2.65	25	1.40	0.98	1.96
C88–C90	Multiple myeloma	15	1.06	0.65	1.63	6	0.60	0.26	1.19
C91–C95	Leukaemia	40	1.39	1.05	1.81	22	1.08	0.73	1.54
C91.0–C91.1	Lymphoid leukaemia	14	1.01	0.61	1.57	6	0.61	0.26	1.20
C91.0	Acute lymphoid leukaemia	5	1.87	0.73	3.92	2	1.17	0.20	3.69
C91.1	Chr. lymphoid leukaemia	9	0.80	0.42	1.40	4	0.49	0.17	1.12
C92.0 –C92.1	Myeloid leukaemia	14	1.40	0.86	2.19	10	1.32	0.72	2.24
C92.0	Acute myeloid leukaemia	4	2.56	0.87	5.86	3	2.32	0.63	5.98
C92.1	Chr. myeloid leukaemia	10	1.18	0.64	2.01	7	1.12	0.52	2.09

Notes: SIR: standardised incidence ratio, 90% CI: 90% confidence interval. Reference: South East Serbia, number of cancer cases: 3,564, area: Bor.

previous analyses (2006–2015), we found 3,564 cancer cases overall (i.e., all malignant cancers except for skin), with 1,008 extra cases in five years. The analysis showed that for all malignant cancers except for skin, there exists a significantly stronger excess of risk in Bor in both males and females (SIR = 1.21 in men and 1.23 in women).

If we look into lung cancer as a cancer site related to exposures from smelters, in the previous analyses, there was no excess of risk (SIR = 1.02 in men and 1.00 in women), although some increasing trends appear from area to area in the Bor District. The new analysis shows a statistically

Table 11.8. Cancer death cases and mortality in Bor, by primary site (ICD-10) and sex, 2001–2015.

ICD-10		Men			Women		
Code	**Primary Site**	**Number**	**CR**	**ASR(W)**	**Number**	**CR**	**ASR(W)**
C00–C96	All sites	1148	303.1	170.8	944	241.6	118.8
C15	Oesophagus	14	3.7	2.1	4	1.0	0.4
C16	Stomach	77	20.3	20.9	8.3	21.2	9.2
C18–C21	Colon, rectum, anus	147	38.8	20.9	83	21.2	9.2
C22	Liver, intrahepatic bile ducts	64	16.9	9.1	49	12.5	5.6
C25	Pancreas	54	14.3	8.1	49	12.5	5.7
C32	Larynx	34	9.0	5.1	8	2.0	0.8
C33–C34	Trachea, bronchi, lungs	351	92.7	52.5	107	27.4	13.5
C43	Melanoma cutis	15	4.0	2.2	13	3.3	1.6
C45	Mesothelioma	1	0.3	0.2	0	0.0	0.0
C47, C49	Connective and soft tissues	2	0.5	1.1	17	0.8	0.7
C50	Breast	4			172	44.0	21.5
C53–C55	Cervix uteri, uterus				157	40.2	21.2
C56, C57	Ovary				70	17.9	8.0
C61	Prostate	106	28.0	13.9			
C62	Testis	4	1.1	0.8			
C64–C66	Kidney, renal pelvis, ureter	25	6.6	3.6	8	2.0	0.8
C67	Bladder	37	9.8	5.4	11	2.8	1.2
C70–C72	Meninges, brain, central nervous system	37	9.8	5.4	41	2.8	1.2
C81–C96	Lymphopoietic tissue	59	9.8	6.8	41	10.5	5.9

(Continued)

Table 11.8. (*Continued*)

ICD-10 Code	Primary Site	Men Number	Men CR	Men ASR(W)	Women Number	Women CR	Women ASR(W)
C81	Hodgkin's disease	8	15.6	9.4	2	10.5	6.4
C82–C85, C96	Non-Hodgkin's lymphoma	21	2.1	1.3	14	0.5	0.5
C88–C90	Multiple myeloma	11	5.5	2.9	3	3.6	2.0
C91–C95	Leukaemia	23	2.9	1.6	28	0.8	0.4
C91.0-C91.1	Lymphoid leukaemia	8	6.1	4.0	5	7.2	4.5
C92.0-C92.1	Myeloid leukaemia	12	2.1	1.7	17	1.3	0.8

Notes: CR: crude rate, ASR(W): standardised incidence rate (world population).

significant lung cancer excess risk, both in men (SIR = 1.15) and in women (SIR = 1.27). This pattern is also observed for specific cancer sites, including cancer of the colon and rectum (SIR = 1.37 in men and 1.13 in women), liver (SIR = 1.02 in men), pancreas (SIR = 1.21 in men and 1.38 in women), kidney (SIR = 1.55 in men and 1.67 in women), bladder (SIR = 1.54 in men and 1.24 in women), thyroid gland (SIR = 1.62 in men and 1.10 in women), lymphopoietic tissue (SIR = 1.58 in men and 1.28 in women), Hodgkin's and non-Hodgkin's lymphoma (SIR = 1.10 in men and 1.87 in women and SIR = 2.08 in men and 1.40 in women, respectively), leukaemia (SIR = 1.39 in men and 1.08 in women) and mesothelioma (SIR = 3.66 in men and 1.98 in women). Also, a similar pattern is observed for cervical cancer (SIR = 1.52) and ovarian cancer (SIR = 1.26) among females and for prostate cancer (SIR = 1.44) among males. This new analysis clearly confirms the criticalities in the cancer profile of the Bor population: Many excesses of risk are stronger than those reported in the previous analyses, with additional cancer sites at risk and risks concerning both the male and female subpopulations (which is an index of environmental exposure, whereas, in the presence of only occupational industrial exposure, the excesses of risk are almost

Table 11.9. Standardised mortality ratio, 90% CI, by primary site (ICD-10) and sex, 2001–2015.

ICD-10 Code	Primary Site	Men				Women			
		Number	SMR	90% CI		Number	SMR	90% CI	
C00–C96	All sites	1148	1.11	1.06	1.17	944	1.24	1.18	1.31
C15	Oesophagus	14	0.83	0.50	1.30	4	0.94	0.32	2.14
C16	Stomach	77	1.10	0.90	1.33	34	0.85	0.63	1.13
C18–C21	Colon, rectum, anus	147	1.26	1.09	1.44	83	1.02	0.85	1.23
C22	Liver and intrahepatic bile ducts	64	1.34	1.08	1.65	49	1.52	1.18	1.93
C25	Pancreas	54	1.13	0.89	1.41	49	1.37	1.06	1.74
C32	Larynx	34	0.82	0.61	1.10	8	2.00	0.99	3.60
C33–C34	Trachea, bronchus and lungs	351	1.11	1.01	1.21	107	1.16	0.98	1.36
C43	Melanoma cutis	15	1.34	0.82	2.06	13	1.57	0.93	2.49
C45	Mesothelioma	1	0.19	0.01	0.88	0			
C47, C49	Connective and soft tissue	2	0.95	0.16	3.00	17	2.24	1.43	3.36
C50	Breast	4	1.51	0.51	3.45	172	1.25	1.10	1.42
C53–C55	Cervix uteri and uterus					157	1.76	1.54	2.01
C56, C57	Ovary					70	1.90	1.54	2.32
C61	Prostate	106	1.41	1.19	1.66				
C62	Testis	4	1.43	0.49	3.26				
C64–C66	Kidney, renal pelvis, ureter	25	1.31	0.91	1.83	8	0.67	0.33	1.21
C67	Bladder	37	1.01	0.76	1.33	11	1.05	0.59	1.73
C70–C72	Meninges, brain and central nervous system	37	0.96	0.72	1.27	41	1.55	1.17	2.01

(*Continued*)

Table 11.9. (*Continued*)

ICD-10 Code	Primary Site	Men				Women			
		Number	SMR	90% CI		Number	SMR	90% CI	
C81 C96	Lymphopoietic system	59	1.11	0.88	1.38	41	1.06	0.80	1.37
C81	Hodgkin's disease	8	1.92	0.95	3.46	2	0.86	0.15	2.70
C82–C85, C96	Non-Hodgkin's myeloma	21	1.45	0.97	2.08	14	1.26	0.76	1.97
C88–C90	Multiple myeloma	11	1.01	0.56	1.66	3	0.43	0.12	1.11
C91–C95	Leukaemia	23	0.87	0.60	1.24	28	1.48	1.05	2.03
C91.0–C91.1	Lymphoid leukaemia	8	0.75	0.37	1.35	5	0.73	0.29	1.54
C92.0–C92.1	Myeloid leukaemia	12	1.10	0.63	1.78	17	1.97	1.25	2.95

Notes: SIR: standardised incidence ratio, 90% CI: 90% confidence interval.

exclusively confined to males). These results might be a combination of two phenomena:

- the more accurate reference rates from Southeast Serbia as compared to all Central Serbia, which allow us to compare populations with more similar characteristics for the exposure of interest;
- the large observational period, which includes relevant years for the residential exposure in the Bor area.

As a next-level approach, in order to describe the population mortality profiles in CSs, we performed an analysis of the standardised mortality ratio (SMR) with 90% CIs for the same selected cancer sites in Bor. Mortality data from the National Statistical Office for the same period was used. Besides cancer mortality, other mortality indicators were analysed: for all causes of death, deaths from cardiovascular, respiratory, digestive, and genitourinary diseases. The observed cases were compared with the expected cases based on standard age classes, gender, geographical area and cancer site-specific rates.

Table 11.10. Death cases, mortality rates, standardised mortality ratio, 90% CI and leading mortality causes (ICD-10) by sex, Bor, 2001–2015.

ICD-10 Code	Diseases and Disorders	Sex	Mortality Indicators					
			Observed	CR	ASR(W)	SMR	90% CI	
C00–D48	Malignant neoplasms	Men	1175	308.3	157.6	1.11	1.06	1.17
		Women	957	216.7	99.4	1.23	1.16	1.30
		All	2132	261.9	126.1	1.16	1.12	1.20
E10–E14	Diabetes mellitus	Men	122	32.2	17.0	1.12	0.96	1.30
		Women	181	46.3	17.6	1.18	1.04	1.34
		All	303	39.4	17.5	1.16	1.05	1.27
I00–I99	Circulatory system diseases	Men	2774	732.4	380.3	1.12	1.08	1.15
		Women	2892	739.8	275.2	1.03	1.00	1.06
		All	5666	736.2	324.7	1.07	1.05	1.09
J00–J99	Respiratory system diseases	Men	298	78.7	40.8	1.19	1.08	1.31
		Women	191	48.9	19.2	1.23	1.08	1.38
		All	489	63.5	28.7	1.19	1.11	1.29
K00–K93	Digestive system diseases	Men	166	43.8	24.0	0.86	0.75	0.98
		Women	158	40.4	17.0	1.19	1.04	1.35
		All	324	42.1	20.3	0.99	0.90	1.08
N00–N99	Genitourinary system diseases	Men	120	31.7	16.4	1.09	0.93	1.97
		Women	99	25.3	10.1	1.09	0.92	1.29
		All	219	28.5	12.9	1.09	0.97	1.21
Q00–Q99	Congenital malformations, deformities, chromosome abnormalities	Men	12	3.2	7.3	1.18	0.68	1.91
		Women	9	2.3	6.1	1.19	0.62	2.08
		All	21	2.7	6.7	1.18	0.79	1.70
R00–R99	Symptoms, signs and abnormal clinical and laboratory findings, not classified elsewhere	Men	403	106.4	57.1	1.01	0.93	1.10
		Women	518	132.5	48.6	1.34	1.24	1.44
		All	921	119.7	53.0	1.17	1.11	1.23

(*Continued*)

Table 11.10. (*Continued*)

ICD-10 Code	Diseases and Disorders	Sex	Mortality Indicators: Observed	CR	ASR(W)	SMR	90% CI	
V01–Y86	External causes of morbidity, mortality	Men	293	77.4	56.4	1.22	1.10	1.34
		Women	131	33.5	19.1	0.90	0.78	1.04
		All	424	55.1	37.5	1.09	1.01	1.19
A00–T98	All diseases and disorders	Men	5559	1467.6	820.3	1.10	1.08	1.13
		Women	5275	1349.5	563.3	1.11	1.08	1.13
		All	10834	1407.6	683.9	1.10	1.08	1.12

Notes: CR: crude rate, ASR(W): standardised mortality rate (world population), SMR: standardised mortality ratio, 90% CI: 90% confidence interval.

The new analysis was concerned with a 15-year period (2001–2015) and used as a reference the cancer mortality rates from Southeast Serbia. Overall, 2,092 cancer death cases (all malignant tumours) and, for all malignant cancers, and excess of mortality risk in Bor in both males and females (SMR = 1.11 in men and 1.24 in women) were observed.

As mentioned earlier, in order to describe the population mortality profiles in Bor, we performed a mortality analysis for all causes of death and deaths from circulatory, respiratory, digestive and genitourinary diseases. The observed death cases were compared with the expected cases based on standard age classes, gender, geographical area and the analysis of SMR with 90% CIs. Mortality and population data from the National Statistical Office for the 15-year period (2001–2015) were used. Furthermore, we calculated reference mortality rates from Southeast Serbia. We observed that for almost all groups of mortality causes, higher mortality risks exist in Bor in both males and females. Higher mortality risks in Bor were found for all diseases and disorders (SMR = 1.10 in men and 1.11 in women), diabetes mellitus (SMR = 1.12 in men and 1.18 in women), diseases of the circulatory system (SMR = 1.12 in men and 1.03 in women), diseases of the respiratory system (SMR = 1.19 in men and 1.23 in women) and congenital malformations, deformities and chromosomal abnormalities (SMR = 1.18 in men and 1.19 in women). Also, some higher mortality risks exclusively in women for symptoms, signs and abnormal clinical and laboratory findings (SMR = 1.34) and for digestive system diseases (SMR = 1.19) were noted. In men, some higher mortality risk for external causes of morbidity and mortality was found (SMR = 1.22).

11.6 Conclusions

In this study, cancer incidence and mortality data for Bor were analysed at the municipality level in the context of the SENTIERI approach. For these causes, in the study, an aetiologic role of environmental exposure is present. While the environmental data were unavailable, further findings can be supported on the basis of *a priori* evaluation of the epidemiological evidence completed in the modified SENTIERI approach in this study. The present results for Bor are based on valid data, at lower potential for bias giving an opportunity for development of similar or more profound future ecological studies. The present findings indicate the need for a more thorough approach to cleaning of the polluted environment in the future.

The fact that most diseases with an increased incidence and mortality risk have multifactorial aetiology should not be disregarded. In the future, the two sectors of environment and health must rebuild a climate of confidence and trust between themselves at the institutional level. With regard to the Bor study results and public health actions, the work of these and other sectors need to be more communicative, objective and transparent.

Note that the Bor study has been implemented using the data prior to the takeover by a new ownership of the Bor copper mining and smelting industrial complex, which happened in 2018. Since then, mining of the copper ore and smelting processes have risen significantly, causing even greater concerns regarding the consequences for the public health of the exposed population.

The Bor study was an outcome of the WHO/UNEP/SAICM project "Strengthening Serbian national capacities and inter-sectorial synergies for safe management of contaminated sites and related hazardous substances to prevent negative impact on human health and the environment" (No.: QSPTF/13/GOV/19) (Matić *et al.*, 2024).

References

Dimitrijević, M. D., Nujkić, M. M., Alagić, S. Č., Milić, S. M., and Tošić, S. B. (2016). Heavy metal contamination of topsoil and parts of peach-tree growing at different distances from a smelter complex. *International Journal of Environmental Science and Technology*, 13(2), 615–630. DOI: 10.1007/c13762-015-0905-3.

Filipović, D. (2015). Rudna Glava in the Foreground of Recent Overviews of the Beginnings of Copper Mining in Europe and of the Development of

Archaeometallurgy. Balcanica XLVI, Annual of the Institute for Balkan Studies, Serbian Academy of Sciences and Arts.

Jelenković, R., Milovanović, D., Koželj, D., and Banješević, M. (2016). The mineral resources of the Bor metallogenic zone: A review. *Geologica Croatica*, 69/1, 143–155. DOI: 10.4154/gc.2016.11.

Matić, B., Jovanović, D., Živković Perišić, S., Miljuš, D., Dejanović, S., *et al.* (2024). The role of Serbian health sector in multisectoral management of chemicals and contaminated sites. *Journal of Earth and Environmental Sciences: JEES-125,* 3(2), ISSN: 2835-7868. https://www.cmjpublishers.com/wp-content/uploads/2024/02/the-role-of-serbian-health-sector-in-multisectoral-management-of-chemicals-and-contaminated-sites.pdf.

Matić, B., Rakić U., Dejanović, S, Jovanović, V., Jevtić, M., and Đonović, N. (2017). Industrially contaminated areas in serbia as a potential public health threat to the exposed population. *Tehnika-Kvalitet IMS, Standardizacija i Metrologija*, 173, 441–447. DOI: 10.5937/tehnika1703441M.

Mudu, P., Terracini, B., and Martuzzi, M. (Eds.) (2014). *Human Health in Areas with Industrial Contamination.* Copenhagen: WHO Regional Office for Europe.

Pirastu, R, Ancona, C., Iavarone, I., Mitis, F., Zona, A., and Comba, P. (2010). SENTIERI Project. Mortality study of residents in Italian polluted sites: Evaluation of the epidemiological evidence. *Epidemiol Prev*. 34 (5–6, supplement 3): 1–2. English, Italian. PMID: 21220828.7.

Pirastu, R., Ancona, C., Iavarone, I., Mitis, F., Zona, A., and Comba, P. (2011). SENTIERI Project. Mortality study of residents in Italian polluted sites: Results. *Epidemiologia e Prevenzione*, 35(5–6, supplement 4), 1–204.

Pirastu, R., Pasetto, R., Zona, A., Ancona, C., Iavarone, I., Martuzzi, M., and Comba, P. (2013). The health profile of populations living in contaminated sites: SENTIERI approach. *Journal of Environmental and Public Health*, 2013, 939267. DOI: 10.1155/2013/939267. Epub 2013 June 18. PMID: 23853611; PMCID: PMC3703355.

Serbula, S. M., Ilic, A. A., Kalinovic, J. V., *et al.* (2014). Assessment of air pollution originating from copper smelter in Bor (Serbia). *Environmental Earth Sciences*, 71, 1651–1661. Available at: https://doi.org/10.1007/s12665-013-2569-7.

Stanković, Đ., Milivojević, D., and Radivojević, D. (2009). POLA VEKA INFORMACIONIH TEHNOLOGIJA U BASENU BOR, Institut za rudarstvo i metalurgiju Bor, „Grafomedtrade“ Bor (in Serbian).

Tasić, V., Kovačević, R., Maluckov, B., Apostolovski-Trujić, T., Cocić, M., Matić, B., and Šteharnik, M. (2017). The content of As and heavy metals in TSP and PM_{10} near copper smelter in Bor, Serbia. *Water Air Soil Pollution,* 228. Article 230. DOI: 10.1007/s11270-017-3393-6.

World Health Organization. Regional Office for Europe. (2013). Contaminated Sites and Health: Report of Two WHO Workshops: Syracuse, Italy, 18 November 2011, Catania, Italy, 21–22 June 2012. World Health Organization. Regional Office for Europe. Available at: https://iris.who.int/handle/10665/108623.

https://doi.org/10.1142/9781800617841_bmatter

Index

www.ingramcontent.com/pod-product-compliance
Lightning Source LLC
LaVergne TN
LVHW010328220926
848484LV00004BB/857
* 9 7 8 1 8 0 0 6 1 7 8 3 4 *